Fundamentals of
Data
Structures in C

for Anna University ECE course

Fundamentals of
Data
Structures in C

for Anna University ECE course

P Sudharsan ME
Program Manager
Infosys Ltd
Chennai

J John Manoj Kumar BE
Program Manager
Standard Chartered Bank
Chennai

CBS

CBS Publishers & Distributors Pvt Ltd

New Delhi • Bengaluru • Chennai • Kochi • Kolkata • Mumbai
Bhopal • Bhubaneswar • Hyderabad • Jharkhand • Nagpur
Patna • Pune • Uttarakhand • Dhaka (Bangladesh)

Fundamentals of
**Data
Structures** in **C**

for Anna University ECE course

ISBN: 978-93-89261-70-7

First Edition: 2020

Published by Satish Kumar Jain and produced by Varun Jain for

CBS Publishers & Distributors Pvt Ltd
4819/XI Prahlad Street, 24 Ansari Road, Daryaganj, New Delhi–110 002, India.
Ph: 23289259, 23266861, 23266867   Website: www.cbspd.com
Fax: 011-23243014   e-mail: delhi@cbspd.com; cbspubs@airtelmail.in
Corporate Office: 204 FIE, Industrial Area, Patparganj, Delhi–110 092
Ph: 4934 4934   Fax: 4934 4935
e-mail: publishing@cbspd.com; publicity@cbspd.com

Branches

- **Bengaluru:** Seema House 2975, 17th Cross, K.R. Road, Banasankari 2nd Stage, Bengaluru-560 070,
  Karnataka, India
  Ph: +91-80-26771678/79   Fax: +91-80-26771680   e-mail: bangalore@cbspd.com

- **Chennai:** 7, Subbaraya Street, Shenoy Nagar, Chennai–600 030, Tamil Nadu, India
  Ph: +91-44-26680620, 26681266   Fax: +91-44-42032115   e-mail: chennai@cbspd.com

- **Kochi:** 42/1325, 1326, Power House Road, Opposite KSEB Power House, Ernakulam–682 018, Kochi,
  Kerala, India
  Ph: +91-484-4059061-65   Fax: +91-484-4059065   e-mail: kochi@cbspd.com

- **Kolkata:** 6/B, Ground Floor, Rameswar Shaw Road, Kolkata–700 014, West Bengal, India
  Ph: +91-33-22891126, 22891127, 22891128 e-mail: kolkata@cbspd.com

- **Mumbai:** 83-C, Dr E Moses Road, Worli, Mumbai–400018, Maharashtra, India.
  Ph: +91-22-24902340/41   Fax: +91-22-24902342   e-mail: mumbai@cbspd.com

Representatives

• Bhopal	0-8319310552	• Bhubaneswar	0-9911037372	• Hyderabad	0-9885175004
• Jharkhand	0-9811541605	• Nagpur	0-9421945513	• Patna	0-9334159340
• Pune	0-9623451994	• Uttarakhand	0-9716462459	• Dhaka (Bangladesh)	01912-003485

Printed at Glorious Printers, Daryaganj, Delhi, India.

PREFACE

We feel immense pleasure in introducing the book *Fundamentals of Data Structures in C* for the students of Anna University–ECE course. Although, a number of textbooks are available on data structures, this text attempts to be different from the rest. It aims at fulfilling the needs of those who want to get acquainted with programming in C and to write programs in C for various data structures with algorithms.

Algorithms in this book are presented for complicated data structures followed by complete executable C programs. These algorithms allow the reader to focus on the approach to solve a problem without concern about declaration of variables and the specifications of the programming language.

This book is organised into 5 chapters. Chapters 1 and 2 explain the basic concepts in C, and Chapters 3–5 implement C language for writing programs using various data structures. All of the concepts in the book are illustrated by diverse examples, wherever required. All the programs in this text have been tested and debugged for two errors. We have strived hard to make sure that all the concepts illustrated in this text have been accompanied by worked out examples.

Although this book is written primarily for examination purposes, the practical aspect has been kept in view and hope that it will appeal to the students and practicing engineers alike. This book has been written particularly focusing on the questions and programs of various university examinations of the past 5 years.

Every effort has been made to present the topics and programs from the fundamentals in an easy and expressive language for the convenience and better understanding of the readers. This book contains more than 100+ programs and 450+ multiple choice questions. A large number of illustrative examples are given throughout the book, to instill the basics of of data structures in C programming language and the logic behind it, in the readers mind.

P Sudharsan
J John Manoj Kumar

ACKNOWLEDGEMENTS

We express our whole-hearted thanks to Mr A Nagoor Kani for all his guidance, support, motivation, encouragement and cooperation, he has extended to us for the past 20+ years. Without him, our careers as authors, would not have taken off.

We acknowledge the contributions of our technical editors, Ms C Mohana Priya and Ms K Uthra, for editing, proof-reading and type-setting of the manuscript and preparing the layout of the book.

Our sincere thanks to all reviewers, for their valuable suggestions and comments, which helped us to explore the subject to greater depth and add more contents with programs in this text.

We are grateful to Mr Satish K Jain, CMD, CBS Publishers & Distributors, for his keen interest in publishing this work in CBS banner. We are thankful to all team members of CBS Publishers & Distributors, for their concern and care in publishing this work.

Finally, a special note of appreciation is due to our relatives, friends, students and the entire teaching community for their overwhelming support and encouragement to our writing.

P Sudharsan
J John Manoj Kumar

CONTENTS

Chapter 2: Functions, Pointers, Structures and Unions   2.1–2.52

Chapter 3: Linear Data Structures   3.1–3.78

Chapter 5: Searching and Sorting Algorithms   5.1–5.98

Chapter 1
C Programming Basics

1.1 Introduction

C is a general-purpose structured programming language developed at Bell Laboratories in 1972. It was designed and written by a systems programmer named **Dennis Ritchie**. C was originally written for programming under an operating system called BASIC.

C is a powerful programming language, which has attracted considerable attention world wide because it is reliable, simple, easy to use, and highly portable (i.e., a program written in C can be transferred easily from one oprating system to another).

C is characterized by the ability to write very concise source programs since it includes a large number of operators within the language. It has a relatively small instruction set but includes extensive library functions, which enhance the basic instructions. Programmers can write additional library functions of their own. Thus, the user can extend the features and capabilities of the C language.

C is often described as a `Middle Level` programming language. This is not a reflection on its lack of programming power but more a reflection on its capability to access the system's low level functions. It was designed to have both relatively good programming efficiency and relatively good machine efficiency. The C compiler combines the capabilities of an assembly language with the features of a high-level language and hence it is well suited for writing both system software like Operating Systems, Interpreters, Editors and Assembly programs.

1.1.1 The C character Set

A character denotes an alphabet, digit, or a special character. These characters can be combined to form variables. C uses constants, variables, operators, keywords, and expressions as building blocks to form a basic program. Valid characters used in C programming are as follows,

ALPHABETS

Lower case	a b c d e f g h i j k l m n o p q r s t u v w x y z
Upper case	A B C D E F G H I J K L M N O P Q R S T U V W X Y Z
Digits	0 1 2 3 4 5 6 7 8 9

SPECIAL CHARACTERS

~	tilde	_	under score	/	slash
!	exclamation mark	+	plus sign	<	less than symbol
@	at symbol	–	minus sign	>	greater than symbol
#	number sign	=	equal to	?	question mark
$	dollar sign	'	apostrophe	{	left flower brace
%	percent sign	"	quotation mark	}	right flower brace
^	caret	:	colon	[	left bracket
&	ampersand	;	semicolon	]	right bracket
*	asterisk	,	comma	.	dot operator
(	left parenthesis	|	vertical bar	\	back slash
)	right parenthesis				

WHITE SPACE CHARACTERS

\b	blank space	\f	form feed	\n	new line
\r	carriage return	\t	horizontal tab	\v	vertical tab

1.1.2 Tokens, Identifiers and Keywords

Tokens

Tokens are the smallest individual unit in a program. Tokens supported in C are as follows,

- Identifiers
- Keywords
- Constants
- String literals
- Operators.

C program is usually written with these tokens, as per the syntax of C language. White spaces are also used to seperate adjacent tokens.

Identifiers

*An **identifier** is a string of alphanumeric characters that begins with an alphabetic character or an under score character that are used to identify or name various programming elements such as variables, function names, array names, tags and members of structures, unions, typedef names, enumeration constants and so on.*

The under score character is considered as a letter in identifiers. The under score character can be used anywhere in an identifier, but usually used in the middle of an identifier. There are 63 alphanumeric characters, i.e., 53 alphabetic characters and 10 digits (i.e., 0-9). There are 53 characters, to represent identifiers. They are 52 alphabetic characters (i.e., both upper case and lower case alphabets) and the under score character. An identifier should be small, informative, memorable and easily pronounceable if possible.

Rules for constructing identifiers

- The first character in an identifier must be an alphabet or an under score and can be followed by any alphabets, or digits or under scores.
- No commas or blank spaces are allowed within an identifier.
- No special character other than under score can be used to represent identifiers.
- The name used to represent an identifier should not be a keyword.
- Identifiers are case sensitive (i.e., AVERAGE is different form average or Average or AVERage).
- Identifiers can be of any length, but only the first 31 characters are recognised.

Table 1.1 : Identifiers

Valid	Invalid
ChennaI	786
PRIYA	R*P
FLOAT	(sum)
average_value	%
sUcHiTrA	100th
WorD	float
_rama	cost$
mark_1	FIVE STAR
One_And_Only	[JeyaBal]
mounTAIN	"Saranya"

Keywords

Keywords are also referred as **reserved words,** *whose meaning has been already defined in the C compiler. All keywords have fixed meanings and their meanings cannot be changed.*

Keywords can be used only for their intended purpose and they cannot be used as identifiers. All keywords must be compulsarily written in lowercase. For using the keywords in a program, no header file is to be included.

The keywords in C are summarized as below:

auto	double	int	struct
break	else	long	switch
case	enum	register	typedef
char	extern	return	union
const	float	short	unsigned
continue	for	signed	void
default	goto	sizeof	volatile
do	if	static	while

1.2 Structure of a C Program

C is a programming language and so an introduction to structure of C program is presented in this section for better understanding the logic of a program by the reader. Every C program contains few basics of building blocks. These building blocks should be written in a correct order and procedure. Let us see a sample program for calculating the sum of two integers.

Program 1.1 :

```
/* Program to calculate the sum of two integers */        /* Title comment */

/* firstpgm.c */

#include<stdio.h>                                          /* Library file access */

void main()                                               /* Function */

  {

    int x, y, sum;                                        /* Declaration of variables */

    printf("Enter the values of X and Y : ");             /* Output statement */

      scanf("%d %d", &x, &y);                             /* Input statement */

    sum = x + y;                                          /* Assignment statement */

    printf("The sum of the variables X and Y = %d", sum); 

                                                          /* Output statement */

  }
```

The program displays the following output

```
Enter the values of X and Y : 25    30
The sum of the variables X and Y = 55
```

Comments are added at each end of the statement in order to understand the overall program organisation easily. Let us see each comment statement in detail.

/* Title comment */

Any comment statement used in a C program must be enclosed within /* */. The comment statement is used to identify the purpose of the program. This statement is usually specified at the start of a program, but can be specified any where in the program. This may or may not be present in the program as per the programmers wish. The comment statement is not an executable statement. It's a good programming practice to include comment statements in a program.

/* Library file access */

The second line in the program #include <stdio.h> refers to a special file which contains, the information about the various input and output operations which must be included in the program when it is compiled. The C compiler will handle the inclusion of the required information automatically. The header file stdio.h contains the information of all the standard input and output functions in C. If you don't include the header file stdio.h in the program, the C compiler may complain about undefined functions and data types used in the program.

/* Function */

The third line in the program `main()` is a function, which may consist of one or more statements. Every C program must contain at least one function which must be `main()`. C program will always begin by executing the `main()` function, which will access other functions used in the program. The empty parenthesis followed by `main()` indicates that this function does not include any arguments. The remaining lines of the program are enclosed within a pair of braces. The statements enclosed within a pair of braces are referred as the body of the function.

/* Declaration of variables */

A declaration of a variable is a statement that gives information to the C compiler about the variables to be used in the program. The general form or the syntax of declaring a variable is

```
data type variable(s);
```

The declaration consists of a data type, followed by one or more variable names, seperated by commas ending with a semicolon. In *Program 1.1,* we have declared

```
int  x, y, sum;
```

where `int` is the data type in C and `x`, `y` and `sum` are the variables to be used in the program. The type of a variable informs the compiler how the variable values are to be stored (i.e., how much of memory space is to be allocated). *Note that all variables used in the program must be declared before all executable statements.*

/* Output statement */

The other lines in the program are statements. The first statement is the `printf` statement, that generates a request to the user to give information (i.e., values) for the variables `x` and `y` for summation. The `printf` statement is usually called as the *output statement* since it displays the message on the screen. The information to be displayed on the screen must be specified within double quotes (" ").

/* Input statement */

The values for the variables `x` and `y` are entered into the computer through the `scanf` statement, which is below the `printf` statement. The `scanf` statement is usually called as the *input statement* because it reads the values to be processed by the computer.

/* Assignment statement */

The fourth line is an important statement for this program since it adds the two values of `x` and `y` entered through the keyboard. This statement is called as an *assignment statement*, since it assigns the value of `x+y` to the variable `sum`.

/* Output statement */

The last line within the flower bracket is the `printf` statement, which displays the output of the program (i.e., the summation of x and y) in the screen.

You may notice that all statements within the flower braces ends with a semicolon. This is compulsarily required for all expression statements. White spaces within `printf`, `scanf` and other expression statements are usually encouraged as a good programming practice.

Important points to be noted while writing programs

- All statements must be written in lower case letters. However, symbolic constants, identifiers can be written in upper case letters.

- All statements must end with a semicolon.

- Comment statements can be included wherever necessary in the program. It's a good programming practice to include comment statements in a program.

- White spaces during variable declaration, within arguments in functions, before and after operators, and other expression statements are usually encouraged as a good programming practice.

- The opening and closing flower braces used in a program should be balanced. For example, if there are three opening braces then there should be three closing braces.

1.2.1 Library Functions

The standard library is not a part of the C language but an environment that supports C language. The standard library provides function declarations, type and macro definition for the standard C language. C language is accompanied by a number of library functions that carries out various commonly used operations. Most of the functions return a value of one or zero stating the condition as true or false respectively. There are many library functions that carries standard input and output operations, (e.g., `scanf()`,`printf()`, `getc()`, `gets()`, etc.,), functions that perform operations on strings (e.g., `strcpy()`, `strcat()`, `strcmp()`, `strlen()`,etc.,), functions that perform mathematical calculations (e.g., `sin()`, `cos()`, `sqrt()`, `pow()`, etc.,) etc., A library function is accessed simply by writing the function name followed by a list of arguments, which represents the information being passed to the function. The arguments must be enclosed in parenthesis and seperated by commas. The arguments passed to the library functions can be constants, variables, or complex expressions. A typical set of C library includes large number of functions, which can be seen in detail in the **APPENDIX**.

1.2.2 Header Files

Library functions that have similar functions are grouped together as programs (with file extensions `.h`) in separate library files. In order to use a library function it is necessary to include a specific information within the main portion of the program, called the ***header*** *files*. The general form or the syntax of including a header file in the program is,

```
#include<filename.h>
```

where `filename` represents the header file and `.h` represents file extension. The different header files in C language are

assert.h	float.h	math.h	stdarg.h	stdlib.h
ctype.h	limits.h	setjmp.h	stddef.h	string.h
errno.h	locale.h	signal.h	stdio.h	time.h

1.2.3 The main( ) function

A function is a self-contained program segment that carries out some specific, well-defined task whenever it is accessed. A program can have one or more functions but there should be at least one function, which must be `main()`. Empty parentheses following the word `main` are necessary. C program will always begin by executing the `main()` function. Some programmers place the `main()` function at the beginning of the file and others use it at the end. Regardless of its location, every C program must contain a `main()` function. The `main()` function consists of two parts namely, the *declaration part* and the *execution part.* The declaration part declares all the variables used in the executable part. Initialization of variables can also be done in the declaration part. Declaration of variables allocates memory spaces according to their data types specified. The execution part contains one or more statements following the declaration part. The execution part uses all the variables declared in the declaration part. These declaration and execution parts must appear between the opening and the closing braces of the `main()` function.

The program begins its execution at the opening flower brace and ends its execution in the closing brace. The closing flower brace of the `main()` function is the logical end of the program. All statements in the declaration and executable part ends with a semicolon compulsarily. A `main()` function may or may not contain a declaration statement but it should have an executable statement. *Program 1.2*, shows a `main()` function with both declaration and executable parts.

Program 1.2 :

```
/* Program to find the area of a rectangle */
/* area.c */
#include<stdio.h>
void main()
 {
   int length = 5, breadth = 3, area;
   area = length * breadth;
   printf("The Area of the rectangle = %d", area);

 }
```

The program displays the following output

```
The Area of the rectangle = 15
```

1.3 Compilation and Linking Processes

Compilation and linking processes of a C program are necessary before executing a C program. Usually, creating and executing a C program for a specific task involves the following processes.

- Typing and editing a program
- Pre-processing a program
- Compilating a program
- Linking a program
- Executing a program.

Typing and Editing a Program

In order to write or create a C program the program codes have to be typed using the editor tool and saved as a file with extension .c. For example, **Program 1.1** in this book is typed and saved as firstpgm.c. The C program editor is basically a text editor for creating a progrm and will have facility to type, modify, save, delete, etc.

Pre-processing a Program

The pre processing is the process performed by the compiler when a program compile command is initiated. In pre-processing, the entire program is scanned to identify the macros, header files and commands. The macros and header files are expanded and commands are removed.

For example, in **Program 1.1,** the header file stdio.h contains the information about various input and output operation which will be included and the command line enclosed with /* */ are removed.

Compilating a Program

The compilation is the process of converting the pre processed program into the object program and save the object program with .obj extension. The **Program 1.1** will be saved as firstprgm.obj after compilation.

Before creation of object program, the source program undergoes debugging. After all the errors are identified and rectified, the analysis and synthesis phase of the compilation are carried. Analysis is the process where the source program is divided into small elements and internal representation of the source program is generated. Synthesis is the phase where the machine code is generated which is the object program.

Linking a Program

Linking is the process of converting the object program in to an executable program. The output of linking process is saved with file extension .exe. When program is developed in modules the object program of various modules are linked to single executable file by the linker.

Executing a Program

Running the .exe file of a program is called execution. Upon executing the .exe file the task for which the program is written will be carried by the system or computer.

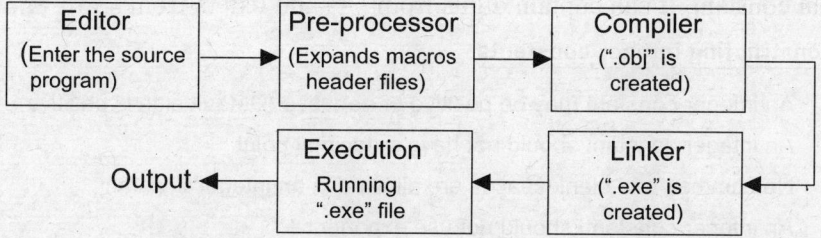

Fig. 1.1 : Compliation and linking process.

1.4 Constants

Constants are tokens in C program representing fixed numeric or character values that do not change during the execution of a program. *Fig. 1.2*, shows the various types of constants in C.

1.4.1 Integer Constants

C deals with several kinds of numbers, one of the most frequently used is the whole number, usually called as an *integer*. *An* **integer constant** *refers to a sequence of digits without a decimal point. Integer constants can be specified in the following three ways.*

- Decimal integer constant
- Octal integer constant
- Hexadecimal integer constant.

The value 35, can be written in any of the three ways as,

35   representing a decimal constant

043   representing an octal constant

0x23   representing a hexadecimal constant.

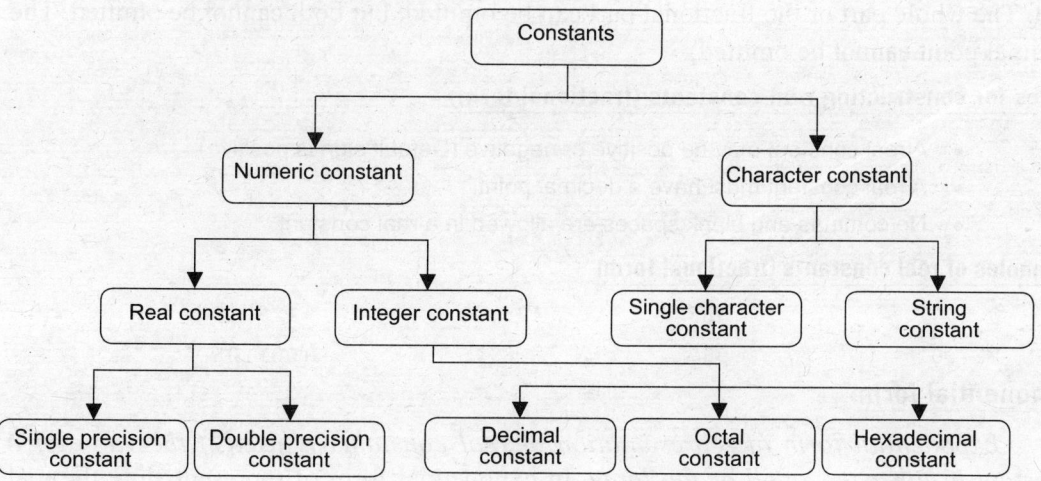

Fig. 1.2 : Types of constants.

The leading 0 (zero) to an integer constant specifies, that it is an octal integer constant. It can contain digits from 0-7. If an octal constant contains digits 8 or 9, an error is reported. The leading 0x or 0X to an integer constant specifies, that it is a hexadecimal constant. It can contain digits from 0-9 and letters from a to f or A to F.

Rules for constructing integer constants

- An integer constant may be positive or negative (Default sign is positive).
- An integer constant should not have a decimal point.
- No commas and blank spaces are allowed in an integer constant.
- An integer constant, should not use exponent e.

The suffix u or U is used for denoting unsigned int constants, l or L is used for denoting long int constants and s or S is used for denoting short int constants. Usually these suffixes are not needed. The compiler automatically considers small integer constants to be of type short and large integer constants to be of type long.

Examples of integer constants

```
12      -5      0     10000    -25000
```

1.4.2 Real Constants

*Real constants are often called as **floating-point constants***. They can be written in two forms. They are,

- Fractional form
- Exponential form.

Fractional Form

The fractional part consists of a series of digits representing the whole part of the number, followed by a decimal point, followed by a series of digits representing the fractional part. The whole part or the fractional part can be omitted, but both cannot be omitted. The decimal point cannot be omitted.

Rules for constructing real constants (fractional form)

- A real constant may be positive or negative (Default sign is positive).
- A real constant must have a decimal point.
- No commas and blank spaces are allowed in a real constant.

Examples of real constants (fractional form)

```
0.0001          -0.963          485.98          +123.456
727.01          0.987           -.8765          +001.05
```

Exponential form

Exponential form of representation of real constants is used if the value of a constant is either too small or too large. In exponential form of representation, the real constant is represented in two parts. The part appearing before e is called **mantissa**, and the part after e is called **exponent**. The exponent consists of an optional plus or minus sign followed by a series of digits. For example 425000 can be expressed as 4.25e5 (or) 42.5e4. Similarly 0.00028 can be expressed as 2.8e–4 (or) 0.28e–3.

Rules for constructing real constants (exponential form)

- The mantissa part and the exponential part should be seperated by a letter e.
- The mantissa part and the exponential part may be positive or negative (Default sign is positive).
- The mantissa and the exponent part must have at least one digit.
- No commas and blank spaces are allowed in a real constant.

Examples of real constants (exponential form)

+3.2e+5	3.4e5	-8.25e-5
-10.9e5	9.72e-10	0.0025e8

1.4.3 Single Character Constants

*A **single character constant** or a character constant contains a single character enclosed within a pair of single quotes.* A single character constant must be either a single alphabet or a single digit or a single special character enclosed in single quotes. The length of a single character constant is only one.

Table 1.2 : Single Character Constants

Valid	Invalid
'A'	"x"
'm'	'xyz'
'2'	'123'
'*'	"pq'
'_'	'$$"

1.4.4 String Constants

*A character string, a **string constant** or a literal consists of a sequence of characters enclosed in double quotes.* A string constant may consist of any combination of digits, letters, escape sequences, and spaces. The end of the string is marked with a single character referred as the NULL character '\0'. ***Note that a single character constant 'A' and the single character string constant "A" are not same.*** The string constant "A" consists of characters A and \0.

'A' - character constant → 'A'

"A" - string constant → 'A' and string endmarker '\0' (NULL)

Further a character constant as an equivalent integer value, where has a single character string constant does not have an equivalent integer value. It occupies two bytes, one for the ASCII code of A and another for the NULL character with a value 0, which is used to terminate all strings.

Rules for constructing string constants

- A string constant may be a single alphabet or a digit, or a special character or sequence of alphabets or digits enclosed in double quotes.
- Every string constant ends up with a NULL character, which is automatically assigned.

Table 1.3 : String Constants

Valid	Invalid
"W"	'Prasad'
"100"	"12-June-1978'
"Madhu"	'10-02-1978"

1.5 Variables

*A quantity, which may vary during program execution, is called as a **variable**.* Variable name is an arbitrary name, an identifier used to refer to the area of memory in which a particular value is stored. These memory locations can contain an integer, real, single character or a string constant.

1.5.1 Declaration of Variables

A data item must be assigned to a variable at some point in the program. The data item can then be accessed later in the program simply by referring to the variable name. The information represented by the variable can change during execution of the program, but the data type associated with the variable cannot change. A variable name used by the programmer can be in a meaningful way to reflect its use or function or nature in the program. All variables used in the program must be declared before all executable statements (i.e., at the start of each function after the opening flower brace ({) of the function). Declaration of variables begins with its type followed by one or more variable names. Declaration assigns the variable to a specific data type and there by allocate memory to the variables depending upon the data type. For example,

```
int a, b, c;
char d1, e2, f3;
float x_1, y_2, z_3;
```

In the first declaration, a, b and c are variables declared as type integer (because of the use of int data type). In the next declaration, d1, e2 and f3 are character variables (because of the use of char data type). In the last declaration, x_1, y_2 and z_3 are floating-point variables (because of the use of float data type).

1.5.2 Initialization of Variables

Variables can also be initialized when they are declared. This is done by adding an equal sign followed by the required value to be initialized after the declaration. For example,

```
int a = 5, b = 10;
char d1 = 'a';
float x_1 = 100.056;
```

The first declaration declares two variables a and b of type integer and thereby initializing the variables with the values 5 and 10 respectively. The next declaration declares a character variable d1 and initializes with a value 'a' (i.e., a single character constant). The last declaration declares a floating-point variable x_1 and initializes with a value 100.056. Variables can also be initialized in the program with an assignment statement of the form,

```
variable = constant;
```

where constant is the value to be assigned to the variable. For example,

```
age = 23;
```

```
salary = 7500.75;
```

The first declaration initializes the value 23 to the variable age, and the next declaration initializes the value 7500.75 to the variable salary. Variables in C can be broadly classified into two types. They are,

- Constant variables
- Volatile variables.

Constant Variables

Variables that cannot be changed by a function in which it is present or by any other function in the same program are called as ***constant variables***. The keyword const is used for this purpose. It protects variables from modification during execution of a program. The following declaration illustrates this,

```
const int x = 34;
```

In the above intialization, x is a variable declared as type integer (because of the use of int data type). The value in the variable (i.e., 34 in x) cannot be modified during execution of the program, because of the use of const keyword before the data type declaration. The C compiler protects constant variable from modification.

Volatile Variables

The variable that can be changed at any time by a function in which it is present or by any other function in the same program are called as ***volatile variables***. The keyword volatile is used for this purpose. The following declaration illustrates this,

```
volatile int z = 6000;
```

1.6 Data Types

The data types supported in a language dictates the type of values, which can be processed by the programming language. C language is rich in its data types. It supports a wide variety of data types each of which may be represented differently within the computer's memory, which allows the programmer to select the appropriate type to the needs of the application. Data types in C can be broadly classified as,

- Primary data types
- User-defined data types
- Empty data type.

1.6.1 Primary Data Types

All C compilers supports four primary data types namely,

- Character data type represented as **char**
- Integer data type represented as **int**
- Floating-point data type represented as **float**
- Double-precision floating-point data type represented as **double**.

Integer Data Types

*Whole numbers, which are used frequently, are referred as **integers**.* It occupies two bytes for their storage. In order to have control over the range of numbers and storage space in memory `short int, int, long int` in both `signed` and `unsigned` forms are used.

Floating-Point Data Types

*Real numbers, which are used frequently, are referred as **floating-point numbers**.* The `float` data type is used to store any number in floating-point format. Floating-point data types store numerical values with a fractional portion. They usually occupy 4 bytes of memory for their storage. These are known as ***single precision numbers***. When the range provided by a `float` data type is not sufficient, `double` data type is used. The `double` data type represents the same data type, that `float` represents but with a greater precision hence it requires 8 bytes of memory for their storage in the memory. These are known as ***double precision numbers***. To increase the precision further we can use `long double` which uses 10 bytes of memory for their storage. These are known as ***extended precision numbers***.

Character Data Types

The `char` data type is used to represent characters that require only one byte of memory for their storage. In order to have control over the range of characters and storage space in memory, `signed` and `unsigned char` data type are used.

Table 1.4, shows the range of values that can be stored in various data types.

Table 1.4 : Size and Range of Data Types in C

Data type	Size (bytes)	Range
char or signed char	1	-128 to 127
unsigned char	1	0 to 255
int	2	-32768 to 32767
short int or signed short int	2	-32,768 to 32,767
int or signed int	2	-32,768 to 32,767
unsigned short int	2	0 to 65,535
unsigned int	2	0 to 65535
long int or signed long int	4	-2,147,483,648 to 2,147,483,647
unsigned long int	4	0 to 4,294,967,295
float	4	3.4e-38 to 3.4e+38
double	8	1.7e-308 to 1.7e+308
long double	10	3.4e-4932 to 1.1e4932

1.6.2 User-defined Data Types

C supports a feature known as type definition that allows the user to define new data types that are equivalent to existing data types. This technique in some situations can help to clarify the source code of a program. The user-defined data type identifier can later be used to declare variables.

Type Definition

The keyword typedef is used to define new data type names. Note that you are not actually creating a new data type, but rather defining a new name for an existing data type. The general form or the syntax of defining a new data type is

```
typedef  data type  identifier;
```

where identifier refers to new name(s) given to the data type. Few examples of type definition are

```
typedef int age;
typedef float average;
```

where age symbolizes int and average symbolizes float. They can be later used to declare variables as

```
age child, adult;
average mark1, mark2;
```

where child and adult are declared as integer variables whereas mark1 and mark2 are declared as floating-point variables. Using type definition you can create meaningful data type names for increasing the readability of the program. They also suggest the purpose of the data type names used in the program.

Enumerated Data Type

The enumerated data type gives an opportunity to invent your own data type with a set of named integer constants represented by identifiers that specifies all the valid values a variable of that type may have. They can be used in any expression where integer constants are valid.

The general form or the syntax of enumerated data type is,

```
enum identifier { enumeration_list } list_of_variables;
```

where `enum` is a keyword, which is used to specify the enumerated data type, `identifier` is an user-defined enumerated data type which can be used to declare the `enumeration_list` also referred as ***enumeration constants***. Both the `identifier` and the `list_of_variables` are optional, but at least one must be present. For example, the enumeration

```
enum month{JANUARY,FEBRUARY,MARCH,APRIL,MAY,JUNE,JULY,
           AUGUST,SEPTEMBER,OCTOBER,NOVEMBER,DECEMBER};
```

creates a new type month in which the `enumeration_list` is JANUARY , . ., DECEMBER. The values for the enumeration list are set automatically to integers 0 to 11. To number the `enumeration_list` from 1 to 12 the `enum` declaration should be

```
enum month{JANUARY = 1,FEBRUARY,MARCH,APRIL,MAY,JUNE,JULY,
           AUGUST,SEPTEMBER,OCTOBER,NOVEMBER,DECEMBER};
```

where the first `enumeration_list` (i.e., JANUARY) is explicitly set to 1 and hence the remaining values are incremented from 1, and result in the values of 1 to 12. To define a variable for the new type month, the statement

```
enum month month_start, month_end;
```

declares `month_start` and `month_end` as variables for the new type month.

1.6.3 Empty Data Type

The keyword `void` *is used to indicate an empty data type.* This data type is not used to declare any variables, but used to indicate nothing. When used as a function return type, `void` means that the function does not return a value. When used in a function parameter list, `void` means the function does not take any parameters.

1.7 Expressions using Operators in C

*An **operator** is a symbol or a special character that tells the computer to perform certain mathematical or logical operations, which is applied to operands to give a result.* Operators are used in programs to manipulate data and variables. The data items that operators act upon to evaluate expressions are called as ***operands***. One of the main reasons for the power of C is its wide range of operators. C not only offers the common arithmetic and logical operators, but also offers operators for bit-level manipulations, pointer operations and so on. The operators used in C are

- Arithmetic operators
- Unary operators
- Increment and Decrement operators
- Relational operators
- Equality operators

- Logical operators
- Bitwise operators
- Conditional operator
- Assignment operators
- Special operators.

1.7.1 Arithmetic Operators

C language has all the basic arithmetic operators. *Table 1.5*, shows various arithmetic operators and their meaning.

Table 1.5 : Arithmetic Operators and its Meaning

Operator	Meaning
+	Addition
−	Subtraction
*	Multiplication
/	Division
%	Modulo operation (remainder after integer division)

The operands that act upon arithmetic operators must represent numeric values, thus the operands can be integer quantities or floating-point quantities. Division of one integer quantity by another is referred to as the integer division. Integer division truncates the fractional part (i.e., the decimal portion of the quotient will be dropped). Modulo division produces the remainder of an integer division as the result. Modulo division is only meaningful between integers.

Suppose when operands x and y are integer variables whose values are 9 and -4 respectively. Several arithmetic expressions and their results are

Expression	Result
x + y	5
x - y	13
x * y	-36
x / y	-2
x % y	1

*When two operands such as x and y are declared as integers, an arithmetic operation performed on these integers are referred as **integer arithmetic***. It always yields an integer value. When two operands such as x and y are declared as real (i.e., float or double or long double) an arithmetic operation performed on these floats are referred to as *floating-point arithmetic*. It always yields a floating-point value. If either of the operands is real, an arithmetic operation performed on these integers is called as *mixed-mode arithmetic*. It always yields a real value (float value). For example,

```
125/25.0 = 5.0
```

Since `25.0` is real constant, `125` is converted to a real (i.e., `125.0`) and the result is also a real.

1.7.2 Unary Operators

The operators that act upon a single operand to produce a new value are called as unary operators. Unary operators usually precede its operands, though some unary operators are written after their operands. *Table 1.6*, shows various unary operators and their meaning.

Table 1.6 : Unary Operators and its Meaning

Operator	Meaning
-	Unary minus
++	Increment by 1
--	Decrement by 1
sizeof	Returns the size of the operand.

When the operator is used before the variable, the operation is applied to the variable first, and then the result is used in the expression. This type of notation is referred as *prefix notation*. When the operator is used after the variable, the value of the variable is used first, and then the operation is applied to the variable. This type of notation is referred as *postfix notation*. If the value of the operand x is 3 then the various expressions and their results are

Expression	Execution
++x	x ← 3 + 1
x++	x ← 3 ; x ← 3 + 1
--x	x ← 3 - 1
x--	x ← 3 ; x ← 3 - 1

The pre-increment operation (`++x`) increments x by 1 and then assigns the value to x. The post-increment operation (`x++`) assigns the value to x and then increments 1. The pre-decrement operation (`--x`) decrements 1 and then assigns to x. The post-decrement operation (`x--`) assigns the value to x and then decrements 1. These operators are usually very efficient, but causes confusion if you try to use too many evaluations in a single statement.

The sizeof Operator

The `sizeof` operator is not a library function but a keyword, which returns the size of the operand in bytes. The `sizeof` operator always, precedes its operand. This operator can be used for dynamic memory allocation. Various expressions and results of `sizeof` operator are

Expression	Result
sizeof(char)	1
sizeof(int)	2
sizeof(float)	4
sizeof(double)	8

1.7.3 Relational Operators

Relational operators are symbols that are used to test the relationship between two variables or between a variable and a constant. These operators are used for checking conditions in control statements. *Table 1.7*, shows various relational operators and their meaning.

Table 1.7 : Relational Operators and its Meaning

Operator	Meaning
<	less than
>	greater than
<=	less than or equal to
>=	greater than or equal to

An expression containing a relational operator is called as an ***relational expression***. The value of the relational expression is either one or zero. If the expression is true, it is represented by an integer value 1. If the expression is false, it is represented by an integer value 0. These operators have left to right associativity.

If the values of x, y, z are 1, 2, 3 respectively then the results and the values returned by the expressions are

Expression	Result	Value
x < y	True	1
(x + y)> = z	True	1
(y + z)>(x +5)	False	0
y >= 10*(x +z)	False	0

1.7.4 Equality Operators

Closely associated with relational operators are two equality operators. *Table 1.8*, shows various equality operators and their meaning.

Table 1.8 : Equality Operators and its Meaning

Operator	Meaning
==	equal to
!=	not equal to

The value of the expression is either one or zero. If the expression is true then the result will be one, if the expression is false then the result will be zero. If the values of x and y are 1 and 2 respectively, then the various expressions and their results are

Expression	Result	Value
y != 2	False	0
x == 2	False	0
x == 1	True	1
y != 3	True	1

1.7.5 Logical Operators

Operators, which are used to combine two or more relational expressions, are called as logical operators. These operators are used to test more than one condition at a time. *Table 1.9*, shows various logical operators and their meaning.

Table 1.9 : Logical Operators and its Meaning

Operator	Meaning
&&	Logical AND
||	Logical OR
!	Logical NOT

1.7.6 Bit wise Operators

C has certain operators known as bit wise operators that are used in applications, which require manipulation of individual bits within a word of memory. These operators permit the programmer to access and manipulate individual bits within a piece of data. Bit wise operators can operate only on integers and characters but not on floats and doubles. *Table 1.10*, shows the various bit wise operators and their meaning.

Table 1.10 : Bit Wise Operators and its Meaning

Operator	Meaning
~	One's complement
<<	Left shift
>>	Right shift
&	Bit wise AND
|	Bit wise OR
^	Bit wise X-OR

Masking

Masking is a process in which a given pattern is transformed into another bit pattern by means of bit wise operators. The original bit pattern is one of the operands in the bitwise operation. The second operand called as the mask is a specially selected bit pattern that brings the desired transformation. Usually logical bitwise AND operator (&) is used for masking.

1.7.7 Conditional Operators (Ternary Operators)

An expression that evaluates condition is called a ***conditional expression***. *The operator used to evaluate conditional expression is called a conditional operator denoted by (?:).* It is also referred as a ***ternary operator***, since it takes three operands as its arguments. The general form or the syntax of the conditional operator is

```
expr1 ? expr2 : expr3;
```

When evaluating a conditional expression, `expr1` is evaluated first. If `expr1` is true, `expr2` is alone evaluated. If `expr1` is false, `expr3` is alone evaluated.

For example,

```
a = ((b > 2)? 8:6);
```

1.7.8 Assignment Operators

Assignment operators are used to assign the result of an expression to a variable. The equals `(=)` sign is used as an assignment operator. It is used to form assignment expressions, which assigns the value of an expression to an identifier, which may be written in the form of

```
identifier = expression;
```

where `expression` represents a constant or a variable or a complex expression. The value evaluated from the expression is stored in the `identifier`, which is a variable.

In addition, C has a set of shorthand assignment operators of the form

```
identifier <operator> = expression;
```

The operator should be an arithmetic operator or a bitwise operator. *Table 1.11*, shows the various shorthand operators and their meaning.

Table 1.11 : Bit Wise Operators and its Meaning

Operator	Meaning
+=	Assign sum
-=	Assign difference
*=	Assign product
/=	Assign quotient
%=	Assign remainder
~=	Assign one's complement
<<=	Assign left shift
>>=	Assign right shift
&=	Assign bitwise AND
|=	Assign bitwise OR
^=	Assign bitwise X-OR

1.7.9 Special Operators

Some of the special operators used in C are listed below. These operators are referred as *seperators* or *punctuators*.

Operator		Meaning
Symbol	Name	
&	Ampersand	Also referred as *address operator* usually precedes the identifier name, which indicates the memory location (address) of the identifier.
*	Asterisk	Also referred as an *indirection operator* and it is an unary operator, which usually precedes the identifier name, which indicates the creation of a pointer operator.
{}	Braces	The opening and closing braces indicate the start and end of a compound statement or a function. A semicolon is not necessary after the closing brace of the statement, except in case of structure declaration.
[]	Brackets	Also referred as *array subscript operator* is used to indicate single and multi dimensional array subscripts.
:	Colon	Used in labels.
,	Comma	Used to link the related expressions together. Comma used expressions are linked from left to right and the value of the right most expression is the value of the combined expression. The comma operator has the lowest precedence of all operators. For example in the expression, `sum = (x = 5,  y = 3,  x + y);` The result will be `sum` = 8. The comma operator is also used to seperate variables during declaration. For example, `int a, b, c;`

Operator		Meaning
Symbol	**Name**	
. . .	Ellipsis	There are three successive periods with no white space in between them. It is used in function prototypes to indicate that this function can have any number of arguments with varying types. For example, `void fun(char c, int n, float f, ...);` The above declaration indicates that `fun()` is a function that takes at least three arguments, a char, an int and a float in the order specified, but can have any number of additional arguments of any type. You can also omit the comma preceding the ellipsis.
#	Hash	Also referred as pound sign which is used to indicate preprocessor directives.
. and ->	Member access operators	Used to access the members of structure or union. For example, if abc is a member of a structure str, the member abc is accessed using the member access operators as `str.abc = 23.45;` `str -> abc = 23.45;` The operator . is referred as the direct member access operator and -> is referred as the indirect member access operator or pointer member access operator.
()	Parenthesis	Also referred as *function call operator* which is used to indicate the opening and closing of function prototypes, function calls, function parameters, etc., Parenthesis are also used to group expressions, and there by changing the order of evaluation of expressions.
;	Semicolon	A statement terminator. It is used to end a C statement. All valid C statements must end with a semicolon, which the C compiler interprets as the end of the statement. For example, `c = a + b;` `b = 5;`

1.8 Managing Input and Output Operations

One of the primary advantages of modern computers is their ability to communicate with the user during program execution. This feature enables the programmer to read the values for the input variables, when the program is in execution without altering the program. Once the required value has been read and the ENTER key is pressed, program execution starts again, and the output is displayed on the screen. To perform input and output operations, C has a number of input and output functions. These functions can be broadly classified into two types. They are,

- Formatted I/O functions
- Unformatted I/O functions.

1.8.1 Formatted I/O Functions

The formatted input/output functions read any type of data as input from the user and displays any type of data as output in the screen. They use *format strings* (also referred as *conversion specifiers*) for identifying the type of the data involved. These functions return an integer value after execution. The return value is equal to the number of values successfully read/written. The return value will be useful for the user to check the correctness of data read/written. The scanf() function is a formatted input function and printf() function is a formatted output function.

The scanf() Function

The scanf() *function permits a program to be interrupted during execution in order to read the required values from the user through the keyboard.* scanf stands for "scan function" or "scan formatted". This function (i.e., scanf()) can be used to read any combination of numeric values, single characters and strings. The general form or the syntax of scanf() function is

```
scanf("format_string", Var1, Var2, ... , Varn);
```

where format_string refers to conversion specification which refers to a string containing certain required formatting information and Var1, Var2,., Varn are arguments that represents individual input data items. Each character group must begin with a % sign followed by a conversion character, which indicates the type of the corresponding data item to be read. A simple form of scanf() function may be

```
scanf("%d", &a);
```

where the variable a has been previously declared as an integer. %d is called as the conversion specifier. Its role is to specify that the number is to be converted to decimal notation before it gets scanned. The ampersand symbol (&) that precedes before the variable name 'a' is compulsary, whenever an integer, a floating-point, or a character value is to be read from the user. Actually ampersand (&) means address of, &a means address of a. The value read by the computer after the user has been entered will be stored in the address (memory location) of `'a'`. However, character strings are an exception to this and doesn't need a leading & before the variable for reading strings. *Table 1.12*, shows different `scanf` format codes for reading input from the user. *Program 1.3*, helps you to understand things better.

Table 1.12 : Format Codes used in scanf() Function

Code	Meaning
%c	Reads a single character
%s	Reads a string
%d	Reads signed short integer
%u	Reads unsigned integer
%ld	Reads signed long integer
%lu	Reads unsigned long integer
%o	Reads octal integer
%x	Reads hexadecimal integer
%e	Reads floating point value
%f	Reads floating point value
%g	Reads floating point value
%lf	Reads double value
%i	Reads decimal,hexadecimal,octal integer
%[..]	Reads string of word(s)

Program 1.3 :

```
/* scanf1.c */
#include<stdio.h>
void main()
{
    char empname[20];                          /* Character with length 20 */
    int empno;
    float salary;
        scanf("%s %d %f", empname, &empno, &salary);
    ...........
}
```

In *Program 1.3*, `scanf()` function uses three conversion specifiers `%s`, `%d` and `%f`. The first conversion specifier `%s` indicates `empname`, which is a string *(Note that for reading strings we don't use & symbol)*. The second conversion specifier `%d` indicates `empno` is an integer. The third conversion specifier `%f` indicates `salary`, which is a floating-point data type.

Important points to be noted while using scanf() function

- Input data items must be seperated by spaces and should match the variables receiving the input in the same order.
- Format specifications contained in the conversion specifications should match the arguments in order.

- The reading will terminate when it encounters an invalid mismatch of data type that is not valid for the values being read.
- When the field width specifier is used, it should be large enough to contain the input data size.

The printf() Function

The printf() *function is used to give output for any data type to a standard output device from the computer .* This function can be used to display any combination of digits, characters, and strings. printf stands for "print function" or "print formatted". The general form or the syntax of printf() function is,

```
printf("format_string", Var1, Var2,..., Varn);
```

where format_string refers to conversion specification which refers to a string containing certain required formatting information and Var1, Var2, ., Varn are arguments that represent the individual output data items. Each character group must begin with a % sign followed by a conversion character, which indicates the type of corresponding data item. A simple form of printf() function may be

```
printf("%d", a);
```

where variable a has been previously declared as integer and %d is called as the conversion specifier. Its role is to specify that the number is to be converted to decimal notation before it is displayed in the screen.

Table 1.13, shows different printf() format codes for displaying the output in the screen. *Program 1.4*, helps you to understand things better.

Table 1.13 : Format Codes used in printf() Function

Code	Meaning
%c	Displays a single character
%s	Displays a string
%d	Displays signed short integer
%u	Displays an unsigned integer
%ld	Displays signed long integer
%lu	Displays unsigned long integer
%i	Displays signed decimal integer
%o	Displays an octal integer, without leading zero
%x	Displays hexadecimal integer without leading 0x.
%hx	Displays hexadecimal integer in small case
%p	Displays hexadecimal integer in upper case
%e	Displays floating point value in exponent form
%f	Displays floating point value without exponent
%g	Displays floating point value either e-type or f-type depending on value
%lf	Displays double value
%n	Aborts program with an error.

Program 1.4 :

```c
/* printf1.c */
# include<stdio.h>
void main()
 {
    char empname[20];
    int empno;
    float salary;
    printf("Enter employee name, no & salary : ");
       scanf("%s %d %f", empname, &empno, &salary);
    printf("\n The employee name is %s", empname);
    printf("\n The employee number is %d", empno);
    printf("\n The employee salary is %f", salary);
 }
```

The program displays the following output

```
Enter employee name, no & salary : Joy    4    8484
The employee name is Joy
The employee number is 4
The employee salary is 8484.000000
```

In *Program 1.4*, the last three printf() functions uses three conversion specifiers %s, %d and %f. The first conversion specifier %s indicates empname which is to be displayed as a string. The second conversion specifier %d indicates empno which is to be displayed as an integer. The third conversion specifier %f indicates salary which is to be displayed as a floating point data. *Note that each output is displayed in new lines because of the use of \n character.*

1.8.2 Unformatted I/O Functions

All unformatted I/O function reads and displays only char data type and doesn't use format strings for identifying the type of the data involved. The getch(), getche(), getchar() are single character unformatted input functions, putch(), putchar() are single character unformatted output functions. The gets(), puts() are unformatted string I/O functions respectively.

The getchar() Function

The simplest of all input operations is reading a single character from the standard input. Reading a single character can be carried out by a function called as getchar(). The getchar() function does not require any arguments, though a pair of empty parenthesis follows the word getchar. Thesyntax of getchar() function is,

```c
variable = getchar();
```

where variable refers to the previously declared character variable. When the above statement is encountered the computer waits until a key is pressed and assigns the character as a value to the getchar() function which is assigned to the character variable on the left hand side of getchar() function. But the getchar() function needs an ENTER key to be pressed to assign the character to the variable.

The getch() Function

The getch() *function is similar to* getchar() *function which reads a single character the instant it is typed without waiting for the* ENTER *key to be pressed.* It will not display the character in the screen, which the user has typed.

The getche() Function

The getche() *function is similar to* getch() *function which reads a single character the instant it is typed without waiting for the* ENTER *key to be pressed.* It echoes or displays the character in the screen, which the user has typed.

The gets() Function

In C the gets() *function is used to receive a string from the standard input device until a new line or eof (end of file) is encountered.* The general form or the syntax of the gets() function is

```
gets(variable);
```

where variable refers to the previously declared character array. The string received by the gets() function may include white space characters. The gets() function will terminate only with a new line character. Note that the gets() function can receive data for only one variable.

The putch() Function

The putch() *function displays a single character the instant it is typed without waiting for the* ENTER *key to be pressed.* It displays the character in the screen, which the user has typed.

The putchar() Function

The simplest of all output operations is displaying a single character in the standard output. Displaying a single character can be carried out by a function called as putchar(). The character variable must be expressed as an argument to the function enclosed in parenthesis in the word putchar. The general form or the syntax of putchar() function is

```
putchar(variable);
```

where variable refers to the previously declared character variable. The putchar() function can be used to display a string in the standard output within a one dimensional, character-type array. Each character can then be stored seperately within a loop.

The puts() Function

In C *the* `puts()` *function is used to display a string at a time in the standard output device followed by a new character.* The general form or the syntax of `puts()` function is

```
puts(argument);
```

The `puts()` function accepts only a single argument. The argument passed within a function must be a data item. It represents a string or a character array. The string displayed by the `puts()` function may include white space characters. The `puts()` function will display only one string at a time.

1.9 Decision Making and Branching Statements

In case of C programs, that we have seen so far, the statements are normally executed sequentially as they appear in the program. By default, the statements in the program are executed sequentially. However, in practice we have situations where we may have to change the order of execution of statements until specified conditions are met. These kinds of situations are dealt in C using decision control statements.

One of several possible actions will be carried out depending upon the outcome of the logical test is referred as **branching***. Some portion of the program has to be executed several number of times or until a particular condition is being met. This is referred as* **looping***.* The different decision control or decision-making statements in C are

- The if statement
- The if-else statement
- The switch statement
- The conditional operator
- The break, continue and goto statements.

1.9.1 The if Statement

C *uses the keyword* `if` *to implement decision control statements.* The if statement can be used to check some conditions and execute one or more statements when specified conditions are met. The general form or the syntax of an `if` statement is

```
if(condition)
{
    executable-X-statement(s);
}
executable-Y-statement(s);
```

Fig. 1.3, shows the operation of an `if` statement. The `condition` used in an `if` statement should be specified within parenthesis. When a program with `if` statement is evaluated the `condition` specified after the `if` statement is executed first. If the `condition` is true then the `executable-X-statement(s)` within the flower braces

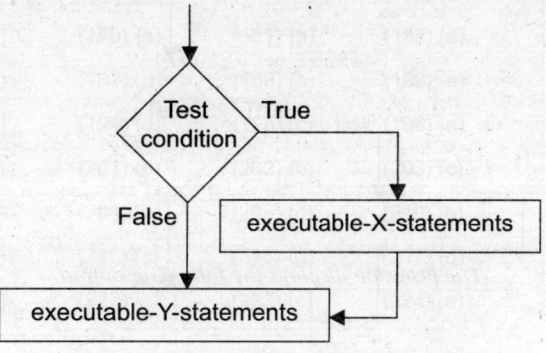

Fig. 1.3 : Flowchart of if statement.

will be executed. If the `condition` is false, the control will be directly transferred to the statement outside the flower braces and executes `executable-Y-statement(s)` without executing `executable-X-statement(s)`. The `executable-X-statement(s)` referred as the body of the `if` statement, should be enclosed within flower braces. ***Note that if the body of the if statement (i.e., executable-X-statement(s)) consists of only one statement the flower braces enclosing them is not necessary.*** If the expression evaluates to zero, it is treated as false, and if the expression evaluates to a non zero value, it is treated as true. *Table 1.14*, shows various expressions that can be used as conditions inside an `if` statement.

Table 1.14 : Single Conditional Expressions and its Meaning

Conditional expression	Meaning
a == b	a is equal to b
a != b	a is not equal to b
a < b	a is less than b
a > b	a is greater than b
a <= b	a is less than or equal to b
a >= b	a is greater than or equal to b

Note that = is used for assignment, where as = = is used for comparison of two quantities. Two or more conditions may be combined in an `if` statement using a logical AND operator (`&&`) or a logical OR operator (`||`). The `if` statement can compare any number of variables in a single `if` statement. ***Program 1.5***, helps you to understand things better.

Program 1.5 :

```
/* Program to find the biggest of two numbers using an if statement */
/* big_num.c */
#include<stdio.h>
void main()
{
   int value1, value2, big = 0;
   printf("Enter two values : ");
     scanf("%d %d", &value1, &value2);
   big = value 1;
   if(value1 < value2)
     {
       big = value2;
     }
   printf("The biggest number is %d", big);
}
```

The program displays the following output

```
Enter two values : 8   21
The biggest number is 21
```

As soon as the user enters 8 and 21 it is assigned to the variables value1 and value2 respectively. The value stored in the variable value1 is assigned to the variable big (i.e., 8 is assigned to the variable big). The if statement checks the condition, whether value1 is lesser than value2 or not (i.e., 8 < 21). Since the condition is true in this case, value2 is assigned to the variable big (i.e., 21 is assigned to the variable big), and the printf() function displays the message.

1.9.2 The if-else Statement

The if-else *statement is an extension of the* if *statement to include execution of one or more statements for false condition also.* The if statement by default will execute a single statement or a group of statements when the condition is true, but if does nothing when the condition is false. *The* else *statement executes a single statement or a group of statements when the condition is false.* The general form or the syntax of an if-else statement is

```
if(condition)
 {
    executable-X-statements(s);
 }
else
    executable-y-statements(s);
 }
```

Fig. 1.4, shows the operation of the if-else statement. The condition used in the if statement should be specified within parenthesis. If the condition is true the statement following if until else will be executed. If the condition is false, the statement(s) following the else will be executed. *Program 1.6*, helps you to understand things better.

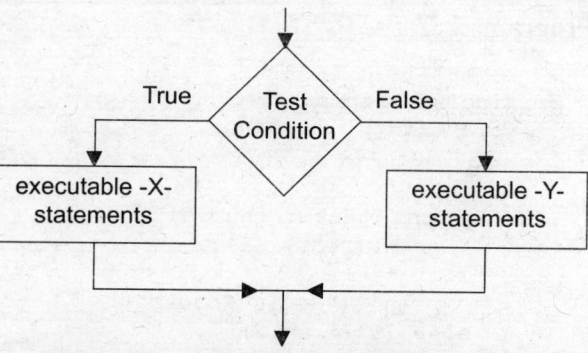

Fig. 1.4 : Flowchart of if-else statement.

Program **1.6** :

```
/* Program to check whether the no is odd or even using an if-else statement */
/* odd_even.c */
#include<stdio.h>
void main()
 {
    int number;
    printf("Enter a number : ");
      scanf("%d", &number);
    if(number % 2 == 0)                        /* if remainder is zero */
```

```
        {
            printf("The number is even");
        }
        else
        {
            printf("The number is odd");
        }
    }
```

The program displays the following output

```
Enter a number : 258
The number is even
```

As soon as the user enters 258 it is assigned to the variable number. The if statement checks the condition, whether the remainder of the variable is zero or not. Since the condition is true in this case, the statement in the if block is executed. The else block is skipped and the program ends.

Multiple if-else Statements

Multiple if-else statement provide a method of testing more than one condition and execute one or more statements for each condition. Multiple if-else statements are much faster than a series of if-else statements, since the if structure is exited when any one of the condition is satisfied.

Program 1.7 :

```
/* mul_if.c */
#include<stdio.h>
void main()
{
    int i;
    printf("Enter the value of i : ");
    scanf("%d", &i);
    if(i == 1)
        printf("Hello Chocks");
    else if (i == 2)
        printf("Hello Girish");
    else if (i == 3)
        printf("Hello Madhu");
    else if (i == 4)
        printf("Hello Mani");
    else
        printf("Hello Moorthy");
}
```

The program displays the following output

RUN 1	Enter the value of i : 2
	Hello Girish
RUN 2	Enter the value of i : 7
	Hello Moorthy

In run1, as soon as the user enters 3 it is assigned to the variable i. The if statement checks the condition, whether i is equal to 1 or not (i.e., if(i==1)). Since the condition is false, control is transferred to next if statement (i.e., else if(i==2)) and the condition, whether i is equal to 2 or not is checked. Since the condition is true, the statement following it is executed. The printf() function displays the message as

```
Hello Girish
```

and all other else if statements are ignored and the program ends.

In run2, since none of the condition is true, all the if statements are ignored and the control is transferred to the else block. The printf() function displays the message

```
Hello Moorthy
```

and the program ends. ***Note that any number of if-else statements can be included in a program, but the conditions specified inside each if statement should be different.***

Nested if-else Statements

One or more if-else *statements inside an* if-else *statement is called as **nested if** statements. **Program 1.8**,* helps you to understand things better.

Program 1.8 :

```
/* nest_if.c */
#include<stdio.h>
void main()
 {
   int choice, yrs_of_exp;
   printf("\nEnter 1 for Project Leader");
   printf("\nEnter 2 for Software Engineer");
   printf("\nEnter your choice: ");
     scanf("%d", &choice);
   printf("Enter Years Of Experience : ");
     scanf("%d", &yrs_of_exp);
   if(choice == 1)
   {
     if(yrs_of_exp > 5)
       printf("\nSalary Hike is 7500");
     else
       printf("\nSalary Hike is 6000");
   }
   else if(choice == 2)
   {
     if(yrs_of_exp > 3)
       printf("\nSalary Hike is 5000");
     else
       printf("\nSalary Hike is 4000");
   }
```

```
        else
           printf("\nInvalid Choice ...");

    }
```

The program displays the following output
```
    Enter 1 for Project Leader
    Enter 2 for Software Engineer
    Enter your Choice : 1
    Enter Years Of Experience : 8
    Salary Hike is 7500
```

When the user enters 1 (which indicates `Project Leader`) and hence the control enters the first `if` statement. The nested if's inside the first `if` statement checks the condition. Since the years of experience (i.e., `yrs_of_exp`) is greater than 5 the `printf()` function displays the message as

```
    Salary Hike is 7500
```

and the control from the nested `if` is exited and simultaneously the control from the main if is also exited and the program ends.

1.9.3 The switch Statement

The `switch` *statement allows us to make a decision from a number of choices. It is usually referred as the* ***switch-case*** *statement.* The general form or the syntax of `switch` statement is

```
    switch(expression)
      {
        case constant1 :
                        statement(s);
        case constant2 :
                        statement(s);


             .                .
             .                .
             .                .


        case constantN :
                        statement(s);
        default        :
                        statement(s);
      }
```

The `switch` statement causes a particular group of statements to be chosen from several available groups. The selection is based upon the current value of an expression, which is included within the `switch` statement. Each `case` is labeled by one or more integer valued expressions. If a `case` matches the expression value, execution starts at that `case`, and ends when it encounters a `break` statement. All `case` expressions must be different. The `case` labeled `default` is executed, if none of the other cases are satisfied. A `default` is optional in a `switch` statement. ***Program 1.9***, helps you to understand things better.

Program 1.9 :

```
/* Program to demonstrate the use of switch statement */
/* switch.c */
#include<stdio.h>
void main()
 {
    int a, b, choice;
    printf(" Enter two numbers : ");
      scanf("%d %d", &a, &b);
    printf("\n Enter 1 for addition : ");
    printf("\n Enter 2 for subtraction : ");
    printf("\n Enter 3 for multiplication : ");
    printf("\n Enter 4 for division : ");
    printf("\n Enter your choice : ");
      scanf("%d", &choice);
    switch(choice)
     {
       case 1:
               printf("Sum is %d", a+b);
               break;
       case 2:
               printf("Difference is %d", a-b);
               break;
       case 3:
               printf("Product is %d", a*b);
               break;
       case 4:
               printf("Quotient is %d", a/b);
               break;
       default:
               printf("Invalid choice");
     }
 }
```

The program displays the following output

```
Enter two numbers : 5  8
Enter 1 for addition
Enter 2 for subtraction
Enter 3 for multiplication
Enter 4 for division
Enter your choice : 1
Sum is 13
```

As soon as the user enters 5 and 8 it is assigned to the variables a and b respectively. The user has entered his choice as 1 and hence the control is transferred directly to the first case statement inside the switch statement, which has the value 1. The printf() function inside the case statement is executed with the sum of a and b as

    Sum is 13

and the break statement below the printf() function transfers the control to the end of switch-case statement.

A *fall through* in a `switch` statement is a case that does not include a `break` statement, there by causing control to the next `case` statement. *Program 1.10*, helps you to understand things better.

Program **1.10** :

/* Program to demonstrate the use of switch without break */

```
/* no_break.c */
#include<stdio.h>
void main()
 {
   int choice = 2;
   switch(choice)
     {
       case 1:  printf("\n This is Case 1 ...");
       case 2:  printf("\n This is Case 2 ...");
       case 3:  printf("\n This is Case 3 ...");
       default: printf("\n This is Default ...");
     }

 }
```

The program displays the following output
```
This is Case 2 ...
This is Case 3 ...
This is Default ...
```

Output of *Program 1.10*, is definitely not what we have expected. We have expected the `printf()` function in `case 2` to be executed but not, other lines. This has happened because we haven't included a `break` statement at the end of each `case` statement and hence the subsequent `case` statements including the `default` statement have been executed.

Limitations of using a switch statement

- Only one variable can be tested with the available case statements with the values stored in them (i.e., you cannot use relational operators and combine two or more conditions as in the case of if or if-else statements).
- Floating-point, double, and long type variables cannot be used as cases in the switch statement.
- Multiple statements can be executed in each case without the use of a pair of braces as in the case of if or if-else statement.

Important points to be noted while using a switch statement

- Character values are also allowed in cases. You can also mix integer and character constants in different cases of switch.
- Integer expression used in different case statements can be specified in any order.

- A switch may occur within another switch, but it is rarely done. Such statements are called as **nested switch** statements.

- A default is optional. If it is not present and if none of the other cases match no action takes place. The default may appear any where in the switch statement, but there should be only one default in a switch-case statement.

- The switch statement is very useful while writing menu driven programs.

1.9.4 The Conditional Operator

*An expression that evaluates conditions is called as **conditional expressions**. The operator used to evaluate conditional expressions is called as a conditional operator (?:).* It's a simple form of an `if-else` statement. The conditional operator is also referred as a ternary operator, since it takes three operands as its arguments. The general form or the syntax of the conditional operator is

```
expr1 ? expr2 : expr3;
```

The `expr1` is the condition for evaluation. When evaluating ternary operator, `expr1` is evaluated first. If `expr1` is true, `expr2` is alone evaluated. If `expr1` is false, `expr3` is alone evaluated. A sample conditional expression is shown below.

```
a = ( (b < 2) ?  8 : 6 );
```

If `b` is lesser than 2, then the value of `a` will be 8. If `b` is greater than 2, then the value of `a` will be 6. *Program 1.11*, helps you to understand things better.

Program 1.11 :

/* **Program to find the biggest of two numbers using a ternary operator** */

```
/* ternary1.c */
#include<stdio.h>
void main()
 {
    int a, b, c;
    printf("Enter the value of a and b : ");
      scanf("%d %d", &a, &b);
    c = ( (a > b) ? a : b );
    printf("The biggest value is %d", c);
 }
```

The program displays the following output

```
Enter the value of a and b : 125   100
The biggest value is 125
```

As soon as the user enters 125 and 100, it is assigned to the variables a and b respectively. The first expression in the conditional statement (i.e., a > b) evaluates to true, (since the value of a is greater than the value of b) and hence, the value of the second expression (i.e., a) is assigned to the variable c and the `printf()` function displays the message and the program ends. *Note that it is not necessary that only arithmetic statements should be evaluated in the conditional operator. They can also evaluate statements depending upon the value returned by the first expression.*

1.9.5 The break, continue and goto Statements

The break Statement

The `break` *statement is used to terminate or to exit at any point of* `switch` *statement.* The general form or the syntax of the `break` statement is

```
break;
```

The `break` statement do not have any expressions or arguments. The `break` statement causes the control to transfer out of the entire `switch` statement. The `break` statement can be also used within a `for`, `while` or `do-while` loops. The keyword `break` breaks the control only from the loop in which it is placed. Refer *Programs 1.9 and 1.10* for the usage of break statement.

The continue Statement

The `continue` *statement is used to transfer the control to the beginning of the loop, there by terminating the current iteration of the loop and starting again from the next iteration of the same loop.* The `continue` statement can be used within a `while` or a `do-while` or a `for` loop. The general form or the syntax of the `continue` statement is

```
continue;
```

The `continue` statement does not have any expressions or arguments. The `continue` statement is usually used within an iteration statement. It causes the control to pass to the loop-continuation portion in which the `continue` statement is enclosed. If a `continue` statement is not enclosed within a loop, a syntax error is indicated by the compiler. The `continue` statement applies only to loops and not for `switch` statement.

The goto Statement

The `goto` *statement is used to transfer the control in a loop or a function from one point to any other portion in that program.* If misused, the `goto` statement can make a program impossible to understand. The general form or the syntax of `goto` statement is

```
goto label;
    statement(s);
    . . .
    . . .
label:
    statement(s);
```

The `goto` statement requires a `label` in order to identify the place where the control is to be transferred. A `label` may be any valid variable name. The `label` must be followed by a colon (`:`). The `label` can be placed anywhere in the program either before or after the `goto` statement. The `goto` statement breaks the normal sequential execution of a program. It merely instructs the computer to resume execution at a specified statement, which is specified as a `label`. The naming rules for labels are the same as those for variables except that `label` is not declared.

1.10 Looping Statements

*A segment of the program that is executed repeatedly is called as a **loop***. Some portion of the program has to be executed several number of times or until a particular condition is satisfied. Such repetitive operation is done through a loop structure. The three loop structures are,

- while loop
- do-while loop
- for loop.

1.10.1 The while Loop

The simplest of all the looping structures in C is the `while` loop. The general form or the syntax of a `while` loop is

```
initialize loop counter;
while(condition)
  {
    statement(s);
    increment or decrement loop counter;
  }
```

Fig. 1.5, shows the operation of the `while` loop. A looping process generally has three steps. They are,

- Initialization of the counter
- Test for a specified condition for execution of the statements in the loop
- Incrementing or decrementing the counter.

In looping, *a sequence of statements is executed until some conditions are satisfied.* A loop consists of two segments. One is referred as the ***body of the loop*** and the other is the ***control statement.*** The control statement tests certain conditions and then directs the repeated execution of the statement(s) contained in the body of the loop structure.

When the `condition` specified inside the parenthesis of `while` loop is satisfied, the control is transferred to the statement(s) inside the loop and executes the body of the

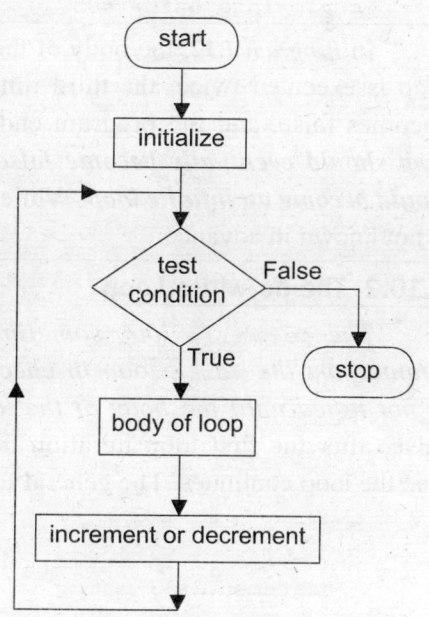

Fig. 1.5 : Flowchart of the while loop.

loop and increments or decrements the loop counter. The loop continues until the condition is violated. *The* while *loop tests the condition before each iteration.* If the condition initially fails, the loop is skipped entirely even in the first iteration itself. *Program 1.12*, helps you to understand things better.

Program 1.12 :

```
/* Program to calculate the simple interest */
/* simple.c */
#include<stdio.h>
void main()
{
    int p, n, count = 1;
    float r, si;
    while(count <= 2)
    {
        printf("\n Enter P, N and R : ");
            scanf("%d %d %f", &p, &n, &r);
        si = (p * n * r) / 100;
        printf("\n Simple interest = %f",si);
        count++;
    }
}
```

The program displays the following output

```
Enter P, N and R : 3000   2    1.5
Simple interest = 90

Enter P, N and R : 4500   4    2.75
Simple Interest = 495
```

In *Program 1.12*, the body of the loop is executed after checking the condition. The loop is executed twice, the third time the while loop tests the condition, it eventually becomes false, and the program ends. *Note that the test condition specified in while loop should eventually become false at one point of the program, otherwise the loop would become an infinite loop.* While loops are generally used when the number of passes is not known in advance.

1.10.2 The do-while Loop

The do-while *loop sometimes referred to as the* do loop *differs from its counterpart the* while *loop in checking the* condition. *The* condition *of the loop is not tested until the body of the loop has been executed once.* If the condition is false, after the first loop iteration the loop terminates. However, if the condition is true the loop continues. The general form or the syntax of the do-while loop is

```
do
{
    statement(s);
}while(condition);
```

In `do-while`, the statements would be executed at least once even if the condition fails for the first time itself. The operation of the `do-while` loop can be stated from the flowchart in *Fig. 1.6. Program 1.13*, helps you to understand things better.

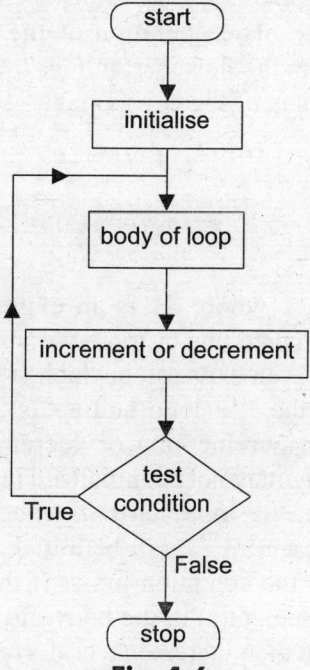

Fig. 1.6 :
Flowchart of do-while loop.

Program 1.13 :

```c
/* do_while.c */
#include<stdio.h>
void main()
{
    char yes;
    int a, b;
    do
     {
        printf("Enter two numbers : ");
          scanf("%d %d", &a, &b);
        printf("SUM = %d", a+b);
        printf("\n Do u want to
                continue say Y/N : ");
          scanf("%c", &yes);
     }while((yes == 'Y')||(yes == 'y'));
}
```

The program displays the following output

```
Enter two numbers : 5  5
SUM = 10
Do u want to continue say Y/N : Y

Enter two numbers :   6  3
SUM = 9
Do u want to continue say Y/N : N
```

The output of *Program 1.13*, clearly states the function of a `do-while` loop. The control enters the loop without testing any condition and the body of the loop is executed once. The `while` statement checks the condition, since it is true (i.e., `Y`) control is transferred again to the starting of the `do` loop and the body of the loop is again executed. This process continues until the condition is false (i.e., `N`). *Note that semicolon is placed after while since it is considered as statement in the do-while loop.*

1.10.3 The for Loop

The `for` loop is most commonly and popularly used loop in C. *The `for` loop allows us to specify three things about the loop in a single line*. They are,

- Initializing the value for the loop
- Condition in the loop counter to determine whether the loop should continue or not
- Incrementing or decrementing the value of loop counter each time the program segment has been executed.

The operation of the `for` loop can be stated from the flowchart in *Fig. 1.7*. The general form or the syntax of the `for` loop is

```
for(e1 ; e2 ; e3)
  {
    statement(s);
  }
```

where `e1` is an expression which is used to initialize one or more variables used in the `for` loop. `e2` is an expression which is used to test the condition in the `for` loop and `e3` is another expression used for incrementing or decrementing the index which may/may not be initialized in the expression `e1`. When the `for` loop starts its execution, the expression(s) present in `e1` will be initialized first and then checks for the condition in `e2`. If the condition satisfies, the statement(s) in the body of the loop is executed once and then increments or decrements the index as in `e3`. The looping action continues as long as the condition

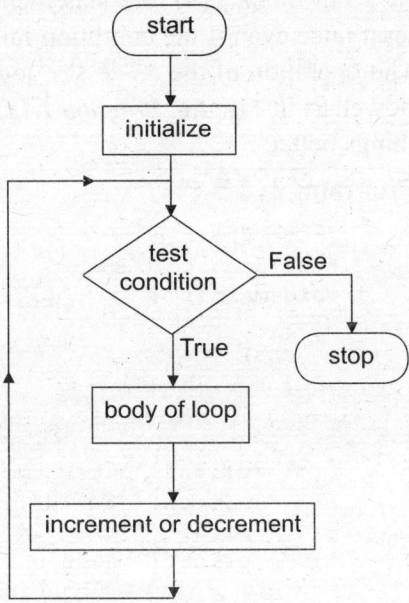

Fig. 1.7 : Flowchart of for loop.

in `e2` is true. *Note that the control goes to e1 only when it enters the loop for the first time only, after that depending upon the condition e2 and e3 are executed alternatively. Note that no semicolon is placed after for since it is a loop and not a statement. Program 1.14,* helps you to understand things better.

Program 1.14 :

```
/* Program to find the sum of first N natural numbers */
/* sum_num.c */
#include<stdio.h>
void main()
  {
    int i, n, sum = 0;
    printf("Enter the limit : ");
      scanf("%d", &n);
    for(i = 1 ; i <= n ; i++)
      {
        sum = sum + i;
      }
    printf("\n The sum of first %d numbers is %d", n, sum);
  }
```

The program displays the following output

```
Enter the limit : 5
The sum of first 5 numbers is 15
```

In *Program 1.14*, the user as entered `5` which is assigned to the variable `n`. The `for` loop is incremented from `1` to `5` (i.e., until `n = 5`) and the sum of `5` natural numbers is added and stored in the variable `sum`. The result is displayed using the `printf()` function.

1.11 Arrays

C *language provides the capability that enables the user to design a set of similar data types called **arrays**.* They are defined in the same manner as ordinary variables except that each array name must be accompanied by a size specification within square braces capable of storing more than one value at a time. An array is defined as a set of homogeneous data items of the same type that share a common name and that are differentiated from one another by their positions within the array. *The individual values in the array are called as **elements**. Each array element is referred by specifying the array name, followed by a number within square braces referred as an **index** or **subscript**.* The dimension of an array is determined by the number of subscripts needed to identify each element. A subscript is always enclosed in square braces [ ], placed after the array name. Each subscript must be expressed as a non-negative integer. The value of each subscript can be expressed as an integer constant, integer variable, or a complex integer expression.

Arrays are structured type of variables, for which, memory is allocated for all the elements declared by the size specification. The ability to use a simple name to represent a collection of items and to refer to an item number enables us to develop efficient programs.

1.11.1 Initialization and Declaration of Arrays

Arrays refer to a group of data items that share a common name with the same data type. Individual data items in an array are referred as elements. Each element in an array is accessed by a subscript, which begins from zero and is always written inside a pair of square brackets. For example,

```
int a[10];
char ab[10][10];
```

The first declaration declares a one dimensional integer array a that can store 10 integer constants (i.e., 0-9). The next declaration declares a two dimensional character array ab that can store 10 string constants of length not more than 10.

Arrays can also be initialized when they are declared. This is done by adding an equal sign followed by the required value to be initialized after the declaration. For example,

```
int a[5] = {1, 2, 3, 4, 5};
char name[6] = {'M', 'a', 'd', 'h', 'u', '\0'};
char name[6] = {"Madhu"};
```

The first declaration declares a one dimensional array a of type integer and initializes the values 1 in a[0], 2 in a[1] and so on up to 5 in a[4] *(Note that the initialization of an array starts from the subscript zero).* The next two initializations are more or less similar, which declares a one dimensional array name of type char. The first declaration initializes a string "Madhu" character by character where as the next declaration directly initializes the string *(Note that a NULL character is compulsary while initializing a string character by character).*

1.12 One Dimensional Arrays

Arrays whose elements are specified by a single subscript are called as **one-dimensional** *or* **single dimension** *or* **single-subscripted** *or* **linear arrays**. The arrays that we have seen so far are all one dimensional arrays. A list of items can be given a variable name using only one subscript and such a variable is called as a one dimensional array also referred as a list. The main purpose of declaration of an array is to reserve space for the elements in memory. The general form or the syntax of an array declaration is

```
storageType dataType arrayName[size];
```

The `storageType` while declaring an array is optional. The `storageType` may be either `auto`, or `register`, or `static`, or `extern`. Refer Section 2.12 for storage classes. The `dataType` specifies the type of element that will be stored in the array. The `arrayName` is the name used to identify the array. The rules for naming arrays are same as identifiers. The `size` is a positive integer constant, indicating the maximum number of elements that can be stored in the array. Each subscript must be expressed as a non-negative integer. In an n-element array, the array elements are stored in $x[0], x[1], \ldots, x[n-2], x[n-1]$ as shown in *Fig. 1.8.*

Fig. 1.8 : Memory allocation of a one dimensional array.

1.12.1 Declaring Arrays

Like ordinary variables, an array variable should also be declared. The general form or the syntax of declaring an array is

```
datatype arrayName[subscript];
```

where `arrayName` is a valid C variable of type `datatype` and `subscript` is an integer constant indicating the maximum number of elements that can be stored in the array. For example the declaration,

```
int days[31];
```

is read as "`days`, an array of 31 `int` values". In the above declaration, `int` specifies the data type of the variable, just as it does with ordinary variables and the word `days` specifies the name of the array. The number 31 is the size of the array, which tells that 31 elements of type `int` is to be stored in the array. This number is referred as the dimension of the array. The square brackets `[]` tell the compiler, that we are dealing with arrays. *Note that an array should not have the same name as an ordinary variable in the same block or function.*

```
int value[50];
char book[102];
extern float amount[20];
```

In the first example, `int` specifies the type of the variable, and the word `value` specifies the name of the array. The number 50 within square braces indicates the maximum number of elements to be stored in the array. The square braces indicate that the variable is an array. Similarly in the second example, `book` is a character array with a size of `102` elements. In the third example, `extern` is used before the data type `float` to indicate that the array `amount` is a global array variable.

Properties of an array

- The type of an array is the data type of its elements.
- The location of an array is the location of its first element.
- The length of an array is the number of data elements in the array.
- The size of the array is the length of the array times the size of the elements, but the size of the array is usually referred as the product of subscripts used.

Important points to be noted while using arrays

- An array name should be a valid C identifier.
- The name of the array should be unique, similar to other variables.
- The values of the elements stored in the array should be of the same type (i.e., if an array is declared as int, it can store elements that are only int).

1.12.2 Accessing Array Elements

Once an array is declared, let us see how individual elements in the array are accessed. This is done with the help of subscripts (i.e., the number specified in the brackets following the array name). The subscript specifies the element's position in the array. All the array elements are numbered, starting from zero. The first item is stored at the address pointed by the array name itself. This can also be referred as position 0. Thus, the first element of the array `days` is referred as `days[0]`. The fourth element of the array is referred as `days[3]`. The ability to use variables as subscripts makes arrays so useful in writing programs. *Program 1.15* helps you to understand arrays better.

Program **1.15** :

/* Program to find the biggest of 10 numbers using arrays. */

```
/* big_10.c */
#include<stdio.h>
void main()
  {
    int value[10], i, big = 0;
    printf("Enter any 10 numbers : ");
      for(i = 0 ; i < 10 ; i++)
        scanf("%d", &value[i]);
    big = value[0];
    for(i = 0 ; i < 10 ; i++)
```

```
        if(big < value[i])
            big = value[i];
        printf("The biggest number is %d", big);
    }
```

The program displays the following output

```
Enter any 10 numbers : 38  26  14  53  62  71  84  11  97  47
The biggest number is 97
```

Program 1.15 we are using an array named `value` to hold `10` numbers to find the greatest of `10` numbers. Without the use of arrays, this program would have used 10 different variables say from a, b, ..., i, j. In ***Program 1.15*** variable i is used as a subscript to refer the elements of the array. The first `for` loop used in the program reads the numbers from the user `10` times. For the first time the subscript value i will be equal to zero (i = 0) and hence the `scanf()` function stores the value entered by the user in the array element `value[0]`, the first element of the array. This process is repeated until i becomes 9. The next `for` loop in the program is used for performing the operation on the body of the loop for finding the greatest of `10` numbers and the result is displayed using the `printf()` function.

Important points to be noted while using subscripts

- A subscript value should not be negative.
- A subscript must always be an integer or an expression that gives integer.
- A subscript must be specified within the square brackets preceding the array name.
- If there are more than one subscript as in the case of multi dimensional arrays each subscript must be specified within seperate square brackets.

1.12.3   Initializing One Dimensional Arrays

So far we have seen, how to read the values of array elements from the user, but we can also initialize an array as an ordinary variable. This is because the reserved storage location for an array contains garbage values. Arrays can be initialized by any one of the following ways.

- Initializing an array during declaration
- Initializing an array using loops.

Initializing an array during declaration

The general form or the syntax of initializing an array during declaration is

```
storageType dataType arrayName[size] = {list of values};
```

where the `storageType` is optional. The `dataType` specifies the type of element that will be stored in the array. The `arrayName` is the name used to identify the array. The `size` is an integer constant, indicating the maximum number of elements that can be stored in the array. In this case `size` is optional. The `list of values` are the values seperated by commas used for initializing the array.

```
int mark[7] = {100, 45, 75, 88, 63};
char name[8] = {'s', 'u', 'd', 'h', 'a', 'r', '\0'};
float amount[] = {12.3, 234.56, 34.56, 5678.90};
```

In the first example, the size of the array is 7 and the values 100, 45, 75, 88, 63 are assigned to the variable mark[0],mark[1],..,mark[4] respectively. In the second example the number of values in the array is less than the size of the array and so the values 's', 'u', 'd', 'h', 'a', 'r' will be assigned to the variable name[0],name[1],.., name[5] and the remaining variables name[6] and name[7] will be set to zero automatically. In the third example, the size of the array is omitted. In such cases, the compiler allocates enough memory space for all the elements given for initialization. Hence, the size of the array amount[] will be 4.

Fig. 1.9 shows the memory occupied by the array mark in the first example. The address occupied by the elements of the array mark are displayed in hexadecimal values. Since the array is of type int each element in the array occupies 2 bytes of memory. Whatever be the initial values, all the array elements would be stored in contiguous memory locations. Whether the array is initialized or not, once it has been declared, memory will be allocated for them continuously depending upon the data type of the array. *Note that the array elements, which are not initialized, are set to zero.*

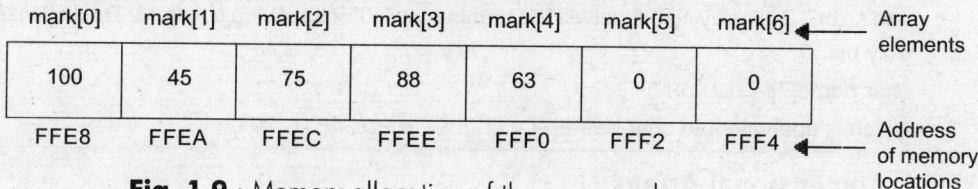

mark[0]	mark[1]	mark[2]	mark[3]	mark[4]	mark[5]	mark[6]	Array elements
100	45	75	88	63	0	0	
FFE8	FFEA	FFEC	FFEE	FFF0	FFF2	FFF4	Address of memory locations

Fig. 1.9 : Memory allocation of the array mark.

Initializing an array using loops

An array can be initialized by using a for loop, a while loop or a do-while loop. The following examples illustrate the use of initializing an array using for loops.

Examples for initializing an array using a for loop

```
1.    int i, mark[50];
      for(i = 0; i < 50; i++)
         {
            mark[i] = 0;
         }
2.    int i, mark[50], x = 5;
      for(i = 0 ; i < 50 ; i++)
         {
            mark[i] = x;
            x += 5;
         }
3.    int i, mark[50];
      for(i = 0 ; i < 50 ; i++)
         {
            scanf("%d", &mark[i]);
         }
```

In the first example, the `for` loop initializes the value of all the elements of the array to zero. In the second example, the `for` loop initializes the value 5 to `mark[0]`, 10 to `mark[1]` and so on. In the third example, the `scanf()` function reads values from the user for the array. Similar to the `for` loop, initialization of an array can be carried out using `while` and `do-while` loops.

Rules for initializing an array

- If the number of initializers is less than the number of elements in the array, the remaining elements are set to zero. For example, int value[5] = { 21, 5, 79 };
- It is an error to specify more initializers than the number of elements in the array. For example, int value[2] = { 4, 3, 2, 1, 5, 7 };
- It is not necessary to specify the length of an array, explicitly in case if initializers are provided for the array during declaration itself. For example, int value[] = { 5, 4 };
- A character array can be initialized by a string constant, resulting in the first element of the array being set to the first character in the string, the second element to the second character and so on. The array also receives the terminating ' \0' in the string constant. The initialization may be

  char name[7] = "GANDHI";

  which is equivalent to char name[7] = {'G', 'A', 'N', 'D', 'H', 'I', '\0'};

1.13 Two-Dimensional Arrays

Arrays whose elements are specified by two subscripts are referred as ***two-dimensional arrays*** *or* ***two dimensional arrays.*** A two-dimensional array is defined in the same manner as a one dimensional array except that it requires two pairs of square brackets for two subscripts. The two-dimensional array is also referred as a ***matrix***. Each subscript must be expressed as a non negative integer. The general form or the syntax of a two-dimensional array is

```
storageType dataType arrayName[rowsize][colsize];
```

where the `storageType` may be either `auto`, or `register`, or `static`, or `extern`. The `storageType` while declaring an array is optional. The `dataType` specifies the type of element that will be stored in the array. The `arrayName` is the name used to identify the array. The rules for naming array names are same as identifiers. The `rowsize` and `colsize` are positive value integer constants. These two subscripts indicate, the number of array elements associated with each subscript. Two square braces are used to indicate the values of two subscripts. For example,

```
int mark[2][2];
```

defines a table as an integer array having 2 rows and 2 columns. An array element starts with an index zero and so the individual elements of the array will be

$$\text{int mark} \begin{bmatrix} [0][1] & [0][1] \\ [1][0] & [1][1] \end{bmatrix}$$

The first element is stored in the memory location [0][0], followed by the second element in [0][1] and so on, and the last element will be stored in the memory location [1][1]. The total number of elements of a two-dimensional array can be calculated from the product of number of rows and the number of columns. An m×n, two-dimensional array can have a table of values having m rows and n columns as shown in **Fig. 1.10.**

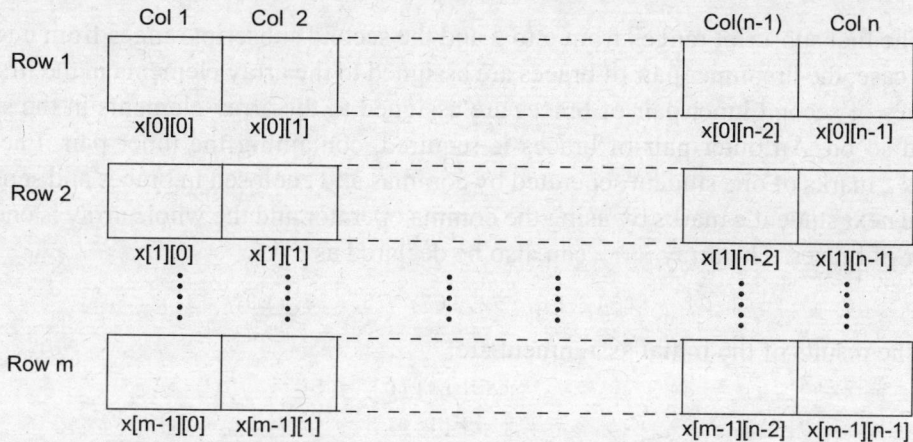

Fig. 1.10 : Memory allocation of a double-dimensional array.

The product of rowsize and colsize indicates the maximum number of elements of the array. It is not compulsary that the rowsize and colsize should be equal. Care must be taken to initialize the values of the array elements. The colsize subscript increases most rapidly and the rowsize subscript increases least rapidly (i.e., all the elements of the first row will be assigned and then all the second row will be assigned and so on).

1.13.1 Initializing Two-Dimensional Arrays

Like one dimensional arrays, a two dimensional array can also be initialized. Arrays can be initialized by the following ways,

- Initializing an array during declaration
- Initializing an array using loops.

Initializing an array during declaration

Similar to one dimensional arrays, a two dimensional array can also be initialized with one or more values during its declaration. The general form or the syntax of initializing an array during declaration is

```
StorageType dataType arrayName[rowsize][colsize] = {List of Values};
```

where the storageType, may be either auto, or register, or static, or extern. The storageType while declaring an array is optional. The dataType specifies the type of elements that will be stored in the array. The arrayName is the name used to identify the array. The rules for naming array names are same as identifiers. *Note that an array should not have the same name as an ordinary variable in the same block or function.* The rowsize and colsize are positive value integer constants, which indicate the number of array elements associated with each subscript.

For example,

```
int mark[4][2]= {   {84,  56},
                    {92,  67},
                    {75,  78},
                    {69,  89} };
```

The first subscript ranges from 0 to 3 and the second subscript ranges from 0 to 1. In the first case, the first inner pair of braces are assigned to the array elements in the first row, the values of second inner pair of braces are assigned to the array elements in the second row and so on. An outer pair of braces is required, containing the inner pair. Each line contains 2 marks of one student seperated by commas and enclosed in braces and seperated from the next student's marks by using the comma operator and the whole array is enclosed in a pair of braces. The array mark can also be declared as

```
int mark[4][2] = {84, 56, 92, 67, 75, 78, 69, 89};
```

The results of the initial assignment are,

```
mark[0][0] = 84              mark[1][0] = 56
mark[0][1] = 92              mark[1][1] = 67
mark[0][2] = 75              mark[1][2] = 78
mark[0][3] = 69              mark[1][3] = 89
```

Consider the following initializations of two-dimensional array.

```
int rank[3][2] = {1, 2, 3, 4, 5, 6};
int rank[][2] = {1, 2, 3, 4, 5, 6};
float amount[2][3] = {25.6, 90.2, 89.4, 72.6};
```

In the first example, the size of the array is 6 (i.e., 3*2) and the values 1, 2, 3, 4, 5, 6 are assigned to the variable rank[0][0], rank[0][1],..., rank[2][1] respectively. In the second example, the rowsize is missing. While initializing an array, the second dimension (i.e., colsize) is necessary, but the first dimension (i.e., rowsize) is optional. If the row size is not assigned, the compiler allocates enough memory space for all the elements given for initialization. The first and second examples are the same. In the third example, the number of values in the array is less than the size of the array and so the values 25.6, 90.2, 89.4, 72.6 will be assigned to the variable amount[0][0], amount[0][1], amount[0][2] and amount[1][0] and the remaining variables amount[1][1] and amount[1][2] will be set to zero automatically.

Program 1.16 :

```
/* Program to calculate the sum of all elements in a matrix. */
/* summat.c */
#include<stdio.h>
void main()
 {
    int a[10][10], i, j, m, n, sum = 0;
```

```
      printf("Enter the order of matrix : ");
        scanf("%d %d", &m, &n);
      printf("Enter the elements of matrix : ");
      for(i = 0 ; i < m ; i++)
        for(j = 0 ; j < n ; j++)
          scanf("%d", &a[i][j]);
      for(i = 0 ; i < m ; i++)
        for(j = 0 ; j < n ; j++)
          sum = sum + a[i][j];
      printf("Sum of elements of the matrix is %d", sum);


    }
```

The program displays the following output

```
Enter the order of matrix : 2    2
Enter the elements of matrix : 1    1    1    1
Sum of elements of the matrix is    4
```

In *Program 1.16* we have used an array "a" and assigned 4 (4x2) elements of type int. The program displays the values and addresses of the memory occupied by the elements of the array matrix. Since the array is of type int each element in the array occupies 2 bytes of memory.

Initializing an array using loops

An array can be initialized by using a **for** loop, a **while** loop or a **do-while** loop. The following examples illustrate the use of initializing an array using for loops.

Examples for initializing an array using a for loop

```
1.    int i, j, x = 5, mark[10][5];
      for(i = 0 ; i < 10 ; i++)
        for(j = 0; j < 5 ; j++)
        {
          mark[i][j] = x;
            x += 5;
        }
2.    int i, j, mark[10][5];
      for(i = 0 ; i < 10 ; i++)
        for(j = 0 ; j < 5 ; j++)
            scanf("%d", &mark[i][j]);
```

In the first example, the for loop initializes the value 5 to the variable mark[0][0], 10 to the variable mark[0][1] and so on. In the second example, the scanf() function reads values from the user for the array. Similarly, arrays can also be initialized using while loop and do-while loops.

1.14 Multidimensional Arrays

There are situations where a table of values of same data type has to be stored. In many cases, we can't use two or double dimensional arrays, in such cases we use *multi dimensional arrays*. *Arrays whose elements are specified by three and four subscripts are called as **three-dimensional** and **four dimensional arrays** respectively.* The general form or the syntax of declaring a multi dimensional array is

```
dataType arrayName[subscript1][subscript2],...,[subscriptN];
```

where the number of subscripts denotes the number of dimensions of the array. For example,

```
int  value[10][10][10];
float sales[2][3][4][5];
```

In the first example, `value` is a three dimensional array that can store `1000` (i.e., `10*10*10`) integer type elements. Similarly in the second example, `sales` is a four dimensional array that can store `120` (i.e., `2*3*4*5`) floating-point type elements. Each element of the matrix is identified by the subscripts used. Multi-dimensional arrays are initialized in a similar fashion like one dimensional arrays. The initializers are listed by rows. Braces are used to seperate the list of initializers from one row to the next, and commas placed after each brace. We don't go in detail of multidimensional arrays because we usually deal with one or two dimensional arrays.

1.15 String Array

A string is any group of characters enclosed in double quotes (except double quotes). A string is an array of characters terminated by a NULL *character and so a string can also be referred to as a string array.* In C language, a string is declared and/or defined by using a character array. For example,

```
"Concept of String Functions"
```

is a string. Whenever a string is stored in memory, the compiler automatically inserts a NULL character (\0) at the end of the string. This NULL character indicates the end of the string. The NULL character is the first ASCII character, which has a value of zero. The NULL character can be used as a signal to let the compiler to know that a string is ended. A string not terminated by a '\0' is not really a string, but merely a collection of characters.

1.15.1 Declaring Strings

The general form or the syntax of declaring a string that can hold an array of characters can be declared as

```
char variable_name[size];
```

where `char` is a data type, `variable_name` is the identifier (i.e., character array) and `size` is the maximum number of characters that the string can store, including NULL character.

For example,

```
char name[10], book[20];
```

where `char` is a data type, `name` is the variable and `10` is the size of the variable that can hold a maximum of ten characters including NULL character (i.e., the character array name can store `9` characters from `name[0]` to `name[8]` and the variable `name[9]` is used for storing the NULL character '\0'). Similarly, `book` is an array of characters that can hold a maximum of `20` characters.

When declaring character arrays, an additional memory space is allocated for the NULL character. Each character within double quotes is treated as an element of an array and stored in the memory. When the compiler encounters a string constant "`Franklin`" it appends the string constant with an additional NULL character. The string constant "`Franklin`" can be read using the `scanf()` function as

```
scanf("%s", name);
```

The `scanf()` function usually accepts an address of the variable, since the name of the array, by itself indicates the starting address of the array and hence we don't use an ampersand (&) before the variable name while reading strings. *It is an error if you use an ampersand before variable names when reading strings.* The `%s` option uses a blank space as a delimiter, since it considers the character seperated by a blank space as two seperate strings. *Note that the array name is used without a subscript in the `scanf()` function, since we are reading the array of characters as a string.*

1.15.2 Initializing Strings

A character array can be initialized by a string constant, resulting in the first element of the array being set to first character in the string, the second element to the second character and so on. The array also receives the terminating '\0' in the string constant. The initialization may be

```
char name[7] = "GANDHI";
```

which is equivalent in declaring as

```
char name[] = {'G','A','N','D','H','I','\0'};
```

In the first method of initialization, '\0' is not necessary, since C inserts the NULL character automatically at the end of the string. But in the next method of initialization, NULL termination must be specified at the end, following the list of character elements. Mentioning the size of the array while initializing its value is optional. As we know that each string is an array of characters, we can also initialize the character of the string individually as we always do when we initialize an array. The same string "`GANDHI`" can also be initialized as

```
name[0] = 'G';
name[1] = 'A';
name[2] = 'N';
name[3] = 'D';
name[4] = 'H';
name[5] = 'I';
name[6] = '\0';
```

1.15.3 Reading Strings

There are three ways to read a string from the user through the keyboard. They are,

- By using the scanf() function
- By using the gets() function
- By using loops.

By using scanf() function

The scanf() *function with format string* %s *can be used to read a string from the user.*
For example,

```
main()
{
    char name[25];
    scanf("%s", name);
}
```

The ***disadvantage of scanf ()*** function is that there is no way to read a multi-word string (i.e., string with spaces) into a single character array variable. The scanf() function terminates its input when it encounters a blank space.

For example, if the input for the scanf() function is

```
United Kingdom
```

the string United will be read in to the variable name, and the string kingdom will not be read since the %s option uses a blank space as a delimiter, it considers the string seperated by a blank space as two strings. *Note that the array name is used without subscripts in the scanf() function while reading strings.*

By using gets() function

The gets() *function with the array* name *can also be used to read a string from the keyboard.* The gets() function overcomes the disadvantage of the scanf() function, since it can read a string of any length with any number of blank spaces and tabs. It gets terminated only when an ENTER key is pressed (i.e., only when it encounters a new line character). For example,

```
main()
{
    char name[20];
    gets(name);                    /* String read from a keyboard */
}
```

If the input to be read is "United Kingdom" then the whole string is stored in the array name. It accepts the blank space between "United" and "Kingdom" as a part of the input string. The disadvantage of gets()function is that, it can be used to read only a single string at a time.

For example, the statement

```
gets(name, designation);
```

is not valid. The advantage of using scanf()function is that it can be used to read more than one string at a time. For example,

```
scanf("%s %s %s", name, designation, address);
```

the `scanf()` function is used to read input for the arrays of `name`, `designation` and `address` at a time, which is not possible by using a `gets()` function.

By using loops

We know that a string is an array of characters, hence a string can also be read, character by character from the user by using loops. You can use either a `for` or a `while` or a `do-while` loop. The loops in the following example reads input from the user until the string is terminated by a new line. The subscript `i` in the program increments the string one by one (i.e., character by character) until a new line character is encountered. For example,

```
main()
{
   int i; char name[20];
   for(i = 0; i != '\n'; i++)          /* String read using a for loop */
      scanf("%c", &name[i]);
}
main()
{
   int i = 0;
   char name[20];
   while(i != '\n')                    /* String read using a while loop */
   {
      scanf("%c", &name[i]);
      i++;
   }
}
main()
{
   int i = 0;
   char name[20];
   do                                  /* String read using a do-while loop */
   {
      scanf("%c", &name[i]);
      i++;
   }while(i != '\n');
}
```

Note that ampersand is used before the array name if you are reading a string, character by character.

1.16 String Library Functions

With every C compiler a large set of useful library functions are provided. *Table 1.15* shows the commonly used string functions along with the purpose of their usage.

Table 1.15 : String Functions

Function	Meaning
stpcpy()	Copies a string into another string
strcat()	Appends one string at the end of another string
strchr()	Finds first occurrence of a given character in a string
strcmp()	Compares two strings
strcmpi()	Compares two strings without regard to case ("i" denotes that this function ignores case)
strcpy()	Copies a string into another string
strcspn()	Finds the initial segment of string s1 consists entirely of characters not from string s2.
strdup()	Duplicates a string
stricmp()	Compares two strings without regard to case (identical to strcmpi)
strlen()	Finds length of a string
strlwr()	Converts a string to lowercase
strncat()	Appends first n characters of a string at the end of another string
strncmp()	Compares first n characters of two strings
strncpy()	Copies first n characters of one string into another string
strnicmp()	Compares two strings without regard to case (identical to strcmpi)
strnset()	Sets first n characters of a string to a given character
strpbrk()	Scans a string, s1, for the first occurrence of any character appearing in s2.
strrchr()	Finds last occurrence of a given character in a string
strrev()	Reverse a string.
strset()	Sets all characters of string to a given character
strspn()	Finds the initial segment of string s1 that consists entirely of characters from string s2.
strstr()	Finds first occurrence of a given string in another string
strtok()	Scans s1 for the first token not contained in s2
strupr()	Converts a string to uppercase
strxfrm()	Transforms the string *s2 into the string *s1 for no more than n characters.

From the above list we shall discuss the functions strlen(), strcpy(), strcat(), strcmp(), strrev(), strlwr(), since these are the most commonly used string library functions.

1.16.1 The strlen() function

The strlen()(**string length**) *function returns the count of number of characters stored in a string.* The general form or the syntax of strlen() function is

```
x = strlen(str);
```

where x is an integer variable which receives the length of the string and str is a valid string variable or a constant. *Program 1.17* helps you to understand things better.

Program 1.17 :

```
/* strlen.c */
#include<stdio.h>
#include<string.h>
void main()
 {
    char str[25];
    printf("Enter a string : ");
      gets(str);
    printf("The length of the string is %d", strlen(str));
 }
```

The program displays the following output

```
Enter a string : Surya
The length of the string is 5
```

In *Program 1.17* the strlen() function takes a single argument (i.e., the string entered by the user), and returns the length of the string. *Note that the length of the string doesn't count the null character.* If a string with spaces is entered by the user the strlen() function returns the length of the string including the number of spaces in between the string. For example,

```
Enter a string : Raj Kumar

The length of the string is 9
```

The strlen() function also counts the space in between the string and returns the length of the string as 9.

1.16.2 The strcpy() function

A string cannot be copied directly using an assignment statement (with the exception of pointers). However, it is possible to copy a string element by element using assignment statements. *The strcpy()(**string copy**) function is used to copy the contents of one string to another.* The general form or the syntax of strcpy() function is

```
strcpy(string1, string2);
```

where string1 and string2 are valid string variables or string constants. The strcpy() function takes two arguments for copying. The string in the second argument (i.e., `string2`) is copied to the first argument (i.e., `string1`). For copying one string into another the base address of the source and target strings should be supplied to this function. *Program 1.18* helps you to understand things better.

Program 1.18 :

```c
/* strcpy.c */
#include<stdio.h>
#include<string.h>
void main()
 {
    char str[25];
    char cpy[25];
    printf("Enter a string : ");
      gets(str);
    strcpy(cpy, str);
    printf("\n The source string is %s", str);
    printf("\n The copied string is %s", cpy);
 }
```

The program displays the following output

```
Enter a string : Krishna
The source string is Krishna
The copied string is Krishna
```

In *Program 1.18* the variable `str` is used to read the string as input and it copies the input string to `cpy`. The string in `str` is copied to `cpy`, character by character until it encounters the NULL character. The NULL character is also copied.

1.16.3 The strcat() function

The `strcat()` (*string concatenate*) *function concatenates the source string at the end of the target string.* The `strcat()` function takes two arguments for concatenating. The string in the second argument (i.e., `string2`) is concatenated with the string present in the first argument (i.e., `string1`). The general form or the syntax of `strcat()` function is

```c
strcat(string1, string2);
```

where `string1` and `string2` are valid string variables or string constants. The content of `string2` is concatenated with the content of `string1`. *Program 1.19* helps you to understand things better.

Program 1.19 :

```
/* strcat.c */
#include<stdio.h>
#include<string.h>
void main()
 {
    char str[25], cat[25];
    printf("Enter a string : ");
      gets(str);
    printf("Enter another string : ");
      gets(cat);
    strcat(str, cat);
    printf("The concatenated string is : %s", str);
 }
```

The program displays the following output

```
Enter a string : Rama
Enter another string : Chandran
The concatenated string is : RamaChandran
```

In *Program 1.19* the first string is stored in the variable str and the second string is stored in the variable cat. The string in cat is concatenated with the string in str and the output is displayed.

1.16.4 The strcmp() function

The strcmp() (*string compare*) *function compares two strings to find whether the strings are equal or not.* The general form or the syntax of strcmp() function is,

```
strcmp(string1, string2);
```

where string1 and string2 are valid string variables or string constants. The strcmp() function takes two arguments for comparing and returns an integer. The string in the first argument (i.e., string1) is compared with the string in the second argument (i.e., string2) character by character until it encounters a mismatch between the two strings or end of one of the two strings. *Program 1.20* helps you to understand things better.

Program 1.20 :

```
/* strcmp.c */
#include<stdio.h>
#include<string.h>
void main()
 {
    char str[25], cmp[25];
    int x;
```

```
        printf("Enter a string : ");
          gets(str);
        printf("Enter another string : ");
          gets(cmp);
        x = strcmp(str, cmp);
        if(x == 0)
          puts("Strings are Equal");
        else if(x > 0)
          printf("String %s is greater than string %s", str, cmp);
        else
          printf("String %s is greater than string %s", cmp, str);
    }
```

The program displays the following output

RUN1   Enter a string : Subramanian

        Enter another string : Subramanian

        Strings are Equal

RUN2   Enter a string : Jaikumar

        Enter another string : Sasikumar

        String Sasikumar is greater than string Jaikumar

RUN3   Enter a string : Yuvaraj

        Enter another string : Jebaraj

        String Yuvaraj is greater than string Jebaraj

In run1, as soon as the user has entered two strings, the strcmp() function compares two strings by moving along character by character, since the end of the string is reached without detecting unequal characters the strings are equal. The strcmp() function returns a value zero and hence the condition in the if statement is true and the printf() function displays the message

```
Strings are Equal
```

In run2, as soon as the user has entered two strings, Jaikumar, Sasikumar and since the string Sasikumar is greater than the string Jaikumar the strcmp() function returns a value greater than zero and the condition in the else if statement becomes true and the printf() function displays the message

```
String Sasikumar is greater than string Jaikumar
```

In run3, as soon as the user has entered two strings, Yuvaraj, Jebaraj and since the string Jebaraj is lesser than the string Yuvaraj the strcmp() function returns a value less than zero and the condition in the `else` block (i.e., the `printf()` function) is executed and displays the message

```
String Yuvaraj is greater than string Jebaraj
```

Note that the value returned by the **`strcmp()`** *function is usually the difference between the ASCII values of the unequal characters of the first mismatch between the two strings.*

1.16.5 The strrev() function

The `strrev()` *(string reverse) function is used to reverse a string.* This function takes only one argument. The general form or the syntax of `strrev()` function is

```
x = strrev(string);
```

where `x` is the string variable to hold the reversed string and `string` is the valid string variable or a string constant which stores the string to be reversed. *Program 1.21* helps you to understand things better.

Program 1.21 :

```c
/* strrev.c */
#include<stdio.h>
#include<string.h>
void main()
 {
   char x[20];
   printf("Enter the string : ");
     gets(x);
   printf("The reversed string is : %s", strrev(x));
 }
```

The program displays the following output

RUN1   Enter the string : Aravind
        The reversed string is : dnivarA

RUN2   Enter the string : Gayathri Devi
        The reversed string is : iveD irhtayaG

In *Program 1.21* the input string is stored in the array `x` and is passed as an argument to the `strrev()` function which reverses the string and displays the output. *Note that the space in between the strings is also reversed in the above output.*

1.16.6 The strlwr() function

The strlwr() (*string lower*) *function is used to convert a string to lowercase.* This function takes only one argument. The general form or the syntax of strlwr() function is

```
x = strlwr(string);
```

where x is the string variable to hold the lowercase converted string and string is the valid string or constant which contains the string (uppercase). *Program 1.22* helps you to understand things better.

Program 1.22 :

```
/* strlwr.c */

#include<stdio.h>

#include<string.h>

void main()

 {

   char x[20];

   printf("Enter the string : ");

     gets(x);

   printf("The case changed string is : %s", strlwr(x));

 }
```

The program displays the following output

RUN1   Enter the string : PRAMILA
        The case changed string is : pramila

RUN2   Enter the string : ZaHiRa
        The case changed string is : zahira

In *Program 1.22* function strlwr() changes the uppercase letters to its lowercase and displays the output. If the input string entered is in lowercase, the compiler gives no error and automatically displays the output. *Note that for each program in this chapter we have included* #include<string.h>, *because all the string functions are pre-defined in the header file string.h.*

1.17 Simple Programs

Program S1 :

Write a program for swapping (interchanging) two numbers (using three variables).

```c
/* swap1.c */
#include<stdio.h>
void main()
 {
   int a, b, c;
   printf("Enter two numbers : ");
     scanf("%d %d", &a, &b);
   printf("Before swapping . . . \n");
   printf("A = %d  B = %d", a, b);
   c = a;
   a = b;
   b = c;
   printf("After swapping . . . \n");
   printf("A = %d  B = %d", a, b);
 }
```

The program displays the following output

```
Enter two numbers : 90   70
Before swapping . . .
A = 90   B = 70
After swapping . . .
A = 70   B = 90
```

Program S2 :

Write a program for swapping (interchanging) two numbers (using two variables).

```c
/* swap2.c */
#include<stdio.h>
void main()
 {
   int a, b;
   printf("Enter two numbers : ");
     scanf("%d %d", &a, &b);
   printf("Before swapping . . . \n");
   printf("A = %d  B = %d", a, b);
   a = a + b;
   b = a - b;
   a = a - b;
   printf("After swapping . . . \n");
   printf("A = %d  B = %d", a, b);
 }
```

The program displays the following output

```
Enter two numbers : 23   54
Before swapping . . .
A = 23   B = 54
After swapping . . .
A = 54   B = 23
```

Program S3:

Write a program for swapping two numbers without using arithmetic operators.

```c
/* swap3.c */
#include<stdio.h>
void main()
  {
    int a, b;
    printf("Enter two numbers : ");
      scanf("%d %d", &a, &b);
    printf("Before swapping . . . \n");
    printf("A = %d  B = %d", a, b);
    a = a ^ b;
    b = a ^ b;
    a = a ^ b;
    printf("\nAfter swapping . . . \n");
    printf("A = %d  B = %d", a, b);
  }
```

The program displays the following output

```
Enter two numbers : 90   70
Before swapping . . .
A = 90   B = 70
After swapping . . .
A = 70  B = 90
```

Program S4:

Write a program to convert temperature in centigrade to fahrenheit.

```c
/* temp_c_f.c */
#include<stdio.h>
void main()
  {
    float cent, faren;
    printf("Enter temperature in centigrade : ");
      scanf("%f", &cent);
    faren = (1.8 * cent + 32);
    /* cent = (faren - 32) / 1.8; for converting temperature in farenheit to centigrade */
    printf("The temperature in fahrenheit is %.2f", faren);
  }
```

The program displays the following output

```
Enter temperature in centigrade : 45
The temperature in fahrenheit is 113.00
```

Program S5 :

Write a program to find the maximum of three numbers using an if statement.

```c
/* max1_if.c */
#include<stdio.h>
void main()
  {
    float a, b, c, max;
    printf("Enter three numbers : ");
      scanf("%f %f %f", &a, &b, &c);
```

```
      max = a;
      if(b > max)
        max = b;
      if(c > max)
        max = c;
      printf("Largest of three numbers is %8.2f", max);
   }
```

The program displays the following output

```
Enter three numbers : 45.75   65.00   34.998
Largest of three numbers is 65.00
```

Program S6 :

Write a program to find whether the given number is odd or even.

```
/* odd_even.c */
#include<stdio.h>
void main()
 {
    int num;
    printf("Enter the number : ");
      scanf("%d", &num);
    if(num % 2 == 1)              /* find the modulus of the number */
      printf("The number is odd");
    else
      printf("The number is even");
 }
```

The program displays the following output

RUN1   Enter the number : 46
        The number is even
RUN2   Enter the number : 73
        The number is odd

Program   S7 :

Write a program to find simple and compound interest.

```
/* sim_com.c */
#include<stdio.h>
void main()
 {
    int n;
    float p, r, dur, simple, compound, total;
    printf("Enter principal,rate,no.of.years,compounding/year \n");
      scanf("%f %f %f %d", &p, &r, &dur, &n);
    simple = (p * dur * r) / 100;
    printf("Simple Interest is %.2f", simple);
    total = p * pow((double)(1+r/(n*100)),(double)(n*dur));
    compound = total - p;
    printf("Compound Interest is %8.2f", compound);
 }
```

The program displays the following output

```
Enter principal,rate,no.of.years,compounding/year
34    124    5    6

Simple Interest is 210.80
Compound Interest is 8636.00
```

Program S8 :

Write a program to find whether the given year is leap or not.

```c
/* leapyear.c */
#include<stdio.h>
void main()
{
    int x;
    printf("Enter the year : ");
        scanf("%d", &x);
    if((x % 4 == 0) && ((x % 100 == 0) || (x % 400 != 0)))
        printf("%d is a leap year", x);
    else
        printf("%d is not a leap year", x);
}
```

The program displays the following output

```
RUN1   Enter the year : 1999
       1999 is not a leap year

RUN2   Enter the year : 2000
       2000 is a leap year
```

Program S9 :

Write a program to find the factorial of a given number.

```c
/* factoria.c */
#include<stdio.h>
void main()
{
    long i, j, x = 1;
    printf("Enter the number : ");
        scanf("%ld", &i);
    for(j = 1 ; j <= i ; j++)
        x = j * x;
    printf("Factorial value of %ld is %ld", i, x);
}
```

The program displays the following output

```
Enter the number : 5
Factorial value of 5 is 120
```

Program S10 :

Write a program to find the sum of digits of an integer.

```c
/* digitsum.c */
#include<stdio.h>
void main()
{
    long x, y = 0, z = 0;
    printf("Enter the number : ");
```

```
    scanf("%ld", &x);
 while(x > 0)
  {
    y = x % 10;
    z = z + y;
    x = x / 10;
  }
 printf("The sum of the digits in the integer is %ld", z);
}
```

The program displays the following output

```
Enter the number: 1234
The sum of the digits in the integer is 10
```

Program S11 :

Write a program to find the average of first 'n' natural numbers using gauss method.

```
/* ave_guas.c */
#include<stdio.h>
void main()
 {
  int n;
  float sum = 0, i, ave;
  printf("Enter a number : ");
   scanf("%d", &n);
  for(i = 1 ; i <= n ; i++)
   {
    sum = sum + i;
   }
  ave = (float)sum / (float)n;       /* type casting */
  printf("Average of %.2d numbers is %.2f \n", n, ave);
  printf("Average by gauss method is %.2f",
             ((float)n*((float)n+1.0)/2.0)/(float)n);
 }
```

The program displays the following output

```
Enter a number : 12
Average of 12 numbers is 6.50
Average by gauss method is 6.50
```

Program S12 :

Write a program using functions to find the square of first N numbers and to calculate its sum.

```
/* square.c */
#include<stdio.h>
int squares(int);
void main()
 {
  int i, n, s, sum = 0;
  printf("Enter the limit : ");
   scanf("%d", &n);

 for(i = 1 ; i <= n ; i++)
   {
```

```
          s = squares(i);
          sum += s;
          printf("Square of %d is %d", i, s);
        }
     printf("Sum of the squares is %d", sum);
   }

   int squares(int b)
   {
      return(b * b);
   }
```

The program displays the following output

```
Enter the limit : 5
Sqaure of 1 is 1
Square of 2 is 4
Square of 3 is 9
Square of 4 is 16
Square of 5 is 25
Sum of the squares is 55
```

Program S13 :

Write a program to check whether the given number is palindrome or not.

```
/* palinnum.c */
#include<stdio.h>
void main()
  {
      long int a = 0, b = 0, c = 0, d = 0, e = 0;
      printf("Enter the number : ");
        scanf("%d", &a);
      e = a;
      while(a > 0)
        {
          b = a % 10;
          c = c + b;
          a = a / 10;
          d = b + d * 10;
        }
      if(e == d)              /* check whether the reverse and the original number are equal */
        printf("%d is a Palindrome number", e);
      else
        printf("%d is not a Palindrome number", e);
  }
```

The program displays the following output

```
RUN1    Enter the number : 234
        234 is not a Palindrome number

RUN2    Enter the number : 989
        989 is a Palindrome number
```

Program S14:

Write a program to check whether the given number is prime or not.

```c
/* prime.c */
#include<stdio.h>
void main()
 {
    int i, flag;  long number;
    printf("Enter the number : ");
      scanf("%ld", &number);
    if(number <= 3)
      printf("The number is prime");
    else
      for(i = 2 ; i < number/2 ; i++)
        if(number % i == 0)
          {
            flag = 1;
            break;
          }
        else
            flag = 0;
    if(flag == 1)
      printf("The number is not prime");
    else
      printf("The number is prime");
 }
```

The program displays the following output

```
RUN1    Enter the number : 7
        The number is prime

RUN2    Enter the number : 21
        The number is not prime
```

Program S15:

Write a program to print the multiplication table of the given number.

```c
/* multable.c */
#include<stdio.h>
void main()
 {
    int i, tables, counter = 1;
    printf("Enter the table number & the counter value : ");
      scanf("%d %d", &tables, &counter);
    printf("The table is . . . \n");
    for(i = 1 ; i <= counter ; i++)
printf("%d x %d = %d \n", tables, i, i * tables);
 }
```

The program displays the following output

```
Enter the table number & the counter value : 5    10
The table is . . .
5 x 1  = 5
5 x 2  = 10
5 x 3  = 15
5 x 4  = 20
5 x 5  = 25
5 x 6  = 30
5 x 7  = 35
5 x 8  = 40
5 x 9  = 45
5 x 10 = 50
```

Program S16 :

Write a program to find the roots of a quadratic equation.

```c
/* quadrat.c */
#include<stdio.h>
void main()
 {
    float a, b, c, d, real, imag, r1, r2, n;
    int k;
    printf("Enter the values of A,B & C : ");
      scanf("%f %f %f", &a, &b, &c);
    if(a != 0)
      {
        d = b * b - 4 * a * c;
        if(d < 0)
           k = 1;
        else if(d == 0)
           k = 2;
        else if(d > 0)
           k = 3;
        switch(k)
          {
          case 1:
                   printf("Roots are Imaginary \n");
                   real = -b / (2 * a);
                   d = -d;
                   n = pow((double)d,(double)0.5);
                   imag = n / (2 * a);
                   printf("r1 = %7.2f + j %7.2f \n",real,imag);
                   printf("r2 = %7.2f - j %7.2f \n",real,imag);
                   break;
          case 2:
                   printf("Roots are Real and Equal \n");
                   r1 = -b / (2 * a);
                   printf("r1 = r2 = %7.2f \n",r1);
                   break;
          case 3:
                   printf("Roots are Real and Unequal \n");
                   r1 = (-b + sqrt((double)d)) / (2 * a);
```

```
                    r2 = (-b - sqrt((double)d)) / (2 * a);
                    printf("r1 = %7.2f \n", r1);
                    printf("r2 = %7.2f \n", r2);
                    break;
            }
        }
    else
        printf("Equation is linear");
}
```

The program displays the following output

```
RUN 1   Enter the values of A,B & C : 1.0   2.0   1.0
        Roots are Real and Equal
        r1 = r2 = -1.00

RUN 2   Enter the values of A,B & C : 1.0   2.0   7.0
        Roots are Imaginary
        r1 = -1.00 + j 8190.50
        r2 = -1.00 - j 8190.50

RUN 3   Enter the values of A,B & C : 0.0   4.0   7.0
        Equation is linear
```

Program S17 :

Write a program to sort an integer.

```c
/* intsort.c */
#include<stdio.h>
void main()
{
    long sort[20], a, i = 0, j, k, number, temp;
    printf("Enter the number : ");
      scanf("%ld", &number);
    while(number > 0)
      {
        a = number % 10;
        sort[i] = a;
        i++;
        number = number / 10;
      }
    for(j = 0 ; j < i-1 ; j++)
      for(k = j + 1 ; k < i ; k++)

        if(sort[j] > sort[k])
          {
            temp = sort[j];
            sort[j] = sort[k];
            sort[k] = temp;
          }
```

```
    printf("The numbers in ascending order .... ");
    for(j = 0 ; j < i ; j++)

        printf("%ld", sort[j]);
  }
```

Program S18:

Write a program to sort the numbers in ascending order.

```
/* ascend.c */
#include<stdio.h>
void main()
 {
   int a[50], i, j, n, temp;
   printf("Enter the limit : ");
     scanf("%d", &n);
   printf("Enter the elements : ");
   for(i = 0 ; i < n ; i++)
     scanf("%d", &a[i]);
   for(i = 0 ; i < n-1; i++)
     for(j = i ; j < n; j++)

        if(a[i] > a[j])
         {
            temp = a[i];
            a[i] = a[j];
            a[j] = temp;

         }

   printf("The sorted list is : ");
   for(i = 0 ; i < n ; i++)
     printf(" %d", a[i]);
 }
```

Program S19 :

Write a program for changing an integer to its equivalent name.

```c
/* numname.c */
#include<stdio.h>
void main()
{
    long num, a[8], i, j, k, c = 0;
    printf("Enter a number : ");
      scanf("%ld", &num);
    while(num > 0)
      {
        j = num % 10;
        a[c] = j;
        num = num / 10;
        c++;
      }
    a[c] = '\0';
    for(k = c-1 ; k >= 0 ; k--)
      switch(a[k])
        {
            case 1 :
                    printf("ONE ");
                    break;
            case 2 :
                    printf("TWO ");
                    break;
            case 3 :
                    printf("THREE ");
                    break;
            case 4 :
                    printf("FOUR ");
                    break;
            case 5 :
                    printf("FIVE ");
                    break;
            case 6 :
                    printf("SIX ");
                    break;
            case 7 :
                    printf("SEVEN ");
                    break;
            case 8 :
                    printf("EIGHT ");
                    break;
```

```
            case 9 :
                        printf("NINE ");
                        break;
            default :
                        printf("ZERO ");
        }
}
```

The program displays the following output

```
Enter a number : 123456
one two three four five SIX
```

Program S20 :

Write a program to find the number of five hundreds, hundreds, fifties, twenties, tens, fives, two's and one's in the amount given.

```
/* rupee.c */
#include<stdio.h>
void main()
  {
    int rs;
    printf("Enter the amount in rupees : ");
      scanf("%d",&rs);
    if(rs >= 500)
      printf("\n The no.of five hundred's are %d", rs/500);
    if(rs >= 100)
      printf("\n The no.of hundred's are %d", rs/100);
    if(rs >= 50)
      printf("\n The no.of fifties are %d", rs/50);
    if(rs >= 20)
      printf("\n The no.of twenties are %d", rs/20);
    if(rs >= 10)
      printf("\n The no.of ten's are %d", rs/10);
    if(rs >= 5)
      printf("\n The no.of fives are %d", rs/5);
    if(rs >= 2)
      printf("\n The no.of two's are %d", rs/2);
    i(rs >= 1)
      printf("\n The no.of one's are %d", rs/1);
  }
```

The program displays the following output

```
Enter the amount in rupees : 243
The no.of hundred's are 2
The no.of fifties are 4
The no.of twenties's are 12
The no.of ten's are 24
The no.of five's are 48
The no.of two's are 121
The no.of one's are 243
```

1.18 Array Programs

Program A1 :

Write a program to find the second largest number in an array.

```c
/* large2.c */
#include<stdio.h>
void main()
 {
    int a[25], temp = 0, i, j, n;
    printf("Enter the limit : ");
       scanf("%d", &n);
    printf("Enter the elements : ");
    for(i = 0 ; i < n ; i++)
       scanf("%d", &a[i]);
    for(i = 0 ; i < n-1 ; i++)
       for(j = i ; j < n ; j++)
          if(a[i] < a[j])
          {
             temp = a[i];
             a[i] = a[j];
             a[j] = temp;
          }
    printf("The second largest number is %d", a[j]);
 }
```

The program displays the following output

```
Enter the limit : 7
Enter the elements : 3   4   2   1   6   8   9
The second largest number is 8
```

Program A2 :

Write a program to read and reverse an array.

```c
/* rev_arr.c */
#include<stdio.h>
void main()
 {
    float a[10], b[10], i, j, n;
    printf("Enter the limit : ");
       scanf("%f", &n);
    printf("Enter the elements : ");
    for(i = 1 ; i <= n ; i++)
       scanf("%f", &a[i]);
```

```
        printf("The reversed array is : ");
        for(i = n ; i >= 1 ; i--)
           printf("%.2f", a[i]);
     }
```

The program displays the following output

```
Enter the limit : 5
Enter the elements : 9  7  6  3  2
The reverse array is : 2.00  3.00  6.00  7.00  9.00
```

Program A3 :

Write a program to find the smallest number & its position in an array.

```
/* smallpos.c */
#include<stdio.h>
void main()
  {
     int a[10], i, n, s = 0, pos = 0;
     printf("Enter the limit : ");
        scanf("%d", &n);
     printf("Enter the elements : ");
     for(i = 0 ; i < n ; i++)
        scanf("%d", &a[i]);
     s = a[0];
     for(i = 0 ; i < n ; i++)
        if(a[i] < s)
        {
           s = a[i];
           pos = i;
        }
     printf("The smallest number is %d", s);
     printf("\n Its position is %d", pos+1);
  }
```

The program displays the following output

```
Enter the limit : 4
Enter the elements :  67  5  74  987
The smallest number is 5
Its position is 2
```

Program A4 :

Write a program to display only the negative elements of an array.

```
/* negative.c */
#include<stdio.h>
void main()
  {
     int i, num, neg[25];
```

```
        printf("Enter the limit : ");
          scanf("%d", &num);
        printf("Enter the elements : ");
        for(i = 0 ; i < num ; i++)
          scanf("%d", &neg[i]);
        printf("The negative elements are : ");
        for( i = 0 ; i < num ; i++)
          if(neg[i] < 0)
          {
            printf(" %d", neg[i]);
            continue;
          }
        printf("\n");
    }
```

The program displays the following output

```
Enter the limit : 5
Enter the elements : -5   4   -3   2   -1
The negative elements are : -5   -3   -1
```

Program A5 :

Write a program to generate fibonacci series using arrays.

```c
/* fiboarr.c */
#include <stdio.h>
void main()
    {
      int i, n, fib[30];
      printf("Enter the limit : ");
        scanf("%d", &n);
      fib[0] = 0;
      fib[1] = 1;
      for(i = 2; i < n; i++)
        fib[i] = fib[i-1] + fib[i-2];
      printf("The fibonacci series is : ");
      for(i = 0; i < n; i++)
        printf("%d  ", fib[i]);
    }
```

The program displays the following output

```
Enter the limit : 8
The fibonacci series is : 0  1  1  2  3  5  8  13
```

1.19 Matrix Programs

Program M1 :

Write a program to calculate the sum of all elements in a matrix.

```c
/* summat.c */
#include<stdio.h>
void main()
 {
   int a[10][10], i, j, m, n, sum = 0;
   printf("Enter the order of matrix : ");
     scanf("%d %d", &m, &n);
   printf("Enter the elements of matrix : ");
   for(i = 0 ; i < m ; i++)
     for(j = 0 ; j < n ; j++)
       scanf("%d", &a[i][j]);
   for(i = 0 ; i < m ; i++)
     for(j = 0 ; j < n ; j++)
       sum = sum + a[i][j];
   printf("Sum of elements of the matrix is %d", sum);
 }
```

The program displays the following output

```
Enter the order of matrix :  2  2
Enter the elements of matrix : 1  1  1  1
Sum of elements of the matrix is  4
```

Program M2 :

Write a program to find the whether the matrix is symmetric or not.Symmetric means the elements in the original matrix and the transpose of the same of the matrix must be same.

```c
/* symetric.c */
#include<stdio.h>
void main()
 {
   int i, j, row, column, mat[10][10], trans[10][10];
   printf("Enter the order of matrix : ");
     scanf("%d %d", &row, &column);
   if(row != column)
     {
       printf("The given matrix cannot be symmetric");
       exit(0);
     }
   printf("Enter the elements in row wise \n");
   for(i = 0 ; i < row ; i++)
```

```
        for(j = 0 ; j < column ; j++)
          {
              scanf("%d", &mat[i][j]);
              trans[j][i] = mat[i][j];
          }
      for(i = 0 ; i  < row ; i++)
        for(j = 0 ; j  < column ; j++)
          if(mat[i][j] != trans[i][j])
            {
                printf("The given matrix is not symmetric");
                exit(0);
            }
      printf("The given matrix is symmetric");
  }
```

The program displays the following output

RUN1   Enter the order of matrix : 3   3
        Enter the elements in row wise
        1 1 1
        1 1 1
        1 1 1
        The given matrix is symmetric

RUN2   Enter the order of matrix : 3   3
        Enter the elements in row wise
        1 2 3
        1 2 3
        1 2 3
        The given matrix is not symmetric

RUN3   Enter the order of matrix : 3   4
        The given matrix cannot be symmetric

Program M3:

Write a program to subtract two matrices.

```c
/* sub_mat.c */
#include<stdio.h>
void main()
 {
    int a[3][3], b[3][3], c[3],[3];
    int i, j;
    printf("Enter the elements of matrix 1 \n");
    for(i = 0 ; i < 3 ; i++)
       for(j = 0 ; j < 3 ; j++)
         scanf("%d", &a[i][j]);
    printf("Enter the elements of matrix 2 \n");
    for(i = 0 ; i < 3 ; i++)
       for(j = 0 ; j < 3 ; j++)
         scanf("%d", &b[i][j]);
    for(i = 0 ; i < 3 ; i++)
```

```
        for(j = 0 ; j < 3 ; j++)
          c[i][j] = a[i][j] - b[i][j];
      printf("\n Result of matrix subtraction is \n");
      for(i = 0 ; i < 3 ; i++)
      {
        for(j = 0 ; j < 3 ; j++)
          printf("%4d  ", c[i][j]);
        printf("\n");
      }
  }
```

The program displays the following output

```
Enter the elements of matrix 1
11      13      15      17      19      21      23      25      27
Enter the elements of matrix 2
 1       3       5       7       9      21      23      25      27
Result of matrix subtraction is
  10      10      10
  10      10       0
   0       0       0
```

Program M4 :

Write a program to multiply two matrices.

```
/* mulmat.c */
#include<stdio.h>
void main()
 {
    int a[5][5], b[5][5], c[5][5], i, j, k, m, n, p, q;
    printf("Enter the order of matrix 1 : ");
    scanf("%d %d", &m, &n);
    printf("Enter the order of matrix 2 : ");
    scanf("%d %d", &p, &q);
    if(n == p)                    /* if column of matrix 1 is equal to the row of matrix 2 */
    {
       printf("Enter the elements of matrix 1 \n");
       for(i = 0 ; i < m ; i++)
         for(j = 0 ; j < n ; j++)
           scanf("%d", &a[i][j]);
       printf("Enter the elements of matrix 2 \n");
       for(i = 0 ; i < p ; i++)
         for(j = 0 ; j < q ; j++)
           scanf("%d", &b[i][j]);
       for(i = 0 ; i < m ; i++)
```

```
            for(j = 0 ; j < q ; j++)
            {
                c[i][j] = 0;
                for(k = 0 ; k < p ; k++)
                    c[i][j] = c[i][j] + a[i][k] * b[k][j];
            }
        }
        else
            printf("Matrix multiplication is invalid");
        printf("Result of matrix multiplication is \n");
        for(i = 0 ; i < m ; i++)
        {
            for(j = 0 ; j < q ; j++)
                printf("%d \t", c[i][j]);
            printf("\n");
        }
}
```

The program displays the following output

RUN1   Enter the order of matrix 1 : 3   3

Enter the order of matrix 2 : 3   3

Enter the elements of matrix 1

1   3   5   7   9   11   21   23   25

Enter the elements of matrix 2

25   23   21   11   9   7   5   3   1

Result of matrix multiplication is

83   65   47

329   275   221

903   765   627

RUN2   Enter the order of matrix 1 : 1   2

Enter the order of matrix 2 : 3   4

Matrix multiplication is invalid

Program M5:

Write a program to evaluate the determinant of a matrix.

```
/* determat.c */
#include<stdio.h>
void main()
{
    float a[10][10], det, ratio;
    int i, j, k, n;
    printf("Enter the order of matrix : ");
        scanf("%d", &n);
    printf("Enter the elements of matrix \n");
    for(i = 1 ; i <= n ; i++)
      for(j = 1 ; j <= n ; j++)
```

```
        scanf("%f", &a[i][j]);
    for(k = 1 ; k <= n-1 ; k++)
      {
        for(i = k+1 ; i <= n ; i++)
          {
              ratio = a[i][k]/a[k][k];
              for(j = 1 ; j <= n ; j++)
                  a[i][j] = a[i][j] - ratio * a[k][j];
          }
        det = 1.0;
      }
    for(i = 1 ; i <= n ; i++)
        det = det * a[i][i];
    printf("The determinant of the matrix is %.3f", det);
  }
```

The program displays the following output

RUN1   Enter the order of matrix : 3
        Enter the elements of matrix

        9   8   7
        6   5   4
        3   2   1
        The determinant of the matrix is 0.000

RUN2   Enter the order of matrix : 3
        Enter the elements of matrix
        9   9   9
        6   5   4
        12   36   4
        The determinant of the matrix is 504.000

1.20 Summary

✍ C is a general-purpose structured programming language developed by Dennis Ritchie.

✍ An identifier is a string of alphanumeric characters that begins with an alphabetic character or an underscore character that are used to identify or name various programming elements such as variables, functions, arrays, structures, unions and so on.

✍ Keywords are the words, which have fixed meanings that has been already defined in the C compiler. They can be used only for their intended purpose and cannot be used as user-defined variables.

✍ Data types supported in a language dictates the type of values, which can be processed by the language. The primary data types are char, int, float and double.

✍ Type definition allows the user to define new data types that are equivalent to existing data types.

✍ The enum data type gives an opportunity to define your own data type with a set of named integer constants.

✍ Constant in C refers to fixed values that do not change during execution of a program.

✍ A character constant contains a single character enclosed within a pair of single quotes.

✍ A string constant or a literal consists of a sequence of characters enclosed in double quotes. It may contain any number of digits, letters, escape sequences, and spaces.

✍ End of a string constant is marked with a single character referred as the NULL character.

✍ A quantity which may vary during execution of a program is called as a variable.

✍ An operator is a symbol or a special character that tells the compiler to perform certain mathematical and/or logical operations.

✍ The sizeof operator is a keyword which returns the size of the operand in bytes.

✍ The conditional operator denoted by (?:) is used to evaluate conditional expressions.

✍ A statement in a program carries out some specific action. Statements used in C are classified as expression, compound and control statements.

✍ An expression statement consists of valid C expressions, ending with a semicolon.

✍ A group of valid C expression statements placed within an opening flower brace '{' and a closing flower brace '}' is referred as a compound statement.

✍ Comment statements used in a program must be enclosed with in /*and */. They may or may not be present in a program.

✍ The preprocessor directive #include<stdio.h> refers to a special file which contains the information about the various standard input and output operations which must be included in the program when it is compiled.

✍ Every C program must contain at least one function, which must be a main. A program will always begin by executing the main() function.

✍ All statements used in both declaration and execution parts must end with a semicolon.

✍ The formatted input/output functions read any type of data as input from the user and displays any type of data as output in the screen. They use format strings for identifying the type of the data involved.

✍ All unformatted input/output function reads and displays only char data type. They don't use format strings for identifying the type of the data involved, since all data are considered in character format.

✍ The scanf() function referred as the input statement is used to read formatted input from the user through the keyboard.

- ✍ The printf() function referred as the output statement is used to display formatted output in the screen.

- ✍ The getchar(), getch(), getche() functions deal with single character input.

- ✍ The putchar() function deals with single character output.

- ✍ The gets() function reads input from the user till it encounters a new line character.

- ✍ The gets() and puts() deals functions with string input and output respectively.

- ✍ The if-else statement is used to execute an action when the condition is true and another action when the condition is false.

- ✍ Any number of if-else statements can be included in a program, but the condition specified inside each group of if statement should be different.

- ✍ A multiple if-else statement can test for many different cases. If more than one condition is true, only the statements after the first true condition will be executed.

- ✍ Switch control statement allows us to make a decision from a no of choices, but it can test for equality.

- ✍ The four keywords used in a switch statement are switch, case, break and default.

- ✍ A default is optional in a switch statement. The default may appear any where inside the switch statement.

- ✍ The break statement is used to terminate or exit from a switch statement.

- ✍ The continue statement is used to transfer the control to the beginning of the loop.

- ✍ The goto statement is used to transfer the control in a loop or function from one point to any other portion in that program where it encounters a label.

- ✍ A segment of the program that is executed repeatedly is called as a loop.

- ✍ Note that the condition specified in the while/do-while loop should eventually become false at one point of the program, otherwise the loop will become an infinite loop.

- ✍ An array is a group of elements of same data type that share a common name and that are differentiated from one another by their positions within the array. The individual values in the array are called as elements.

- ✍ The array name should be unique and should be a valid C variable name.

- ✍ Each element in an array is referred by specifying the array name followed by one or more subscripts enclosed in square braces.

- ✍ A subscript must always be a positive integer or an integer expression.

- ✍ Arrays can be initialized to the same type in which they are declared.

- ✍ It is an error to specify more initializers than the number of elements in the array.

- ✍ It is not necessary to specify the length of an array, explicitly in case if initializers are provided for the array during declaration itself.

✍ A character array can be initialized by a string constant, resulting in the first element of the array being set to the first character in the string, the second element to the second character and so on. The array also receives the terminating '\0' in the string constant.

✍ Arrays whose elements are specified by one subscript are called as one dimensional or single dimension or single subscripted or linear arrays. It is also called as a list.

✍ Arrays whose elements are specified by two subscripts are called as two dimensional arrays or double dimensional arrays. It is also called as a matrix.

✍ The number of elements in a multi-dimensional array is the product of its subscripts.

✍ Any copy of characters specified within double quotes is a string constant.

✍ Strings can be read through scanf() and gets() functions.

✍ Strings can also be read character by character using the functions scanf(), getchar(), etc.

✍ The strlen() function returns the number of characters present in a string.

✍ The strcpy() function copies the contents of one string into another string.

✍ The strcat() function concatenates the second argument with the content of first argument.

✍ The strcmp() function compares two strings to find whether the strings (passed as arguments) are equal or not.

✍ The strrev() function reverses the string passed as arguments.

✍ The strlwr() function converts the string to lowercase.

1.21 Short-answer Questions

1. *What is an identifier ?*

Identifiers are names for various programming elements in a C program, such as variables, arrays, functions, structures, unions, labels, etc., An identifier can be composed only of uppercase, lowercase letters, underscore and digits, but should start only with an alphabet or an underscore.

2. *What are unary operators ?*

The operators that act upon a single operand are called as unary operators. The unary operators used in C are –, ++, – – and sizeof operators.

3. *Discuss about the increment operator in C ?*

Increment operator (++) is an unary operator which causes its operand to be increased by 1. If x = 3, ++x is equal to 4.

4. *Discuss about the decrement operator in C ?*

Decrement operator (– –) is an unary operator which causes its operand to be decreased by 1. If x = 3, – –x is equal to 2.

5.   What are binary operators ?

The operators that act upon two operands are called as binary operators. The binary operators used in C are +, −, *, /, %, = etc.,

6.   What are ternary operators ?

The operators that act upon three operands are called as ternary operators. The ternary operator available in C is (?:). This operator is also referred as conditional operator.

7.   What are relational operators ?

The operators that are used to test the relationship between two variables or between a variable and a constant are called as relational operators. The relational operators in C are greater than (>), greater than or equal to (>=), less than (<), less than or equal to (<=) and equal to (==).

8.   What are logical operators ?

The operators that are used to combine two or more relational expressions are called as logical operators. The logical operators used in C are logical AND (&&), logical OR (||) and logical NOT (!).

9.   What are bit wise operators ?

The operators that are used in applications that require manipulation of individual bits within a word of memory are called as bit wise operators. The bit wise operators used in C are complement (~), left shift (<<), right shift (>>), bitwise AND (&), bitwise NOT (!) and bitwise OR (^).

10.   Define masking.

Masking is a process in which a given pattern is transformed into another bit pattern by means of bit wise operators.

11.   What is an assignment operator ?

The operator that is used to assign the result of an expression to a variable is called as an assignment operator. The most commonly used assignment operator is equal to (=).

12.   State the purpose of a comma operator.

The comma operator is used to link related expressions together.

13.   What is meant by an expression ?

An expression is a combination of constants, variables, operators and function calls written in any form as per the syntax of the C language.

14.   Define control statements.

Control statements are statements that alter the execution of statements depending upon the conditions specified inside the parenthesis. For example,

```
If (a > b)
 {
    printf (" a is greater")
 }
```

15.   What is the restriction related to putchar() function compared to puts() function ?

It can output only one character at a time, whereas a puts() function can output a string (of any number of character(s) at a time.

16. ***Distinguish between getchar() and gets() functions.***

getchar()	gets()
a) Used to receive a single character.	a) Used to receive a single string with white spaces and blanks.
b) Does not require any argument.	b) It requires a single argument.

17. ***Distinguish between scanf() and gets() functions.***

scanf()	gets()
a) Strings with spaces cannot be accessed until ENTER key is pressed.	a) Strings with any number of spaces can be accessed.
b) All data types can be accessed.	b) Only character data type can be accessed.
c) Spaces and tabs are not acceptable as a part of the input string.	c) Spaces and tabs are perfectly acceptable as a part of the input string.
d) Any number of characters, integers, strings, floats can be received at a time.	d) Only one string can be received at a time.

18. ***Distinguish between printf() and puts() functions.***

puts()	printf()
a) They can display only one string at a time.	a) They can display any number of characters, integers or strings a time.
b) All data types are considered as characters.	b) Each data type is considered seperately depending upon the conversion specifications.

19. ***What is a label in goto statement ?***

A label may be any valid variable name and must be followed by a colon. The label is placed immediately before the statement where the control is transferred from goto statement. The label may be any where in the program either before or after the goto statement.

20. ***Define a loop.***

A segment of the program that is executed repeatedly is called as a loop. Examples are while, do-while and for loop.

21. ***State the three parts of the loop expression in for loop.***

The three parts of the loop expression are initialization, condition and incrementation or decrementation.

Example :

for(i = 0 ; i < 5 ; i ++)

Here, i = 0 is initialization

i < 5 is condition

i ++ is incrementation

22. ***How can you initialize more than one variable in a for loop ?***

More than one variable can be initialized in a for loop by seperating the initializations with commas. Example:

for (a = 0, b = 3, c = 5 ; a < 50 ; a++, b++, c++)

23.   What is the difference between a pre-increment and a post-increment operation ?

A pre-increment operation such as ++a, increments the value of a by 1, before a is used for computation, while a post-increment operation such as a++, uses the current value or present value of a in the calculation and then increments the value of a by 1.

24.   Distinguish between break and continue statement.

break	continue
a) Used to terminate the loops or to exit loop from a switch.	a) Used to transfer the control to the start of loop.
b) The break statement when executed causes immediate termination of loop containing it.	b) The continue statement when executed caused immediate termination of the current iteration of the loop.

25.   Distinguish between while and do-while loops.

while loop	do-while loop
a) The while loop tests the condition before each iteration.	a) The do-while loop tests the condition after the first iteration.
b) If the condition fails initially the loop is skipped entirely even in the first iteration.	b) Even if the condition fails initially the loop is executed once.

26.   What is an array?

An array is a group of related data items that share a common name and type.

27.   Define a string constant.

Any group of characters enclosed in double quotes is a string constant. It is terminated by a null character (\0).

28.   How is a string stored in an array ?

Each character of the string is stored in separate, but continuous memory locations with its last character as the null character (\0), which is automatically inserted at the end of the string.

29.   State the rules for using subscript variables.

- They must be an integer.
- The subscript value cannot be negative.
- The subscript must be with square braces after the array name.
- If there are more than one subscript each should be given in seperate square braces continuously without any commas. Example : arr[10][10][10];

30.   State some of the rules for initializing an array.

- If the number of initializers is less than the number of elements in the array, the remaining elements are set to zero.
- It is an error to specify more initializers than the number of elements in the array.
- It is not necessary to specify the length of an array, explicitly in case if initializers are provided for the array during declaration itself.

Chapter 2
Functions, Pointers, Structures and Unions

2.1 Introduction

*A **function** is a self-contained program segment (block of segments) created to perform some specific well-defined task that can be called in any part of a program . Functions break large computing tasks into smaller ones.* Almost all programming languages have some equivalent of the function, such as a subroutine or a procedure. C allows programmers to define their own functions for carrying out various individual tasks. C has been designed to make functions efficient and easy to use. C programs generally consist of many small functions rather than few big ones. Every C program consists of one or more functions, but there should be at least one function which must be `main()`. Each time when a new program is written, `main()` function must be defined. The `main()` function calls another function to share the work.

C functions may be classified into two categories namely ***user-defined functions*** and ***library functions***. The `main()` function that we have seen so far in the programs is an example of an user-defined function. These functions are defined by the user according to his requirements. User can implement their ideas in developing user-defined functions. `printf()` and `scanf()` are the examples of library functions. The main difference between these two categories is that, an user-defined function has to be developed by the user at the time of writing a program, where as library functions are previously defined functions. C language has a standard library which provides a rich collection of functions for performing input, output, string manipulations, common mathematical calculations, and many other functions for performing useful operations.

> **Note : Commonly used library functions are listed in Appendix IV.**

A function can be called by simply using the function name as a statement. A function will carry out its intended action whenever it is called from some other portion of the program. The same function can be accessed (called) from different places within a program. Once the function has carried out its intended action, control will be returned to the next statement from which the function is accessed. Usage of function is explained in detail with the help of ***Program 2.1.***

Program 2.1 :

```
/* function.c */
#include<stdio.h>
void func();                                    /* Function prototype */
void main()
 {
   printf("\n Inside main function");
   func();                                      /* Function call statement */
   printf("\n Again inside main function");
 }
void func()                                     /* Function definition */
 {
   printf("\n Inside func function");
 }
```

The program displays the following output

```
Inside main function
Inside func function
Again inside main function
```

The execution of a C program starts from main() function, the first printf() function executes and displays the message.

Inside main function

And the control is transferred to the next statement, the function call statement (i.e., func()). When the compiler encounters a function call, control is transferred to that function. The function is executed and displays the message.

Inside func function

Once the function func() has completed its execution, control is transferred or returned to the next statement from which the function is called (i.e., to the second printf() function inside the main() function). The printf() function displays the message

Again inside main function

and the program ends. ***Note that a semicolon is present in the statement, when the function is called within the main() function***. This is compulsary when the function is called from another function. Generally, a function will process information that is passed to it from the calling part of the program. Information is passed to the function via special identifiers called ***arguments*** or ***parameters*** and returned via the return statement.

There are many advantages in using functions for writing programs. For example, many programs may require a particular group of statements to be executed repeatedly from different places within the program. These repeated statements can be placed within a single function, which can be executed whenever it is needed. Thus, the use of a function avoids the need for same statements to be repeated in a program and thus saves memory.

The use of functions also enables a programmer to build user-defined library functions of frequently used routines containing system dependent features. Each routine can be programmed as a seperate function and stored within a special library file. These user-defined library functions can be added as a later part of the C programming library.

2.1.1 Defining a Function

A function definition consists of two parts. They are,

- Argument declaration
- Body of the function.

The first part of the function specification consists of type specification of the value returned by the function followed by the function name, a set of arguments (may or may not be present) seperated by commas and enclosed in parenthesis. If the function definition does not include any arguments, an empty pair of parenthes must follow the function name. The general form or the syntax of defining a function is,

```
returnType functionName (argument list)
{
    declaration(s);
    statement(s);                    } Body of the function
    return(expression);
}
```

where the `returnType` specifies the data type of the value to be returned by the function. C assumes that every function will return a value. The `returnType` is assumed to be of type `int` by default if it is not specified. The `functionName` is used to identify the function. The rules for naming a function are same as identifiers. The `argument list` specified inside the parenthesis after the function name is optional. It contains valid variable names with their data types preceding them. Semicolons are not allowed after the closing parenthesis, while defining a function.

Most functions have a list of parameters and a return value that provides means for communication between functions. *The arguments in the function reference, which defines the data items that are actually transferred, are referred as **actual arguments** or **actual parameters**. The arguments that represent the names of data items that are transferred to the function from the calling portion of the program are referred as **formal arguments** or **formal parameters**.* All variables declared in function definitions are local variables. Their scope is visible known only in the function in which they are defined. Function arguments are also local variables.

The second part of the function definition contains a statement or a group of statements that defines the action to be done by the function. These statements starting from the opening brace of the function to the last statement before the closing brace of the function is called as the ***body of the function***. The variables used in the function must be declared in the start of the function following the opening brace of the function or before all expression statements.

For example,

```
int add(int x, int y)           /* Argument declaration */
{
    int z;
    z = x + y;              } Body of the function
    return(z);
}
```

The function named `add()` takes up two integers namely `x` and `y`. The function returns an integer value, since `int` is specified before the function name `add`. The data type of the arguments passed inside a function may be specified in the declaration or before the opening brace of the function body as shown in the following example.

```
int add(x, y)
int x, y;                    /* Argument declaration */
 {
    int z;
    z = x + y;    } Body of the function
    return(z);
 }
```

On executing a `return` statement, control of the program returns to the calling function. The `expression` within parenthesis of the `return` statement is the value, which is returned to the called function. If the function reaches the final closing braces without any `return` statement, control will be transferred to the calling function without passing any value. The `returnType` specified in this case is `void`.

2.1.2 The return Statement

The `return` *statement is used to return the control from the called function to the next statement of the calling function.* The `return` statement also causes the program logically to `return` to the point from where the function is accessed (called). The `return` statement returns one value per call. The `return` statement can be any one of the types as shown below,

```
return;
return();
return(constant);
return(variable);
return(expression);
return(conditional expression);
return(function);
```

The first and second `return` statements, does not return any value, and are just equal to the closing brace of the function. If the function reaches the end without using a `return` statement, the control is simply transferred back to the calling portion of the program without returning any value. The presence of an empty `return` statement (without any expression or constant or variable) is recommended in such situations. These `return` statements return a value 1 to the calling function.

The third `return` statement returns a constant to the calling function.

For example,

```
    if(x <= 1)
       return(1);
```

The `return` statement returns a constant 1 when the condition specified inside the `if` statement evaluates to true.

The fourth `return` statement returns a variable to the calling function.

For example,

```
if(x <= 15)
    return(x);
```

The `return` statement returns a variable (which may return any value) to the calling function depending upon the value of the variable `x`.

The fifth `return` statement returns a value depending upon the result of the expression specified inside the parenthesis.

For example,

```
return(a + b * c);
```

Returns a value depending upon the values of `a`, `b`, and `c`.

The sixth `return` statement returns a value depending upon the result of the conditional expression specified inside the parenthesis.

For example,

```
return(a > b ? a : b);
```

The above `return` statement returns a value depending upon the value of the variables `a` and `b`.

The last `return` statement calls the function specified inside the parenthesis and collects the result obtained from that function, and returns it to the calling function.

For example,

```
return(power(3,2));
```

Returns a value, obtained after evaluating the `power()` function.

Important points to be noted while using return statement

- The limitation of a return statement is that it can return only one value from the called function to the calling function.
- The return statement can be present anywhere in the function, not necessarily at the end of the function.
- Number of return statements used in a function are not restricted, since the return statement which executes first will return the value from the called function to the calling function and the other return statements are left unexecuted.
- If the called function does not return any value, then the keyword void must be used as the return type specifier.
- Parenthesis used around the expression in a return statement is optional.

2.1.3 Function Prototypes

C function returns an integer value by default. When a function has to return a value other than `int`, then it is necessary to mention the first line of the function in the program, before it is used, which is called as the *function prototype* also referred as the *function declaration*. The function prototype not only identifies the return type of the function, but also the name of the function, the number of parameters in the function, the data types of the parameters passed and the order in which they are expected.

Function prototypes are usually written at the beginning of the program explicitly before all user-defined functions including the `main()` *function.* The general form or the syntax of declaring a function prototype is,

```
returnType functionName(dt1 arg1,dt2 arg2,...,dtn argn);
```

where `returnType` represents the data type of the value that is returned by the function, and `dt1,dt2,...,dtn` represents the data types of the arguments `arg1, arg2,...,argn`. The data types of the arguments should be specified compulsarily in the function prototype, but specifying the arguments of the data types is optional. The function prototype resembles the first line of the function definition (but the first line of the function definition does not end with a semicolon). The names of the arguments within the function prototype need not be declared elsewhere in the program. The data types of the actual arguments must confirm to the data types of the arguments within the prototype. Prototypes are desirable because they facilitate error checking between the parameters of a function, the function definition and the function call statement. In *Program 2.5*, we are using a function prototype,

```
int sum(int x);
```

where `sum` is the name of the function, `int` before the function name `sum()` indicates that the function returns a value of type `integer`. The variable `x` inside the parenthesis is the parameter passed to the called function. The data type `int` before the parameter `x` indicates that it is of type integer.

2.1.4 Types of Function

A function depending upon the arguments present or not and whether a value is returned or not, may be classified as

- Function with no arguments and no return values
- Function with no arguments and return values
- Function with arguments and no return values
- Function with arguments and return values.

Function with no Arguments and no Return Values

When a function has no arguments, it does not receive any data from the calling function, similarly when a function has no return values, the calling function does not receive any data from the called function. Hence, there is no data transfer between the calling function and the called function. If such functions are used to perform any operation, they read the input and display the output in the same block. *Program 2.2*, helps you to understand things better.

Program 2.2 :

```
/* nargnret.c */
#include<stdio.h>
void name();                        /* Function prototype */
void main()
 {
    name();
 }
```

```
void name()
{
    char empname[25];
    printf("Enter the employee name : ");
        scanf("%s", empname);
    printf("The employee name is %s", empname);
}
```

The program displays the following output

```
Enter the employee name : Rama
The employee name is Rama
```

In *Program 2.2*, `main()` is the calling function which calls the function `name()`. The function `name()` contains no arguments and hence, there are no argument declarations. Note that the called function (i.e., `name`) receives its data (i.e., name of the employee) directly from the input terminal (i.e., keyboard) and displays the contents of `empname` to the output terminal (i.e., screen) in the called function itself. No `return` statement is employed since there is nothing to be returned. The closing brace of the function indicates the end of execution of the function, thus returning the control, back to the calling function. The keyword `void` is used before the function `name()` to indicate that there are no return values.

Function with no Arguments and Return Values

When a function has no arguments, it does not receive any data from the calling function, but it can do some process and then return the result to the called function. Hence, there is data transfer between the calling function and the called function. *Program 2.3*, helps you to understand things better.

Program 2.3 :

```
/* Program to find the sum of even numbers within the limit. */
/* retnoarg.c */
#include<stdio.h>
int sum();                              /* Function prototype */
void main()
{
    printf(" %d", sum());
}
int sum()
{
    int i, n, result = 0;
    printf("Enter the limit : ");
        scanf("%d", &n);
    printf("Sum of Even numbers within %d is ", n);
    for(i = 2 ; i <= n ; i += 2)
      {
        result += i;
      }
    return(result);
}
```

The program displays the following output

```
Enter the limit : 10
Sum of Even numbers within 10 is 30
```

In *Program 2.3,* `main()` is the calling function which calls the function `sum()`. The function `sum()` contains no arguments and hence, there are no argument declarations. *Note that the called function (i.e., `sum()`) receives its data (i.e., the limit for calculating the even numbers) directly from the input terminal (i.e., keyboard) in the called function (i.e., `sum()`).* The `return` statement is employed in this function to return the sum of first n even numbers calculated within the limit and the result is displayed from the `main()` function to the standard output device (i.e., screen). *Note that `int` is used before the function name `sum()` instead of `void` since it returns a value of type `int` to the called function.*

Function with Arguments and no Return Values

When a function has arguments, it receives data from the calling function. The `main()` function will not have any control over the way in which the function receives its input data. We can also make the calling function to read data from the input terminal and pass it to the called function. This approach seems better because the calling function can check the validity of data, before it is passed over to the called function. *Program 2.4*, helps you to understand things better.

Program 2.4 :

```
/* Program to calculate simple interest. */
/* argnoret.c */
#include<stdio.h>
void simple(float, float, int);        /* Function prototype or declaring functions */
float principal, rate;
int year;
void main()
  {
     printf("Enter principal,rate of interest & no.of years \n");
       scanf("%f %f %d", &principal, &rate, &year);
     simple(principal, rate, year);
  }
void simple(float principal, float rate, int year)
  {
     float simple interest = 0.0;
     simple_interest = ( principal * rate * year ) / 100;
     printf("The simple interest is %.2f", simple_interest);
  }
```

The program displays the following output

```
Enter principal,rate of interest & no.of years
5000    1.5     2
The simple interest is 150.00
```

In *Program 2.4,* `main()` is the calling function which calls the function `simple()`. The function receives three arguments (i.e., `principal`, `rate` and `year`). The first two arguments are of floating type and the third argument is of integer type. *Note that the called function (i.e.,* `simple()`*) receives its data from the calling function (i.e.,* `main()`*).* No `return` statement is employed since there is nothing to be returned. The closing brace of the function signals the end of the function thus returning the control back to the calling function. The keyword `void` is used before the function name `simple()` to indicate that there are no return values.

Function with Arguments and Return Values

When a function has arguments it receives data from the calling function and does some process and then returns the result to the called function. In this way the `main()` function will have control over the function. This approach seems better because the calling function can check the validity of data before it is passed to the called function and to check the validity of the result before it is sent to the standard output device (i.e., screen). *Note that when a function is called, a copy of the values of actual arguments is passed to the called function.* *Program 2.5,* helps you to understand things better.

Program 2.5 :

```
/* Program to find the sum of first N natural numbers. */
/* arg_ret.c */
#include<stdio.h>
int sum(int x);                        /* Function prototype */
void main()
{
    int n;
    printf("Enter the limit : ");
       scanf("%d", &n);
    printf("Sum of first %d natural numbers is %d", n, sum(n));
}

int sum(int x)
{
    int i,result = 0;
    for(i = 1 ; i <= x ; i++)
       result += i;
    return(result);
}
```

The program displays the following output

```
Enter the limit : 5
Sum of first 5 natural numbers is 15
```

In *Program 2.5,* `main()` is the calling function which calls the function `sum()`. The function `sum()` receives a single argument. *Note that the called function (i.e.,* `sum()`*) receives its data from the calling function (i.e.,* `main()`*).* The `return` statement is employed in this function to return the sum of the `n` natural numbers entered from the standard input device and the result is displayed from the `main()` function to the standard output device. *Note that* int *is used before the function name* sum() *instead of* void *since it returns the value of type* int *to the called function.*

Important points to be noted while calling a function

- Parenthesis are compulsary after the function name.
- The function name in the function call and the function definition must be same.
- The type, number and sequence of actual and formal arguments must be same.
- A semicolon must be used at the end of the statement when a function is called.

2.2 Passing Values to Functions

Arguments to a function are usually passed in two ways. They are,

- Call by value
- Call by reference.

In call by value, a copy of the variable is made and passed to the function as argument. In this method, changes made to the parameters of the function have no effect on the variables, which calls the function, because the changes are made only inside the function. In call by reference, the address of each argument is passed to the function. In this method, changes made to the parameters of the function will affect the variables, which calls the function.

2.2.1 Call by Value

*The process of passing the actual value of variables is called as **call by value**.* In this case, the values from the calling function are passed as arguments to the called function. When a value is passed to a function via an actual argument, the value of the actual argument is copied to the variable into the called function, hence the value of the corresponding formal arguments can be altered within the function, but the value of the actual argument within the calling function will not be altered. *Program 2.6* helps you to understand things better.

Program 2.6 :

```
/* Program to demonstrate call by value. */
/* call_val.c */
#include<stdio.h>
void add(int);
void main()
 {
    int x = 5;
    printf("\n Value of x before calling the function is %d", x);
    add(x);
    printf("\n Value of x after calling the function is %d", x);
 }
void add(int y)
 {
    y = y + 5;
    printf("\n Value of x in the called function is %d", y);
 }
```

> *The program displays the following output*
>
> ```
> Value of x before calling the function is 5
> Value of x in the called function is 10
> Value of x after calling the function is 5
> ```

The first `printf()` function displays the original value of `x` and is then passed to the function `add()`. In the `add()` function, the value of `x` is copied to `y`. This value is added with 5 and the new value (i.e., 10) is displayed in the second `printf()` function. ***Note that the altered value of the formal argument is displayed as*** `y`. Finally, the value of `x` within the `main()` function is again displayed in the last `printf()` function. The output of the program shows that the value of `x` is not altered within the `main()` function, but the value of `x` is copied to the variable `y` in the function `add()` and the value of `y` is added by 5 and is displayed in the output and once again when the control transfers from the `add()` function to the `main()` function the value of `x` is retained as 5 and displayed in the output.

The advantages of passing arguments to a function using call by value method are

- Expressions can be passed as arguments.
- Unwanted changes to the variables in the calling function can be avoided.

Its disadvantage is

- Information cannot be passed back from the calling function to the called function through arguments.

2.2.2 Call by Reference

*The process of calling a function using pointers to pass the address of variable is called as **call by reference**.* In this case, instead of passing the values to the called function, a reference of the argument is passed to the called function by passing the address of the variable. The address operator (&) is placed after the argument type in the function definition. ***Program 2.7*** helps you to understand things better.

Program 2.7 :

```
/* Program to demonstrate call by reference. */
/* call_ref.c */
#include<stdio.h>
void add(int *a);
void main()
 {
    int x = 5;
    printf("\n Value of x before calling the function is %d", x);
```

```
    add(&x);
    printf("\n Value of x after calling the function is %d", x);
}
void add(int *y)
{
    *y = *y + 2;                                    /* Increments the value of a by 2 */
    printf("\n Value of x in the called function is %d", *y);
}
```

The program displays the following output

```
Value of x before calling the function is 5
Value of x in the called function is 7
Value of x after calling the function is 7
```

When the control enters the `main()` function, the value of x is initialized to 5 and is displayed in the output by the first `printf()` function. The next line of the program

```
add(&x);
```

calls the `add()` function, which receives a pointer to x (since the address operator is placed after the argument type). ***Note that a copy of address of*** x ***is passed to the*** `add()` ***function.*** In this case, the changes made to the parameters of the function will affect the variables, which called the function. Hence, the value of `*y` incremented in the `add()` function increments the value of x by 2. Hence, using a pointer as an argument, the called function can access and modify the object pointed to the pointer in the calling function. ***Note that the address of the parameter is not the address of the argument, but the address of the copy of the argument.*** Hence, the address of the variable x remains unchanged, and only its value is incremented by 2, in the `add()` function.

In case of call by reference, original values of the arguments in the calling function may get altered if modification of these values is made in the function. However, in the case of call by value, the original values of the arguments in the calling function can be retained.

2.3 Recursion

Recursion *is a process by which a function calls itself repeatedly until some specified condition has been satisfied.* A function is called recursive if a statement within the body of a function calls the same function. Recursion is a process of defining something in terms of itself. When a recursive program is executed, the recursive function calls are not executed immediately. They are placed on a stack (Last In-First-Out) until the condition that terminates the recursive function. The function calls are then executed in reverse order, as they are popped off the stack. ***Program 2.8,*** helps you to understand things better.

Program 2.8 :

```
/* Program to find factorial of a number using recursive function. */
/* fact_rec.c */
#include<stdio.h>
long fact(long);                        /* Function prototype */
void main()
 {
    long num, fac;
    printf("Enter a number : ");
      scanf("%ld", &num);
    fac = fact(num);
    printf("The factorial of %ld is %ld", num, fac);
 }
long fact(long x)
 {
    if(x <= 1)
      return(1);
    else
      return(x * fact(x - 1));
 }
```

The program displays the following output

RUN1   Enter a number : 0
        The factorial of 0 is 1

RUN2   Enter a number : 3
        The factorial of 3 is 6

In run1, when the number entered is 0, let us see what action the function fact() does. The value of the parameter num (i.e.,1) is copied into x. Since x turns out to be 0 the condition (x <= 1) is satisfied and hence 1 is returned through the return statement (i.e., the factorial value of 0 is 1) and the printf() function displays the message. The factorial of 0 is 1.

In run2, when the number entered is 3, the condition (x <= 1) fails, and hence the control is transferred to the else block.

```
x * fact(x - 1);
```

Since the current value of x is 3, the above statement would be

```
3 * fact(2);
```

will be evaluated. The expression on the right-hand side includes a call to the function fact() with x equal to 2, which returns

```
2 * fact(1);
```

Once again the function `fact()` is called with `x` equal to 1. This time, the function returns 1. The sequence of operations can be summarized as

```
fact = 3 * fact(2)
     = 3 * 2 * fact(1)
     = 3 * 2 * 1
     = 6
```

Recursive functions can be effectively used to solve problems where the solution is expressed in terms of successfully applying the same solution to the subsets of the problem. Don't be frightened by the apparent complexity of recursive functions. They are sometimes the simplest answer for evaluating certain mathematical functions. However, there is always an alternative non-recursive solution available. This will normally use loops and may lack elegance of the recursive solution.

Important points to be noted while using functions

- C program may contain one or more functions.
- A function can be called any number of times.
- A function can be called from any other function.
- The called function returns only one value per call.
- A function declaration must end with a semicolon.
- A function definition should not end with a semicolon.
- A function cannot be defined within another function definition.
- A function that can call itself until a condition is satisfied is called recursion.
- A function cannot be declared and defined more than once in a program.
- A function definition may appear in any order.
- The function name in the function call and the function definition should be the same.
- The order in which functions are defined in a program and the order in which they are called need not be the same.
- If a function takes no arguments, a parenthesis after the function name is a must.
- The type, order, and number of actual arguments must be the same as that of the formal arguments.

Advantages of Recursive Functions

- Used to create clearer and simpler versions of several algorithms.
- Memory occupied when the recursive function is called is very less compared to an ordinary function.
- Recursion is a very useful way of creating and accessing certain dynamic data structures such as linked lists, stacks, queues, etc.

2.4 Pointers

We know that variables are stored in memory. Each memory location has a numeric address. Variable names in C have the capability to enable the programmer to refer to the memory location by its name, but the compiler must translate these names into their respective memory addresses. This process is automatically performed by compiler. The address or name of a memory location points to whatever it contains within that memory location. Passing addresses does not violate the rules of programming. The addresses are passed only to the functions that need to access those memory locations. At any moment, in a program execution, each existing variable has a unique address associated with it. The memory location formerly occupied by such a local variable is freed when the program execution ends and the memory location is allocated to another variable later in some other program. *A pointer is a variable, which represents the location (not the value) of a data item, such as a variable or an array element.*

2.5 Pointer Operators

Like all variables, a pointer variable should also be declared. The general form or the syntax of a pointer variable is

```
dataType *variableName;
```

where `dataType` is the type of the `variableName`, which may be of any valid type. `*` is called as the ***indirection operator*** also referred to as the ***dereferencing operator***, which states that the variable is not an ordinary variable but a pointer variable. The `variableName` is the name of the pointer variable. The rules for naming pointer variables are same as identifiers. When an `*` is placed before the `variableName`, the value returned by it, is not the value of the pointer variable, but the address of the value stored in the memory location. The content of any pointer variable can be accessed with the help of indirection operator. If `ptr` is a pointer variable, then `*ptr` is used to access the content of the pointer variable `ptr`. The indirection operator (`*`) can only act upon operands, which are pointer variables.

All the variables declared in a program occupy a specific address in memory. It is possible in C, to obtain the address of the variable using an operator referred as the ***address operator***. This operator may precede a variable name or an array element, but should not precede a constant, an expression, or an unsubscripted name of an array. Let `ptr` be a variable that represents a data item. The compiler automatically assigns memory cells for the data item. The data item can then be accessed if we know the location (address), of the memory cell. The address of the variable `ptr` can be determined by the expression `&ptr` where `&` evaluates the address of the operand (i.e., `ptr`) as shown in *Fig. 2.1*.

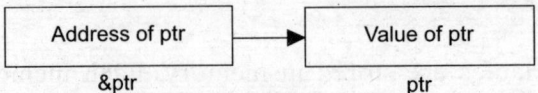

→

&ptr   ptr

Fig. 2.1 : Representation of a pointer variable.

For example,

```
main()
  {
      int *ptr, i = 25;
      ptr = &i;
  }
```

where ptr is a pointer variable for storing the address of an integer value, hence ptr is an integer pointer. In the above program segment, ptr is made to point i by storing the address of i (i.e., &i) in ptr.

Program 2.9 :

```
/* Program for accessing a variable through a pointer. */
/* pointer.c */
#include<stdio.h>
void main()
  {
    int a = 5, b = 10, *ptr;
    ptr = &a;
    printf("\n Initial values a = %d & b = %d", a, b);
    b = *ptr;
    printf("\n Changed values a = %d & b = %d", a, b);
    printf("\n Value of ptr is %d", ptr);
    printf("\n Value of the address of a is %d", &a);
    printf("\n Value of *ptr is %d", *ptr);
  }
```

The program displays the following output

```
Initial values a = 5 & b = 10
Changed values a = 5 & b = 5
Value of ptr is 4094
Value of the address of a is 4094
Value of *ptr is 5
```

In *Program 2.9* we are using a pointer variable *ptr. The pointer variable is initialized by assigning the address of the variable a to ptr as (ptr = &a). In the expression b = *ptr, the symbol * is placed before a pointer to access the content of the variable which is assigned to it. One can be surprised when you see the output of the program particularly the value of the address of a, when you execute this program in different computers at different times you will get different values of address which will be allocated to the variable a at different times.

2.6 Initialization of Pointers

Like ordinary variables, a pointer variable can be initialized. Static and external (global) pointer variables are initialized with NULL by default. Automatic (local) pointer variables can be initialized either with NULL or with the address of some other variable, which is already defined.

For example the statement,

```
ptr = &i;
```

is called as *pointer initialization*, since the address of i is initialized to the pointer variable ptr. A pointer variable can be also initialized as an ordinary variable in declaration itself. For example,

```
int i, *ptr = &i;
```

is also valid, since ptr points to the value stored in the address of i.

2.7 Pointer Arithmetic

An interesting use of pointers is pointer arithmetic. The arithmetic operations that can be performed on pointers are addition and subtraction. You can use the increment (++) and decrement (--) operators on the size of the variable pointed by the pointer. For example if

```
ptr++;
```

is an integer pointer, the increment operation moves the pointer by two bytes (since an integer is two bytes long) from where it is previously pointed. Similarly, if ptr is a character pointer, the increment operation moves the pointer by one byte (since a character is one byte long) from where it is previously pointed. If a decrement operation is carried out on a pointer, the pointer moves backwards, depending upon the size of the variable pointed by the pointer.

For example, if

```
ptr--;
```

is a character pointer. The decrement operation moves the pointer backwards by one byte from where it is previously pointed. The expression,

```
ptr += 6;
```

increments the pointer by 6 elements of its type. Similarly, a pointer can be subtracted from an integer. Besides incrementation and decrementation of a pointer to an integer, you may also subtract one pointer from another pointer in order to find the number of elements of their type which seperates the two pointers. Other operations like multiplication and division of pointers are prohibited. You cannot add or subtract float or double type pointers from each other.

2.8 Structure

*A **structure** is a derived type usually representing a collection of variables of same or different data types grouped together under a single name. The variables or data items in a structure are called as members of the structure.* A structure may contain one or more integer variables, floating-point variables, character variables, arrays, pointers, and even other structures can also be included as members. Structures help to organize data, particularly in large programs, because they allow a group of related variables to be treated as a single unit.

There are two fundamental differences between an array and a structure. One is, an array demands a homogeneous data type, i.e., the elements of an array must be of the same data type, where as a structure is a heterogeneous data type, since it can have any data type as its member. The second difference is that elements in an array are referred by their positions, where as members in a structure are referred by their unique names.

2.8.1 Declaring a Structure

Structure declarations are somewhat more complex than array declarations since a structure must be defined in terms of its individual members. The general form or the syntax of declaring a structure is

```
struct <structure_name>
{
    data type   member1;
    data type   member2;
    data type   member2;
        .           .
        .           .
        .           .
    data type   memberN;
};
```

In the above declaration, `struct` is a keyword, followed by an optional user-defined structure name usually referred as a tag, which is used to identify the structure, and then the list of members with its data types. The list of member declarations is enclosed in a pair of flower braces. The closing brace of the structure and the semicolon ends the structure declaration. A typical example for declaring a structure is

```
struct employee
{
    int empno;
    char empname[15];
    float salary;
};
```

where `employee` is the name of the structure (i.e., tag). The structure `employee` contains three members of type `int`, `char` and `float` representing `empno`, `empname` and `salary` respectively.

2.8.2 Defining a Structure

Declaring a structure merely describes the template of the structure. Declaration does not reserve any memory space but it creates a new data type. Defining a structure means creating variables to access the members in the structure. Creating a structure variable allocates sufficient memory space to hold all the members of the structure. Structure variables can be created during structure declaration or by explicitly using the structure name. The general syntax for defining the structure is,

```
struct <structure_name>
{
    data type   member1;
    data type   member2;
    data type   member2;
       ⋮          ⋮         ⋮
    data type   memberN;
}structure_variable(s);
```

The structure can be defined declaration itself and an example for defining the structure during declaration is,

```
struct employee
{
    int empno;
    char empname[15];
    float salary;
}emp1;
```

where the structure `employee` declares a variable `emp1` of its type. The `structure_variable(s)` can be declared like ordinary variables and used to access the members of the structure. More than one `structure_variable` can also be declared by placing a comma in between the variables. *Fig. 2.2* shows the memory allocation for the structure `employee`.

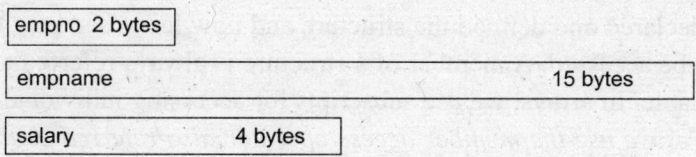

Fig. 2.2 : Memory occupied by structure employee.

The syntax for defining the structure using the structure name is,

```
struct   structure_name   structure_variable(s)
```
or
```
structure_name   structure_variable(s);
```

Note that the keyword struct is optional when defining the structure using the structure name.

A typical example for declaring a structure using the structure name is

```
struct employee emp1;        or        employee emp1;
```

If you need only one structure variable, the structure name or the tag is not necessary. For example,

```
struct
{
    int empno;
    char empname[15];
    float salary;
}emp1;
```

where the structure declares and defines a variable named `emp1`. *Note that the structure name or the structure variables may be omitted, but both, should not be omitted.* The member names within a particular structure must be distinct from one another though a member name can be same as the name of the variable, which is defined outside the structure. For example,

```
struct employee
{
    int empno;
    char empname[15];
    float salary;
}emp;
char empname[15];
```

where character variable `empname` inside the structure shares the same name with the variable `empname` declare outside the structure. *Note that a structure declaration that is not followed by a list of variables reserves no storage space, it merely describes the template (i.e., the structure is only defined not declared).*

2.8.3 Accessing Structure Members

We have declared and defined the structure and now let us see how the elements of the structure can be accessed. A member of a structure is always referred and accessed by the structure variable. In arrays, we use subscripts for accessing individual elements of an array. *In structures we use the member access operator also referred as the dot operator "." to access the members independently.* The general form or the syntax for accessing a member of the structure is

```
structure_variable.member_name
```

The structure variable followed by a period operator and the member name refers an individual member. For example, for accessing the structure elements, you can use

```
emp1.empname  similarly  emp1.empno
```

In the above example, `emp1` is the structure variable, which is used to access the

members `empname` and `empno` using the member access operator.

To assign a value to the member of the structure (say `empname`) use,

```
emp1.empname = "suresh";
```

This assigns the name "`suresh`" to the structure member `empname`. Similarly, to display the value of structure member `empname` on the screen use,

```
puts(emp1.empname);
```

You can also display the contents of `empname`, character by character by the following program segment,

```
for(i = 0 ; emp1.empname[i] != '\0' ; i++)
    putchar(emp1.empname[i]);
```

2.8.4 Initializing a Structure

Defining a structure is a way of creating a new data type. Like any other data type, the members in the structure cannot be initialized. The syntax of C language prevents the programmer from initializing individual structure members within the structure template. Structure members can be initialized only by using structure variables during structure declarations or by explicitly using the structure name. The syntax for initializing the structure during structure declaration is,

```
struct <structure_name>
  {
    data type   member1;

    data type   memberN;
  }structure_variable = {value1, value2,...., valueN};
```

For example,

```
struct employee
  {
    int height;
    float weight;
  }emp1 = {175, 77.5};
```

The above initialization initializes 175 to the structure member `height` and 77.5 to the structure member `weight`. The values to be initialized for the structure members must be enclosed within a pair of braces. ***Note that the constants to be initialized to the structure members must be in the same order in which the members are declared in the structure.***

Program **2.10** :

```
/* Program to initialize the structure using the structure name. */
#include<stdio.h>
struct class_room_1
  {
    char name[15];
    long int reg_no;
  }s3;
```

```
    void main()
    {
      struct class_room_1 s1 = {"Jeyabal", 876543};
      struct class_room_1 s2 = {"Vanitha"};
      printf("Name is %s", s1.name);
      printf("\nReg no is %d", s1.reg_no);
      printf("\n\nName is %s", s2.name);
      printf("\nReg no is %d", s2.reg_no);
      s3 = s1;
      printf("\n\nName is %s", s3.name);
      printf("\nReg no is %d", s3.reg_no);

    }
```

The program displays the following output

```
Name is Jeyabal
Reg no is 876543

Name is Vanitha
Reg no is 0

Name is Jeyabal
Reg no is 876543
```

Program 2.10 shows an example for initializing the structure using the structure name. The values to be initialized for the members of a structure must be enclosed within a pair of braces. *Note that the constants to be initialized to the members of the structure must be in the same order in which the members are declared in the structure.* If some of the structure members are not initialized, then the C compiler will automatically initialize them to zero. This is what you get for the structure variable s2 in *Program 2.10.*

The information contained in one structure variable can also be assigned to another structure variable using a single assignment statement. In *Program 2.10* we have used,

```
    s3 = s1;
```

to copy the values in structure variable s1 to s3. This type of assignment statements will be necessary when many members of the structure are to be initialized. *Note that one structure variable can be assigned to another only when they both are of the same structure type.*

2.9 Structure within a Structure

Nested structures are nothing but a structure within another structure (i.e., a structure can contain one or more structures as its members). A structure may be defined and/or declared inside another structure. For example,

```
struct employee
  {
    int empno;
    char name[15];
    struct employ_add
      {
        int no;
        char street[15];
        char area[15];
        long pincode;
      }address;
    char deptname[15];
    float salary;
  }emp1, emp2;
```

In the above structure declaration `employee` is the main structure. It gets additional information about the employee address from the structure `employ_add`. The member of a nested structure is accessed as

main_structure_variable.sub_structure_variable.sub_structure_member

For example,

`emp1.address.no`

refers to the sub structure member, `no` which is accessed using the structure variable of the main structure `emp1`, followed by the structure variable of the inner structure `address`. It is not necessary that, the sub structure should be specified inside the main structure. Instead the sub structure variable can be defined inside and declared outside the main structure as follows,

```
struct employ_add          /* sub structure */
  {
    int no;
    char street[15], area[15];
    long pincode;
  } address;
struct employee            /* main structure */
  {
    int empno;
```

```
        char name[15], deptname[15];
        float salary;
        struct employ_add address;
    }
```

where `address` is the sub structure variable defined inside the main structure `struct employee`. *Program 2.11* illustrates the concept of nested structures.

Program 2.11 :

```
/* nestruct.c */
#include<stdio.h>
void main()
 {
   struct employee
     {
        int empno;
        char empname[20];
        struct employ_add
          {
             int no;
             char *street;
             char *area;
             char *city;
             long pincode;
          }address;
        char *deptname;
        float salary;
     }emp1;
   printf("Enter empno,empname,deptname,salary of the employee \n");
   scanf("%d%s%f%f",&emp1.empno,emp1.empname,emp1.deptname,&emp1.
                         salary);
   printf("Enter the address of the employee");
   printf("\n (no,street name,area name,city,pincode)\n");
      scanf("%d %s %s %s %ld",&emp1.address.no,
                         emp1.address.street,emp1.address.area,
                         emp1.address.city,&emp1.address.pincode);
   printf("\n Employee No = %d", emp1.empno);
   printf("\n Employee Name = %s", emp1.empname);
   printf("\n Department Name = %s", emp1.deptname);
   printf("\n Employee Salary = %.2f", emp1.salary);
```

```
        printf("\n Employee Address....\n %d, %s, \n %s, \n %s-%d.",
                            emp1.address.no, emp1.address.street,
                            emp1.address.area, emp1.address.city,
                            emp1.address.pincode);
}
```

The program displays the following output
```
Enter empno, empname, deptname, salary of the employee
1  Raaji  Management  8000
Enter the address of the Employee
(no, street name, area name, city, pincode)
25  First_main_road  Adyar  Chennai  600020
Employee No = 1
Employee Name = Raaji
Department Name = Management
Employee Salary = 8000.00
Employee Address ....
25, First_main_road,
Adyar,
Chennai-600020.
```

Note that a structure cannot be nested inside the same structure. For example,
```
struct employee
{
    int empno;
    char empname[15];
    struct employee emp;            /* Error */
};
```
is not a valid nested structure.

2.10 Unions

*A **union** is another compound data type like a structure that may hold objects of different types and sizes, with the compiler keeping track of the size. Unions provide a way to manipulate different kinds of data in a single area of storage.* The general form or the syntax of defining a `union` is

```
union <union name>
{
    data type member1:
        ⋮        ⋮        ⋮
    data type memberN;
}union_variable(s);
```

The syntax of a `union` is identical to that of a structure except that the keyword `struct` is replaced with the keyword `union`. An example for declaring a `union` is

```
union exam
  {
    int roll_no;
    char name[15];
    int mark1, mark2, mark3;
  };
```

In the above example, `union exam` has 5 members. Second member is a character array name having 15 characters (i.e., 15 bytes). Second member is of type `int` that requires 2 bytes for their storage. All the other members `mark1`, `mark2`, `mark3` are integers which requires 2 bytes for their storage. *Fig. 2.3* shows the memory allocation of members of `union exam`.

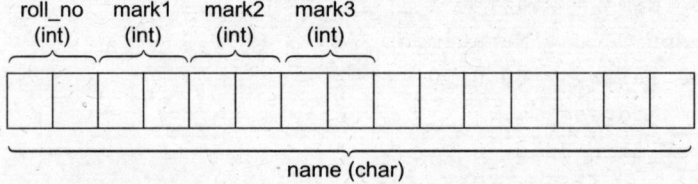

Fig. 2.3 : Memory occupied by union exam.

In the union, all these 5 members are allocated in a common place of memory (i.e., all the members share the same memory location). A union shares the memory space instead of wasting storage on variables that are not being used. The compiler allocates a piece of storage that is large enough to hold the largest type in the `union`. In the above declaration, the member `name` requires, 15 characters, which is the maximum of all members, hence a total memory space of 15 bytes is allocated. In case of structures, the total memory space allocated will be 23 (i.e., 15+2+2+2+2) bytes.

Program 2.12 :

```
/* size.c */
union exam                          /* Union declaration */
  {
    int roll_no;
    char name[15];
    int mark1, mark2, mark3;
  }u1;

struct exam1                        /* Structure declaration */
  {
    int roll_no;
    char name[15];
    int mark1, mark2, mark3;
  }s1;
```

```
void main()
{

   printf("The size of the union is %d\n", sizeof(u1));
   printf("The size of the structure is %d", sizeof(s1));
}
```

The program displays the following output

```
The size of the union is 15
The size of the structure is 23
```

2.10.1 Initializing a Union

There is a major difference between union and structures in terms of storage. In structures, each member has its own storage location where as all the members of a union occupy the same memory location (i.e., The union stores all the values of its members in a single location). This states that, although a union may contain many members of different data types, it can handle only one member at a time. Hence, you can initialize only one member in a union. If you try to initialize more than one member, the last initialized value will be assigned to all of its members.

Program 2.13 :

```
/* Program to demonstrate initialization of a union. */
/* initial.c */
union exam
{
   int roll_no, mark1, mark2, mark3;
}u1;
void main()
{
   u1.roll_no = 252;
   u1.mark1 = 89;
   printf("Roll No : %d\n", u1.roll_no);
   printf("Mark 1  : %d\n", u1.mark1);
   printf("Mark 2  : %d\n", u1.mark2);
   printf("Mark 3  : %d\n", u1.mark3);
}
```

The program displays the following output

```
Roll No : 89

Mark 1  : 89

Mark 2  : 89

Mark 3  : 89
```

Note that in the program even the uninitialized members (i.e., mark2 & mark3) take the same value as its other members.

2.11 Storage Classes

In C, all variables have data types and storage classes. Data type refers to the type of information represented by a variable (i.e., int, char, etc.,). *Storage classes determine the storage of the variable, initial value of the variable if not initialized, life of the variable (i.e., how long the variable would be active in the program) and the scope of variables within the program.* **Scope** of a variable is defined as the region over which the variable is visible or valid. Scope is usually established by a set of flower braces. Any variable declared in C have only one the four storage classes. They are,

- Automatic storage class
- Register storage class
- Static storage class
- External storage class.

2.11.1 Automatic Storage Class

*Variables defined inside a block and are local to the block in which they are declared, are referred as **automatic variables** or **local variables**.* Automatic variables can only be accessed by the block in which they are declared. Any another block cannot access its value.

When declaring a local variable inside a block, it is more precise in C to use the keyword auto before the declaration of the variable. The following statement is an example for local variable declaration,

```
auto int number;
```

These variables are created as new variables each time the block is executed, and they are destroyed when the block encounters a return statement or the closing brace of the block. The life of the variable will be inside the block in which it is declared. The C compiler treats any variable declared inside a block as an automatic variable by default, so the keyword auto is not necessary along with the variable declaration. Automatic variables are stored in memory. When an automatic variable is not initialized, it takes an unpredictable value referred as *garbage value*. *Program 2.14*, illustrates the scope and life of automatic variables.

Program 2.14 :

```
/* auto.c */
#include<stdio.h>
void main()
  {
    auto int x = 5;
      {
        auto int x = 4;
          {
            auto int x = 3;
```

```
            printf("%d \t",x);
        }
      printf("%d \t",x);
    }
  printf("%d \t",x);
}
```

The program displays the following output

3   4   5

Since all `printf()` functions occur within the block in which it is defined, x is local to the block. ***Note that the default storage class of a variable defined inside a function is auto. Also note that values passed to the function as arguments are also considered as local variables.***

2.11.2 Register Storage Class

The register storage class is similar to the auto storage class since the variables defined inside a block are local to the block in which it is specified. Register variables can only be accessed by the block in which it is declared. Its value cannot be accessed by any other function. When declaring a register variable inside a block, it is more precise in C to use the keyword `register` before the definition of the variable.

The following statement is an example for register variable definition,

```
register int number;
```

Register variables are stored in CPU registers, and hence they can be accessed faster than the automatic variables, which are stored in memory. Hence, if a variable is used in many places in a program it is better to declare its storage class as `register`. The initial value by default in case of register variables is also a garbage value. *Program 2.15*, illustrates the scope and life of register variables.

Program **2.15** :

```
/* register.c */
#include<stdio.h>
void main()
 {
   register int x;
   for(x = 1; x <= 10; x++)
     printf("  %d",x);
 }
```

The program displays the following output

1   2   3   4   5   6   7   8   9   10

We cannot use register storage class for all types of variables. For example,

```
register double x;
register float y;
register long z;
```

The above declarations are wrong because the CPU registers in the micro computers are usually 16 bit registers and therefore it cannot hold a float or a double value, which require 32 and 64 bytes respectively for storing a value. If the above declarations are used you won't get error messages, but the compiler would treat these variables as automatic variables.

2.11.3 Static Storage Class

The word static in general refers to anything that is inert to change. In a program if static variables are declared within individual blocks they are local to the block in which they are declared. *The static variables retain their values throughout the life of the program.* Hence, if a block is exited and then re-entered at a later time, the static variables declared in that block will retain their values (i.e., the static variable is not destroyed on exit from the block, instead its value is preserved, and becomes available again when the block is executed next time. Hence, the static variables are initialized only once, when the program begins its execution). When declaring a static variable inside a block, it is more precise in C to use the keyword static before the definition of the variable. The following statement is an example for static variable declaration,

```
static int number;
```

Static variables are stored in the memory. When a static variable is not initialized, it takes a value of zero. The value of the variable persists between different function calls in the program. *Program 2.16*, illustrates the scope and life of the static variables.

Program 2.16 :

```
/* static.c */
#include<stdio.h>
void sum();                          /* Function prototype */
void main()
{
   sum();                            /* Function call */
   sum();
   sum();
   sum();
}

void sum()
{
   static int x = 1;
   printf("%d  ", x);
   x = x + 1;
}
```

The program displays the following output

1   2   3   4

We get the output as above because the value of the variable persists (does not change) between different function calls. If the variable is not defined as static then the output of the program will be

1   1   1   1

2.11.4 External Storage Class

*Variables that are both alive and active throughout the entire program are called as **external variables** also referred as **global variables**.* Global variables can be accessed by any function in the program. Global variables do not belong to any particular function (i.e., they are global in nature) and is usually declared outside all functions, usually at the beginning of the program, but after all preprocessor directives. Global variables can be referred by any function that follows the variable declaration in the program. Any function in the program can access the value of the global variable and can change the value of the variable if desired. When declaring an external variable or a global variable in a program, it is more precise in C to use the keyword `extern` before the declaration of the variable. The following statement is an example for external variable declaration.

```
extern int number;
```

Global variables are also stored in the memory. When an external variable is not initialized it takes a value of zero. The value of the variable persists between different function calls as long as the program execution ends.

Note that initialization of an extern variable takes place only once. (i.e., when the program starts up, before the control enters to the main() function). Program 2.17, illustrates the scope and life of the global variables.

Program 2.17 :

```
/* extern.c */
#include<stdio.h>
void test1();
void test2();
void test3();
void main()
 {
    extern int value;
    value = 100;
    test1();
    test2();
    test3();
    printf("%d  ", value);
 }
void test1()
 {
    extern int value;
    value += 100;
    printf("%d  ", value);
 }
void test2()
 {
    extern int value;
    value = 1000;
    printf("%d  ", value);
 }
```

```
void test3()
{
   extern int value;
   value += 100;
   printf("%d ", value);
}
```

The program displays the following output

```
200   1000   1100   1100
```

The keyword `extern` is not used in the program to declare the variable `value`, since the variable is declared above all functions (including the `main()` function). If you want to declare a variable inside a function, and make that variable a global one, you have to precisely use the keyword `extern` before the definition of the variable. *Table 2.1*, shows the differences between the various storage classes.

Table 2.1 : Differences Between Various Storage Classes

Storage Class	Keyword	Storage Place	Scope of the Variable	Initializer Value	Type of the Variable
automatic	auto	memory	Inside the block in which it is defined	garbage value	local
register	register	CPU	Inside the block in which it is defined	garbage value	local
static	static	memory	The value of the variable persists as long as the program execution comes to an end	garbage value	local
external	extern	memory	Variables are both active and alive in the entire program	garbage value	global

2.12 The Preprocessor

The C preprocessor is a collection of special statements called directives that are executed at the beginning of the compilation process, which permits the insertion of files in a C program. The commonly used preprocessors are

- File inclusion
- Macro definition
- Conditional compilation.

These preprocessors contains the following directives,

#define	#endif	#ifdef	#line
#elif	#error	#ufndef	#pragma
#else	#if	#include	#undef

All preprocessor directives begins with a # sign. This sign is followed by a directive word. Preprocessor directives usually appear at the beginning of a program, but it may appear any where in the program. However, the directive will apply only to the portion of the program following its appearance.

2.12.1 File Inclusion

File inclusion is the first type of preprocessor directive. *The* #include *preprocessor directive is used for including one or more files in the program. The* #include *directive instructs the compiler to make a copy of a specified file to be included in place of the directive.* The general form or the syntax of #include directive is

```
# include <filename>
# include "filename"
```

The first form of the file inclusion (i.e.,filename enclosed in angle brackets <>) causes the preprocessor to look for a file with the specified filename in special directories set aside for files. The name of these files are specified by each individual implementation such as stdio.h, math.h, string.h, etc., The letter 'h' generally designates a header file which is incuded usually at the beginning of the program. The #include directive is also used with programs consisting of several source files that are compiled together. This is referred to as nested includes. The number of levels of nesting allowed varies for different C compilers (Usually eight nested file inclusions are used in a standard C compiler).

The second form of the file inclusion (i.e., filename enclosed in double quotes " ") causes the preprocessor to look for a file with the specified filename in the current working directory. The use of quotes is generally reserved for including files specifically related to the program that is currently handled. Usually these files (general purpose files) are created by the user for its use in the program.

2.12.2 Macro Definition

Macro definition is the second type of preprocessor directive. *The* #define *preprocessor directive is used to define a symbolic constant and assign it a value. The* #define *directive defines an identifier and a set of characters, which will be substituted for the identifier each time it is encountered in the source file.* The general form or the syntax of #define directive is

```
#define identifier set-of-characters
```

where identifier also referred as macro name or macro template substitutes the set-of-characters each time it encounters in the source file (program). *Program 2.18*, illustrates the use of #define directive.

Program **2.18** :

```
/* macro1.c */
#define FIVE 5
#define SIX FIVE+1
#define SEVEN SIX+1
void main()
  {
    printf("%d \t %d \t %d", FIVE, SIX, SEVEN);
  }
```

> ***The program displays the following output***
> ```
> 5  6  7
> ```

The first three statements in ***Program 2.18*** i.e.,

```
#define FIVE 5
#define SIX FIVE+1
#define SEVEN SIX+1
```

are macro definitions. They comprise of identifiers (i.e., FIVE, SIX and SEVEN) and their corresponding set-of-characters (i.e., 5, FIVE+1 and SIX+1). There may be any number of spaces between the identifier and the set-of-characters, but once the set-of-characters begins, there should not be any space in between them. They can be terminated by a new line.

During preprocessing, the identifiers (also referred to as the macro templates) are replaced by their respective set-of-characters (also referred to as the macro expansion), and assigns 5 for FIVE, 6 for SIX and so on.

Similarly, you can display messages using the #define directives. If the identifiers FIVE, SIX and SEVEN are encountered in the program following the macro definition, they are replaced by 5, 6 and 7 respectively. However, if the identifiers are encountered before the macro definition they are not replaced.

Program 2.19 :

```
/* macro2.c */
#define A "X is Big"
#define B "Y is Big"
void main()
 {
    int X, Y;
    printf("\n Enter value for X and Y : ");
    scanf("%d %d", &X, &Y);
    if(X > Y)
       printf(A);
    else
       printf(B);
 }
```

> ***The program displays the following output***
> ```
> Enter value for X and Y : 2  1
> X is Big
> ```

From the output we see that X is greater than Y and the program displays the message.

Before displaying the message the compiler preprocesses the identifier A and substitutes the string X is Big. Note that the string (set-of-characters) is enclosed in double quotes, hence for the compiler, the printf() function will actually appear to be

```
printf("X is Big");
```

Note that if the string is not enclosed in double quotes, the compiler will not display the message. If the `set-of-characters` for a macro is longer than one line, a backslash (\) character must be placed at the end of the line indicating that the `set-of-characters` continues on the next line. For example,

```
#define Two_line_string "This is a two line\
                        string"
```

The `identifier` or the macro name can have arguments. Each time the identifier is encountered, the arguments used in the macro name are replaced by the actual arguments found in the program. This type of macro is called as the ***function macro***. *Program 2.20*, helps you to understand things better.

Program 2.20 :

```
/* macro3.c */
#define PI 3.14159
#define AREA_OF_CIRCLE(R) (PI*R*R)
void main()
 {
   float radius, area;
   printf("Enter the radius of the circle : ");
     scanf("%f", &radius);
   area = AREA_OF_CIRCLE(radius);
   printf("The area of the circle is %4f", area);
 }
```

The program displays the following output
```
Enter the radius of the circle : 4
The area of the circle is 50.2654
```

wherever `AREA_OF_CIRCLE(radius)` appears in the program, the value of `radius` is substituted for `R` in the `set-of-characters`, the symbolic constant `PI` is replaced by its value (i.e., `PI` is replaced by `3.14159`) and the macro is expanded in the program. For example, the statements

```
area = AREA_OF_CIRCLE(4);
```

is expanded as

```
area = (3.14159*4*4);
```

and the program evaluates the area of the circle and displays the message. Symbolic constants and macros can be undefined using the `#undef` preprocessor directive. The general form or the syntax of `#undef` directive is

```
#undef identifier
```

where `identifier` is the already defined macro name.

The scope of a symbolic constant or macro is from its definition (i.e., from `#define`) upto or until it is undefined using `#undef` directive. After the `#undef` directives, any references made to the corresponding identifiers will be ignored. For example,

```
#define BLUE 5
#define BLACK 10
void main()
 {
    . . . . .
    #undef BLUE
    . . . . .
    #undef BLACK
 }
```

where the identifiers BLUE and BLACK inside the program represents 5 and 10 respectively, until it is undefined by using the directive #undef.

2.12.3 Conditional Compilation

Conditional compilation is the third type of preprocessor directive, that enables the programmer to control the execution of preprocessor directives and the compilation directives are #if, #else, #elif *and* #endif. The general form or the syntax of #if, and #endif directives are

```
#if expression
   statement(s);

#endif
```

If the expression following the #if directive is true the statement(s) following them will be executed, otherwise the statement(s) are skipped. The #endif directive marks the end of an #if block. *Program 2.21*, helps you to understand things better.

Program 2.21 :

```
/* compile.c */
#define A 500
void main()
 {
    #if A == 500
       printf("A has a constant value of 500");
    #endif
 }
```

The program displays the following output

```
A has a constant value of 500
```

Because the value of A is equal to 500. Note that the expression that follows the #if directive is evaluated at compile time, and hence, it must contain only previously defined identifiers and no variables may be used. Similarly #elif, #else directive are also used in programs for compiling the directives based on conditions.

Important points to be noted while using preprocessor directives

- All preprocessor directives should begin with an # sign.
- Preprocessor directives should not end with a semicolon.
- There may be any number of spaces between the identifier and set-of-characters.
- No spaces should be left in between the set-of-characters.
- An #if directive must accompany an #endif directive.

2.13 Simple Programs

Program S1 :

Write a program to read an integer of N digits and reduce the integer to a single digit number.

```c
/* digitcnt.c */
#include<stdio.h>
long DigitCount(long n);
void main()
 {
   long no;
   printf("Enter an integer : ");
     scanf("%ld", &no);
   printf("Its single digit is %ld", DigitCount(no));
 }
long DigitCount(long n)
 {
    return(n / 10 == 0 ? n : DigitCount(n / 10 + n % 10) );
 }
```

The program displays the following output

```
Enter an integer : 987456
Its single digit is 3
```

Program S2 :

Write a program to generate fibonacci series using recursion.

```c
/* genfibo.c */
#include <stdio.h>
void fibo(int, int, int);              /* function prototype */
void main()
 {
   int x = -1, y = 1, z = 0;
   printf("Enter any number : ");
     scanf("%d", &z);
   printf("The fibonacci series is : ");
   fibo(z, x, y);                       /* calling the function fibo() */
 }
```

```c
void fibo(int x, int f, int f1)          /* function definition */
{
    if(x == 0)
        exit(0);
    printf("%d  ", f+f1);
    fibo(--x, f1, f+f1);                 /* calling the recursive function fibo() */
}
```

The program displays the following output

```
Enter any number : 5
The fibonacci series is : 0  1  1  2  3
```

Program S3 :

Write a program to generate tribonacci series using recursion.

```c
/* gentribo.c */

#include <stdio.h>

void tribo(int, int, int, int);

void main()
{
    int x = -1, y = 1, z = 0, n;
    printf("Enter any number : ");
    scanf("%d", &n);
    printf("The tribonacci series is : ");
    tribo(n, x, y, z);                   /* calling the function tribo() */
}

void tribo(int x, int f, int f1, int f2)
{
    if(x == 0)
        exit(0);
    printf("%d  ", f + f1 + f2);
    tribo(--x, f1, f2, f + f1 + f2);     /* calling the recursive function tribo() */
}
```

The program displays the following output

```
Enter any number : 6
The tribonacci series is : 0  1  1  2  4  7
```

Program S4 :

Write a program using recursion to calculate the NCR of a given number.

```c
/* ncr.c */
#include<stdio.h>
float ncr(int n, int r);
void main()
{
    int n, r, result;
    printf("Enter the value of N and R : ");
    scanf("%d %d", &n, &r);
    result = ncr(n, r);
    printf("The NCR value is %.3f", result);
}

float ncr(int n, int r)
{
    if(r == 0)
        return 1;
    else
        return(n * 1.0 / r * ncr(n - 1, r - 1) );
}
```

The program displays the following output

```
Enter the value of N and R : 5   2
The NCR value is : 10.00
```

Program S5 :

Write a program to calculate NPR of a given number.

```c
/* npr.c */
#include<stdio.h>
int npr(int n, int r);
void main()
{
    int n, r, result;
    printf("Enter the value of N and R : ");
    scanf("%d %d", &n, &r);
    result = npr(n, r);
    printf("The NPR value is : %d", result);
}
```

```
    int npr(int n, int r)
    {
        if(r == 0)
            return 1;
        else
            return(n * npr(n-1, r-1));
    }
```

The program displays the following output

```
Enter the value of N and R : 5  2
The NPR value is :  20
```

Program S6 :

Write a program to calculate the least common multiple of a given number using recursion.

```
/* lcm.c */
#include<stdio.h>
int allone(int a[], int n);
long int lcm(int a[], int n, int prime);
void main()
{
    int a[20], status, i, n, prime;
    printf("Enter the limit : ");
        scanf("%d", &n);
    printf("Enter the numbers : ");
    for(i = 0; i < n; i++)
        scanf("%d", &a[i]);
    printf("The least common multiple is %ld", lcm(a, n, 2));
}

int allone(int a[], int n)
{
    int k;
    for(k = 0; k < n; k++)
        if(a[k] != 1)
            return 0;
    return 1;
}
```

```
long int lcm(int a[], int n, int prime)
 {
   int i, status = 0;
   if(allone(a, n))
     return 1;
   for(i = 0; i < n; i++)
     if((a[i] % prime) == 0)
       {
           status = 1;
           a[i] = a[i] / prime;
       }
   if(status == 1)
     return(prime * lcm(a, n, prime));
   else
     return(lcm(a, n, prime=(prime == 2)?prime+1 : prime+2));
 }
```

The program displays the following output

```
Enter the limit :   6
Enter the numbers :   6   5   4   3   2   1
The least common multiple is 60
```

Program S7 :

Write a program to calculate the greatest common divisor of a given number using recursion.

```
/* gcd.c */
#include<stdio.h>
int check_limit(int a[], int n, int prime);
int check_all(int a[], int n, int prime);
long int gcd(int a[], int n, int prime);
void main()
 {
   int a[20],stat, i, n, prime;
   printf("Enter the limit : ");
     scanf("%d", &n);
   printf("Enter the numbers : ");
   for(i = 0; i < n; i++)
     scanf("%d", &a[i]);
```

```
        printf("The greatest common divisor is %ld", gcd(a, n, 2) );

}

int check_limit(int a[], int n, int prime)
{
    int i;
    for(i = 0; i < n; i++)
        if(prime > a[i])
            return 1;
    return 0;
}

int check_all(int a[], int n, int prime)
{
    int i;
    for(i = 0; i < n; i++)
        if((a[i] % prime) != 0)
            return 0;
    for(i = 0; i < n; i++)
        a[i] = a[i] / prime;
    return 1;
}

long int gcd(int a[], int n, int prime)
{
    int i;
    if(check_limit(a, n, prime))
        return 1;
    if(check_all(a, n, prime))
        return(prime * gcd(a, n, prime));
    else
        return(gcd(a, n, prime = (prime == 2) ? prime+1 : prime+2));
}
```

The program displays the following output

```
Enter the limit :  5
Enter the numbers :  99   55   22   77   121
The greatest common divisor is 11
```

Program S8 :

Write a program to find the value of X^N using recursion.

```c
/* powerxn.c */
#include<stdio.h>
double power(double, int);
void main()
{
    double x;
    int n;
    printf("Enter the values of X and N : ");
    scanf("%lf %d", &x, &n);
    printf("The power of %.3lf to %d is %.3lf", x, n, power(x,n));
}
double power(double x, int n)
{
    return((n == 0) ? 1 : x * power(x, n-1));
}
```

> *The program displays the following output*
>
> ```
> Enter the values of X and N : 3.250      6
> The power of 3.250 to 6 is 1178.420
> ```

Program S9 :

Write a program using recursion to find the smallest and biggest number in an array.

```c
/* smallbig.c */
#include<stdio.h>
int min(int a[], int n);
int max(int a[], int n);
void main()
{
    int i, n, a[20];
    printf("Enter the value of N : ");
    scanf("%d", &n);
    printf("Enter the elements of the array : ");
    for(i = 0; i < n; i++)
        scanf("%d", &a[i]);
    printf("\nThe smallest number in the array is %d", min(a, n-1));
    printf("\nThe biggest number in the array is %d", max(a, n-1));
}
```

```c
int min(int a[], int n)
{
  if(n == 0)
    return a[n];
  else
    return(a[n] < min(a, n-1) ? a[n] : min(a, n-1));
}
int max(int a[], int n)
{
  if(n == 0)
    return a[n];
  else
    return(a[n] > max(a, n-1) ? a[n] : max(a, n-1));
}
```

The program displays the following output

```
Enter the value of N : 5
Enter the elements of the array : 9    5    4    77    55
The smallest number in the array is 4
The biggest number in the array is 77
```

Program S10 :

Write a program to reverse a string using recursion.

```c
/* rev_str.c */
#include<stdio.h>
#include<string.h>            /* strlen() */
char* revstr(char*);
char* reverse(char* str, int n);
void main()
{
  char str[20];
  printf("Enter the string : ");
    scanf("%s", str);
  printf("Its reversal is %s", revstr(str));
}
char* revstr(char *str)
{
  return reverse(str, strlen(str));
}
```

```
char* reverse(char* str, int n)
{
  char t;
  if((*str == '\0') || (str > (str + n - 1)))
    return str;
  t = *(str + n - 1);
  *(str + n - 1) = *str;
  *str = t;
  reverse(str + 1, n-2);
  return str;
}
```

The program displays the following output

```
Enter the string : SoftwarE
Its reversal is ErawtfoS
```

2.14 Summary

✍ A function is a self-contained program segment that performs some specific well defined task.

✍ Three steps in using a function are declaring a function, defining a function and calling the function.

✍ A function is called by specifying its name, followed by a pair of parenthesis which contains parameters if needed.

✍ The argument that represents the names of data items that are transformed into the function from the calling portion of the program are called as formal arguments or formal parameters.

✍ The corresponding arguments in the function reference which define the data items that are actually transferred are called as actual arguments or actual parameters.

✍ The return statement is used to return the information from the function to the calling function of the program.

✍ A function may be declared anywhere as long as its declaration is above all references to the function. So, a function should be declared before it is called.

✍ The execution of a function is terminated when it executes the return statement or when the last statement of the function is executed or when it encounters the closing brace of the function.

✍ A function prototype declares the return type of the function and declares the number type and the sequence of the parameters that a function will receive.

✍ Recursion is a process by which a function calls itself repeatedly until some specified condition has been satisfied.

✍ Scope of a variable is defined as the region over which the variable is visible or valid. Scope is usually established by a set of flower braces.

✍   A pointer is a variable, that represents the memory location (not the value) of a data item, such as a variable or an array element.

✍   Like all variables a pointer variable should also be declared before they are used.

✍   The ampersand operator (&) gives the address of a variable.

✍   The three values that can be used to initialize a pointer are zero, null and address.

✍   A pointer that has not been initialized is referred to as the dangling pointer.

✍   The arithmetic operations that are performed on pointers are addition and subtraction.

✍   Arguments to a function are passed in two ways. They are, call by value and call by reference.

✍   Call by value is the process of passing the actual value of variables.

✍   Call by reference is the process of calling a function using pointers to pass the address as argument must include a pointer as its parameter.

✍   A structure is a derived type usually representing a collection of variables of same or different data types grouped together under a single name.

✍   The keyword struct begins the structure declaration followed by the structure name, the structure members are enclosed in braces, followed by the semicolon which ends the structure declaration.

✍   A structure definition creates a new data type that can be used to declare variables.

✍   The variables or data items in a structure are called as members of the structure.

✍   Usually the structure type declaration appears at the top of the source code file, before any variables or functions are defined.

✍   The tag name for a structure is optional. If the structure is defined without a tag name, the variables of the derived data type must be declared in the structure definition, and no other variables of the new structure can be declared.

✍   A structure variable may be initialized with a structure variable of the same type.

✍   A member of a structure is always referred and accessed using the member access operator via the structure variable name.

✍   A structure that contains another structure as its member is called as nesting of structures or structure within a structure.

✍   Array of structures are defined as a group of data of different data types stored in consecutive memory locations with a common variable name. You can also use arrays as structure members.

✍   A structure containing a member that is a pointer to the same structure type is called as a self-referential structure.

✍   The structure pointer operator (->) is used for accessing a member of a structure via a pointer to the structure.

✍   A bit field is a built-in feature that allows you to access a single bit. They must be defined as a member of a structure and declared as type unsigned or int.

✍ Each bit field can then be accessed individually similar to other members of a structure. They enable better memory utilization for storing the data in the minimum number of bits required.

✍ A union is another compound data type like a structure that may hold the variables of different types and sizes, with the compiler keeping track of the size.

✍ The keyword union begins the union declaration followed by the tag (i.e., union name), within the braces, union members are declared.

✍ A union may contain many members of different types, but the memory space reserved for a union is large enough to store its largest member.

✍ Storage classes determine the lifetime of the storage, associated with the variable. The different storage classes in C are auto, register, static, and extern.

✍ Variables defined inside a function, that are local to the function in which they are declared are called as auto variables or local variables.

✍ Variables defined inside a function and declared as register are called register variables.

✍ Variables, which retain their values throughout the life of the program, are called as static variables.

✍ The variables that are both alive and active throughout the program are called as external or global variables. They can be accessed from any part of the program.

✍ The default value of an auto and register variable is a garbage value. The default value of a static and extern variable is zero.

2.15 Short-answer Questions

1. **Define a function**

   A function is a self-contained program segment (group of segments) that performs some specific well-defined task.

2. **How is a function invoked ?**

   A function is called by specifying its name, followed by a pair of parenthesis, which contains parameters or arguments if required.

3. **What are the information, we can get from the function declaration ?**

   The information that can be obtained from a function declaration are
   - Name of the function
   - Number, order and the type of arguments supplied to the function
   - Type of the value returned by the function.

4. **What are function prototypes? Where they are placed ?**

   Defining a function before it is accessed is called as function prototype. Function prototypes are generally placed at the start of the program before the main() function.

5.   When is the execution of a function terminated ?

The execution of a function is terminated when it executes the return statement or when the last statement of the function is executed.

6.   State the function of a return statement.

A function of a return statement is to return a value from the function. Since it is usually placed at the end of a function, return statement also terminates the function.

7.   What are the types of passing variables to a function ?

There are two types of passing variables to functions. They are,

- Call by value
- Call by reference.

8.   Define scope of a variable.

Scope of a variable is defined as the region over which the variable is visible or valid. Scope is usually established by a set of flower braces.

9.   Define auto variables.

Variables defined inside a function and are local to the function in which they are declared are called as auto variables. They are also called as local variables.

10.   Define register variables.

Variables defined inside a function and declared as register are called register variables. They are stored in CPU registers and can be accessed faster than other variables which are stored in memory.

11.   Define static variables.

The variables, which retain their values throughout the life of the program, are called as static variables.

12.   Define external variables.

Variables that are both alive and active throughout the entire program are called as external variables. They are also called as global variables. These variables are usually declared above the main() function.

13.   Distinguish between local and global variables.

Local variables	Global variables
a) These are declared within the body of the function.	a) These are declared outside the function.
b) These variables can be referred only within the function in which it is declared.	b) These variables can be referred from any part of the program.
c) The value of the variables disappear once the function finishes its execution.	c) The value of the variables disappear only after the entire execution of the program.

14.   State the differences between the function prototype and function definition.

Function prototype	Function definition
a) It declares the function.	a) It defines the function.
b) It ends with a semicolon.	b) It doesn't ends with a semicolon.
c) The declaration needn't include parameters.	c) It should include names for the parameters.

15.   What is a recursive function ?

Recursion is a process by which a function calls itself repeatedly until some specified condition has been satisfied. A function is called recursive if a statement within the body of a function calls the same function.

16.   Compare and contrast recursion and iteration.

Comparison of recursion and iteration

- Both are based on a control structure.
- Both involves repetition.
- Both involves a termination test.
- Both can occur infinitely.

Iteration	Recursion
a) Iteration explicitly user a repetition structure.	a) Recursion achieves repetition through repeated function calls.
b) Iteration terminates when the loop condition fails.	b) Recursion terminates when a base case is recognised.
c) Iteration keeps modifying the counter until the loop continuation condition fails.	c) Recursion keeps producing simple versions of the original problem until the base case is reached.
d) An infinite loop occurs when the loop continuation test never becomes false.	d) An infinite loop occurs if the recursion step doses not reduce the problem each time in a manner that converges on the base case.
e) Iteration normally occurs within a loop so the extra memory assignment is omitted.	e) Recursion causes another copy of the function and hence a considerable memory space's occupied.

17.   Distinguish between a user-defined function and a library function.

The statements of a user-defined function are written as a part of the program, according to the programmer's requirements but library functions are predefined.

18.   Define a pointer.

A pointer is nothing but a variable that contains the address of another variable.

19.   How is a pointer initialized ?

int *a, b;

a = &b;

a now contains the address of b.

The second declaration a = &b; is called as the pointer initialization.

20.   Define an address.

An address is an integer having fixed number of bits labeling a byte in the memory. Addresses are positive integer values that range from zero to the last location in the memory.

21.   State the uses of pointers.

The uses of pointers are,

*   Pointers provide an easy way to represent multidimensional arrays.
*   Pointers increase the execution speed.
*   Pointers reduce the length and complexity of program.

22.   What is the difference between address stored in a pointer and a value at that address ?

The address stored in the pointer is the address of a variable. The value of the variable is stored in another location. The indirection operator (*) returns the value stored at the address.

23.   What is the difference between indirection operator and address of operator ?

The indirection operator (*) returns the value of the address stored in a pointer. The address of operator (&) returns the memory address of the variable.

24.   State the difference between call by value & call by reference.

Call by value	Call by reference
a) int a;	a) int &a ;
b) Formal parameter a is a local variable.	b) Formal parameter a is a local reference.
c) It cannot change the actual parameter.	c) It can change the actual parameter.
d) Actual parameter may be a constant, a variable, or an expression.	d) Actual parameter must be a variable.

25.   What is a structure template ?

A template is a list of the members of a structure along with their type specifications, enclosed within flower braces.

26.   State the use of period (.) operator.

The period operator with an individual member of a structure followed by the member name is used to access the structure variable. Example, emp.empno;

where, emp is the structure variable and empno is the structure member.

27. **What is meant by nested structures ?**

A structure that contains another structure as its member is referred as nested structures.

28. **State the restrictions laid on bit-fields.**

The restrictions laid on bit-fields are

- Bit-fields cannot be created.
- Bit-fields are machine dependant.
- Address of a bit-field cannot be read.
- Bit-fields cannot be declared as static.

29. **State the advantages of using bit-fields.**

The advantages of using bit-fields are

- To store boolean variables (true/false) in one byte.
- To encrypt certain routines to access the bits within a byte.
- To transmit status information of devices encoded into one or more bits within a byte.

30. **State the difference between arrays and structures.**

Arrays	Structures
a) An array is an single entity representing a collection of data items of same data types. Individual entries in an array are called elements.	a) A structure is a single entity representing a collection of data items of different data types. Individual entries in a structure are called members.
b) An array declaration reserves enough memory space for its elements.	b) The structure definition reserves enough memory space for its members.
c) There is no keyword to represent arrays but the square braces [ ] preceding the variable name tells us that we are dealing with arrays.	c) The keyword struct tells us that we are dealing with structures.
d) Initialization of elements can be done during array declaration.	d) Initialization of members can be done only during structure definition.
e) The elements of an array are stored in sequence of memory locations.	e) The members of a structure are not stored in sequences of memory locations.
f) The array elements are accessed by its followed by the square braces [ ] within which the index is placed.	f) The members of a structure are accessed by the dot operator (also called as the period operator).

31. *State the difference between structures and unions.*

Structures	Unions
a) Each member in a structure occupies and uses its own memory space.	a) All the members of a union use the same memory space.
b) The keyword struct tells us that we are dealing with structures.	b) The keyword union tells us that we are dealing with unions.
c) All the members of a structure can be initialized.	c) Only the first member of an union can be initialized.
d) The members of a structure are not stored in sequence of memory locations.	d) The members of an union are stored in sequence of memory locations.
e) More memory space is required since each member is stored in a seperate memory locations.	e) Less memory space is required since all members are stored in the same memory locations.

Chapter 3
Linear Data Structures

3.1 Introduction

Programs consist of algorithms and data structures. An ***algorithm*** is a step-by-step recipe for solving an instance of a problem. An algorithm states the actions to be executed and the order in which the actions are to be executed. A ***data structure*** represents the logical relationship that exists between individual elements of data to carry out certain tasks. In other words, a data structure defines a way of organising all data items that consider not only the elements stored but also stores the relationship between the elements. To develop a program for an algorithm, we should first decide the steps and select an appropriate data structure for that algorithm. The choice and implementation of a data structure is as important for easier manipulation of data. Therefore, algorithms and data structures are closely related to each other for developing a program.

Arrays let you access lots of data fast with the ability to access its individual elements. You can have arrays of any data type. However, you cannot make arrays bigger if your program decides it needs more space, but we can build our own set data type out of arrays to solve this problem if necessary. A data structure is an actual implementation of an array.

Primary Data Structures

Primary data structures are the basic data structures that directly operate upon the machine instructions. They have different representations on different computers. All the basic constants (integers, floating-point numbers character constants, string constants) and pointers are considered as primary data structures.

Secondary Data Structures

Secondary data structures are more complicated data structures derived from primary data structures. They emphasize on grouping same or different data items with relationship between each data item. Secondary data structures can be broadly classified as static data structures and dynamic data structures. If a data structure is created using static memory allocation, (i.e., a data structure formed when the number of data items are known in advance) it is known as ***static data structure*** or ***fixed size data structure.*** If a data structure is created, using dynamic memory allocation (i.e., a data structure formed when the number of data items are not known in advance) is known as ***dynamic data structure*** or ***variable size data structure.***

Dynamic data structures can be broadly classified as linear data structures and non-linear data structures. *Linear data structures*, have a linear relationship between its adjacent elements. Linked lists are examples of linear data structures. A *linked list* is a linear dynamic data structure that can grow and shrink during its execution time. Elements can be inserted and deleted from any end of the list. If the elements are inserted and deleted in the same end, then the list is referred to as a *Stack*. If the elements are inserted in one end and deleted in another end, then the list is referred to as a *Queue*. A *circular linked list* is similar to a linked list except that the first and last nodes are interconnected.

Non-linear data structures does not have a linear relationship between its adjacent elements. In a linear data structure, each node has a link which points to another node, where as in a non-linear data structure, each node may point to several other nodes. A *tree* is a non-linear dynamic data structure that may point to one or more nodes at a time. A *graph* is similar to tree except that it has no hierarchical relationship between its adjacent elements.

3.1.1 Stacks

A stack is an ordered collection of elements in which insertions and deletions are restricted to one end. The end from which elements are added and/or removed is referred to as *top* of the stack. Stacks are also referred as "*piles*" and "*push-down lists*". The first element placed in the stack will be at the bottom of the stack. The last element placed in the stack will be at the top of the stack. The last element added to stack is the first element to be removed. Hence stacks are referred to as Last-In-First-Out (LIFO) lists.

Consider *Fig. 3.1* where five discs are placed one over the other through a shaft. If we want to insert a new disc say 6. It is possible to insert the disc on the top of disc 5. Because of the presence of shaft, it is not possible to insert the disc 6 in the middle of the discs. Similarly, if we want to remove a disc say 4, it is only possible to remove the disc 5 first and then the disc 4 and so on. Because of the presence of the shaft, it is not possible to remove the disc from the middle of the discs. This similar operation is carried out in a stack.

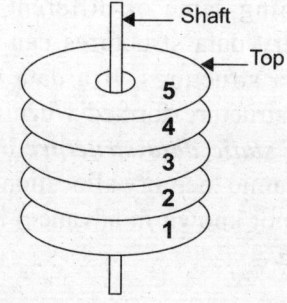

Fig. 3.1 : Shaft with five discs.

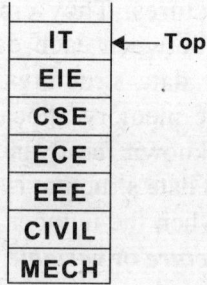

Fig. 3.2 : Stack operation.

A stack is referenced via a pointer to the top element of the stack referred as **top pointer**. The top pointer keeps track of the top element in the stack. Initially, when the stack is empty, the top pointer has a value zero and when the stack contains a single element, the top pointer has a value one and so on. *Fig. 3.1* shows the representation of a stack.

The primary operations that can be carried on a stack are insertions and deletions. These operations on a stack are referred to as *PUSH* and *POP* operations respectively. A PUSH operation adds a new element to the stack. Each time a new element is inserted in the stack, the top pointer is incremented by one before the element is placed on the stack. A POP operation deletes the top most element in the stack. Each time an element is removed from the stack the top pointer is decremented by one. The stack shown in *Fig. 3.2* consists of seven elements. We know that insertions and deletions can occur only at the top of the stack, which means EIE cannot be deleted until IT is deleted, CSE cannot be deleted until IT and EIE are deleted and so on. The elements from the stack may be popped (i.e., removed/deleted) only in the reverse order in which they are pushed (i.e., inserted/added) on to the stack.

3.2 Array and its Representation using Stack

One of the simplest ways of representing a stack is by means of a single dimensional array. Both stacks and arrays are ordered collection of elements, but an array and a stack are two different things. The number of elements in the array is fixed and it is not possible to change the number of elements, but the size of the stack varies continuously when elements are pushed and popped. The stack is stored in a part of the array, so an array can be declared large enough to hold the maximum number of elements of the stack.

During execution of a program, the stack size can be varied within the space reserved for it. One end of the array (i.e., bottom of the stack) is fixed and the other end of the array (i.e., top of the stack) is constantly changed depending upon the elements pushed and/or popped in the stack. Care must be taken in handling extreme cases, like not to push an element into a fully occupied stack and not to pop an element from an empty stack.

To represent a stack using arrays, we need,

- An array to hold the elements of the stack, which can be of any data type such as int, char, float, etc.,
- An integer variable to indicate the top position of the stack within the array.

3.2.1 Push Operation

PUSH is an operation used to add a new element in to a stack. The PUSH operation of a stack is implemented using arrays. When implementing the PUSH operation, overflow condition of a stack is to be checked (i.e., you have to check whether the stack is full or not). If the size of the stack is defined as 15, then it is possible to insert (i.e., add) only 15 elements into the stack. It is not possible to add any more elements to the stack, since there is no space to accommodate the elements in the array.

The following procedure helps you to understand things better.

- Create a function **PUSH()** with one argument, the ELEMENT to be added.
- Assign the new element in the array by incrementing the TOP variable by 1.

3.2.2 Pop Operation

POP is an operation used to remove an element from the TOP of the stack. POP operation of a stack is also implemented using arrays. When implementing the POP operation, underflow condition of a stack is to be checked (i.e., You have to check whether the stack is empty or not). The user should not POP an element from an empty stack. This type of an attempt is illegal and should be avoided. If such an attempt is made, the user should be informed of the underflow condition.

The following procedure can help you to understand things better.

- Create a function **POP()** with one argument, the address of the element to store the popped value.
- Decrement the TOP variable by 1.

3.2.3 PEEK Operation

PEEK is an operation used to display the element from the TOP of the stack, pointed by the TOP pointer without actually removing it. PEEK operation of a stack is implemented using arrays. When implementing the PEEK operation, underflow condition of a stack is to be checked (i.e., You have to check whether the stack is empty or not). The user should not try to PEEK an element from an empty stack. This type of an attempt is illegal and should be avoided. If such an attempt is made, the user should be informed of the underflow condition.

The following procedure can help you to understand things better.

- Create a function **PEEK()** with one argument, the address of the element to store the top value.

3.2.4 Empty Stack

If a stack contains no elements, it is referred to as an empty stack. If an element is to be popped from a stack, or to view the contents of the stack, you have to check whether the stack is empty or not, since elements cannot be popped from an empty stack nor contents cannot be displayed from an empty stack. This type of an attempt is illegal and should be avoided. If such an attempt is made, the user should be informed of the underflow condition. To check whether the stack is empty or not, we are using a variable top in *Program 3.1.* Assume that you are initializing the variable top to -1. If the variable top is, -1 during execution the output message "Stack underflow on POP" is displayed. This condition is referred as empty condition of a stack.

The following procedure can help you to understand things better.

- Create a function **isEmpty()** with no arguments.
- Check whether the TOP pointer is equal to -1 (initialization of TOP is -1).
- If the condition is true, it means, that the stack is empty.
  Otherwise, the stack is not empty.

3.2.5 Fully Occupied Stack

If a stack contains elements equal to its size (i.e., STACK_SIZE as in ***Program 3.1***), *we say that the stack is full.* If an element is to be pushed into a stack, you have to check whether the stack is full or not, since elements cannot be pushed into a stack when it is fully occupied. This type of an attempt is illegal and should be avoided. If such an attempt is made, the user should be informed of the overflow condition. To check whether the stack is full or not, we are using a pointer top in ***Program 3.1.*** We are checking whether the variable top is equal to the maximum size of the array minus one (Since an array starts from zero, if the size of the array is 5, and if the stack is full the last element of the stack is stored in the array location 4). If the condition is satisfied the output message, "Stack Overflow on PUSH" is displayed. This condition is referred as fully occupied stack.

The following procedure can help you to understand things better.

- Create a function is **Full()** with no arguments.
- Check whether the TOP pointer is equal to STACK_SIZE – 1.
- If the condition is true, it means, that the stack is full. Otherwise, the stack is not full.

Program 3.1 :

```
/* Program to implement a stack using Arrays. */
/* arrstack.c */
#include<stdio.h>
#include<conio.h>
#define STACK_SIZE 5
int push(int value);
int pop(int *value);
int peek(int *value);
int size();
int isEmpty();
int isFull();
void view();
void displayMenu();
int top = -1, stack[STACK_SIZE];
void main()
{
        int data, status, choice;
        displayMenu();
```

```c
    while(1)
     {
      printf("\n\n ? ");
        scanf("%d", &choice);
      switch(choice)
      {
           case 1 :
               printf("\n Enter the element : ");
               fflush(stdin);
                 scanf("%d", &data);
               status = push(data);
                if(status == -1)
                  printf("\n Stack Overflow on PUSH !!!");
               break;
           case 2 :
               status = pop(&data);
                if(status == -1)
                  printf("\n Stack Underflow on POP !!!");
               else
                  printf("\n The popped value is %d", data);
               break;
           case 3 :
               status = peek(&data);
               if(status == -1)
                  printf("\n Stack is Empty !!!");
               else
                  printf("\n The top most value is %d", data);
               break;
           case 4 :
               printf("\n Stack Size is = %d", STACK_SIZE);
               printf("\n Current Stack Elements = %d", size());
               break;
           case 5 :
               view();
               break;
           default:
               printf("\n End of Run of your Program . . .");
               exit(0);
      }
     }
}
```

```
void displayMenu()
 {
          printf("\n Representation of Stack Using Arrays ...");
          printf("\n\t 1. Push ");
          printf("\n\t 2. Pop  ");
          printf("\n\t 3. Peek  ");
          printf("\n\t 4. Size  ");
          printf("\n\t 5. View ");
          printf("\n\t 6. Exit ");
 }
int push(int value)
 {
          extern int top, stack[];
          if(isFull())
             return -1;
          stack[++top] = value;
          return 0;
 }
int pop(int *value)
 {
          extern int top, stack[];
          if(isEmpty())
             return -1;
          *value = stack[top--];
          return 0;
 }
int peek(int *value)
 {
          extern int top, stack[];
          if(isEmpty())
             return -1;
          *value = stack[top];
          return 0;
 }
int isEmpty()
 {
          extern int top, stack[];
          if(top == -1)
             return 1;
           else
              return 0;
 }
```

```
int isFull()
{
        extern int top, stack[];
        if(top == STACK_SIZE - 1)
            return 1;
        else
            return 0;
}
int size()
{
        extern int top, stack[];
        return top+1;
}
void view()
{
        extern int top, stack[];
        int i;
        if(isEmpty())
         {
           printf("\n The Stack is Empty !!!");
           return;
         }
        printf("\n The content of the stack is ... TOP");
        for(i = top; i >= 0; i--)
            printf(" --> %d", stack[i]);
        if(isFull())
            printf("\n The Stack is Full !!!");
}
```

The program displays the following output

```
Representation of Stack Using Arrays ...
        1. Push
        2. Pop
        3. Peek
        4. Size
        5. View
        6. Exit
```

```
? 2

Stack Underflow on POP !!!

? 3

Stack is Empty

? 4

Stack Size is = 5

Current Stack Elements = 0

? 5

The Stack is Empty !!!

? 1

Enter the element : 1

? 1

Enter the element : 2

? 1

Enter the element : 3

? 1

Enter the element : 4

? 1

Enter the element : 5

? 1

Enter the element : 6

Stack Overflow on PUSH !!!

? 5

The content of the stack is ... TOP --> 5 --> 4 --> 3 --> 2 --> 1

The Stack is Full !!!

? 4

Stack Size is = 5

Current Stack Elements = 5

? 3

The top most value is 5

? 2

The popped value is 5

? 2

The popped value is 4

? 2

The popped value is 3
```

```
? 3

The top most value is 2

? 2

The popped value is 2

? 2

The popped value is 1

? 2

Stack Underfl ow on POP !!!

? 6

End of Run of your Program . . .
```

3.3 Queues

A ***queue*** *is an ordered collection of elements in which insertions are made at one end and deletions are made at the other end.* The end at which insertions are made is referred to as the rear end, and the end from which deletions are made is referred to as the front end. The first element placed in a queue will be at the start of the queue. The last element placed in a queue will be at the end of the queue. In a queue, the first element inserted will be the first element to be removed. So a queue is sometimes referred to as ***First-In-First-Out*** (FIFO) lists.

Consider five persons waiting in front of a ticket counter in a line for buying their tickets. The person who is standing in front of the line will get the first ticket, the second person will get the next ticket, and so on. If a new person wants to buy a ticket (say the sixth person) he should stand after the fifth person. These similar operations are carried out in a queue.

Queues have many applications in computer systems. Most computers have only a single processor, so only one user may be served at a time. Entries from other users are placed in a queue. Each entry gradually advances to the front of the queue as users receives their service. Queues are also used to support print spooling. *Fig. 3.3* shows the representation of a queue. Queue shown in *Fig. 3.3* consists of 5 elements 1, 2, 3, 4 and 5.

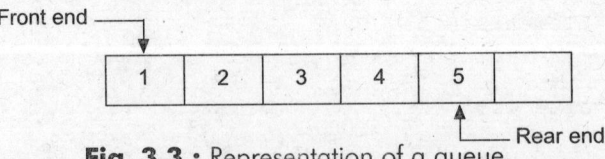

Fig. 3.3 : Representation of a queue.

Where front end is the pointer pointing to the first element in the queue, rear end is the pointer pointing to the last element in the queue. Element 1 is the first element of the queue, and 5 is the last element in the queue. If you want to delete an element say 3, you have to first delete element 1 and then element 2 and then the element 3. The front end is shifted from 1ˢᵗ element 1 to 4.

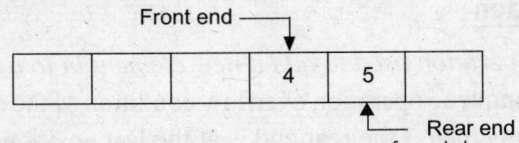

Fig. 3.4 : Queue representation after deletion.

Similarly, if you want to add a new element (say 6) it is added after 5 because it is in the rear end. After inserting a new element, the rear end is shifted from element 5 to element 6, which is the last position of the queue. Beyond this position, you cannot insert any elements in the queue.

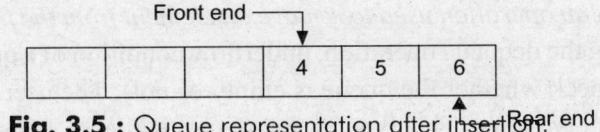

Fig. 3.5 : Queue representation after insertion.

The primary operations that can be done on a queue are insertions and deletions. These operations on a queue are referred as ***enqueue*** and ***dequeue*** operations respectively. An enqueue operation adds a new element in a queue. This process is carried out by incrementing the rear end and adding a new element at the rear end position. A dequeue operation is carried out by incrementing the front end and deleting the first element at the front end position.

3.4 Arrays and its Representation using Queue

One of the simplest ways of representing a queue is by means of a single dimensional array. Both queues and arrays are ordered collection of elements. However, an array and a queue are two different things. The number of elements in the array is fixed and it is not possible to change the number of elements stored in the array. But the size of a queue is constantly changed when elements are enqueued and dequeued. The queue is stored in a part of the array, so an array can be declared large enough to hold the maximum number of elements of the queue. During execution of a program, the queue size can be varied within the space reserved for it. One end of the array (i.e., bottom of the queue) is referred as the rear end and the other end of the array (i.e., top of the queue) is referred as the front end. Care must be taken in handling extreme cases, like not to insert an element into a fully occupied queue and not to delete an element from an empty queue. To represent a linear queue using arrays, we need,

- An array to hold the elements of a queue, which can be of any data type such as int, float or char etc.,

- An integer (i.e., FRONT) to store the position of the first element of the queue.

- Another integer (i.e., REAR) to store the position of the last element of the queue.

3.4.1 Enqueue Operation

ENQUEUE is an operation used to add a new element in to a queue at the rear end. When implementing the enqueue operation overflow condition of the queue is to be checked. (Since we cannot insert any data if the rear end is at the last position of the queue).

The following procedure helps you to understand things better.

- Create a function **ENQUEUE()** with two arguments, the array and element to be added.

- Assign the new element in the array by incrementing the REAR variable.

3.4.2 Dequeue Operation

DEQUEUE is an operation used to remove an element from the front end of the queue. When implementing the dequeue operation, underflow condition of a queue is to be checked (i.e., You have to check whether the queue is empty or not). The user should not delete an element from an empty queue. This type of an attempt is illegal and should be avoided. If such an attempt is made, the user should be informed of the underflow condition.

The following procedure can help you to understand things better.

- Create a function **DEQUEUE()** with two arguments, the array and the address of the element to store the dequeued value.

- Increment the FRONT variable by 1.

3.4.3 PEEK Operation

PEEK is an operation used to display the element from the front of the queue, pointed by the FRONT pointer without actually removing it. PEEK operation of a queue is implemented using arrays. When implementing the PEEK operation, underflow condition of a queue is to be checked (i.e., You have to check whether the queue is empty or not). The user should not try to PEEK an element from an empty queue. This type of an attempt is illegal and should be avoided. If such an attempt is made, the user should be informed of the underflow condition. The following procedure can help you to understand things better.

- Create a function **PEEK()** with one argument, the address of the element to store the front value.

3.4.4 Empty Queue

*If a queue contains no elements, it is referred as an **empty queue**.* If an element is to be dequeued from a queue, you have to check whether the queue is empty or not, because elements cannot be dequeued from an empty queue. The following procedure can be used to create and check an empty queue.

- Create a function **ISEMPTY()** with no arguments.
- Check whether the FRONT pointer and REAR pointer is equal to -1.
- If the condition is true, it means, that the queue is empty.
  Otherwise, the queue is not empty.

3.4.5 Fully Occupied Queue

If a queue contains elements equal to its size (i.e., QSIZE), *we say that the queue is full.* If an element is to be inserted into a queue, you have to check whether the queue is full or not, since elements cannot be inserted into a queue when it is fully occupied. This type of an attempt is illegal and should be avoided. If such an attempt is made, the user should be informed of the overflow condition.

To check whether the queue is full or not, we are using a pointer REAR in *Program 3.1*. We are checking whether the pointer REAR is equal to the maximum size of the array minus one (Since an array starts from zero, if the size of the array is 5, and if the queue is full the last element of the queue is stored in the array location 4). If the condition is satisfied the output message, "Queue Overflow on ENQUEUE" is displayed. This condition is referred as *fully occupied queue.* The following procedure can help you to understand things better.

- Create a function **isFull()** with no arguments.
- Check whether the REAR pointer is equal to QSIZE – 1.
- If the condition is true, it means, that the queue is full. Otherwise, the queue is not full.

Program 3.2 :

/* Program to implement a linear queue using arrays */

```
/* lqueue.c */
#define QSIZE 4              /* Maximum size of the Queue */
void displayMenu();
int isEmpty();
int isFull();
int size();
int enqueue(int value);
int dequeue(int *value);
int peek(int *value);
void view();
int queue[QSIZE], front = -1, rear = -1;
void main()
 {
    int status, choice, data;
    displayMenu();
    while(1)
     {
        printf("\n ? ");
          scanf("%d", &choice);
```

```
        switch(choice)
        {
            case 1 :
                    printf("Enter the element : ");
                    fflush(stdin);
                      scanf("%d", &data);
                    status = enqueue(data);
                    if(status == -1)
                      printf("Queue Overflow on ENQUEUE...");
                    break;
            case 2 :
                    status = dequeue(&data);
                    if(status == -1)
                      printf("Queue Underflow on DEQUEUE ...");
                    else
                      printf("The Dequeued Value is %d", data);
                    break;
            case 3 :

                    status = peek(&data);
                    if(status == -1)
                      printf("\n Queue is Empty");
                    else
                      printf("\n The Top most value is %d", data);
                    break;
            case 4 :
                    printf("\n Queue Size is = %d", QSIZE);
                    printf("\n Current Queue Elements = %d", size());
                    break;
            case 5 :
                    view();
                    break;

            default:
                    printf("\n End of Run of your Program . . .");
                    exit(0);

        }
    }
}
```

```c
void displayMenu()
{
    printf("\n Representation of Linear Queue using Arrays ...");
    printf("\n\t 1. Enqueue ");
    printf("\n\t 2. Dequeue ");
    printf("\n\t 3. Peek ");
    printf("\n\t 4. Size ");
    printf("\n\t 5. View ");
    printf("\n\t 6. Exit ");
}
int isEmpty()
{
    extern int queue[], front, rear;
    if(front == -1 && rear == -1)            /* Check for Empty Queue */
        return 1;
    else
        return 0;
}
int isFull()
{
    extern int queue[], front, rear;
    if(rear == (QSIZE - 1))                  /* Check for Occupied Queue */
        return 1;
    else
        return 0;
}
int enqueue(int value)
{
    extern int queue[], front, rear;
    if(isFull())                             /* Check for Rear at last position */
        return -1;
    else if(isEmpty())                       /* Check for Empty Queue */
        front = rear = 0;
    else
        rear = rear + 1;
    queue[rear] = value;
    return 0;
}
int dequeue(int *value)
{
    extern int queue[], front, rear;
```

```
        if(isEmpty())                                    /* Check for Empty Queue */
          return -1;
        *value = queue[front];
        if(front == rear)          /*Check for Queue contains only one element*/
          front = rear = -1;
        else
          front = front + 1;
        return 0;
    }
int peek(int *value)
    {
        extern int queue[], front, rear;
        if(isEmpty())                                /* Check for Empty Queue */
          return -1;
        *value = queue[front];
        return 0;
    }
int size()
    {
        extern int queue[], front, rear;
        return (rear - front + 1);
    }
void view()
    {
        extern int queue[], front, rear;
        int f;
        if(isEmpty())
          {
            printf("Queue is EMPTY !!!");
            return;
          }
        printf("\nContent of the Queue is ... \n FRONT -->");
        for(f = front; f != rear+1; f = f + 1)
          {
            printf(" %d --> ", queue[f]);
          }
        printf(" REAR");
        if(isFull())
          printf("\n Queue is FULL");
    }
```

The program displays the following output

```
Representation of Linear Queue using Arrays ...
    1. Enqueue
    2. Dequeue
    3. Peek
    4. Size
    5. View
    6. Exit
? 2
Queue Underflow on DEQUEUE ...
? 3
Queue is Empty
? 4
Queue Size is = 5
Current Queue Elements = 1
? 5
Queue is EMPTY !!!
? 1
Enter the element : 1001
? 1
Enter the element : 1002
? 1
Enter the element : 1003
? 1
Enter the element : 1004
? 1
Enter the element : 1005
Queue Overflow on ENQUEUE...
? 5
Content of the Queue is ...
FRONT -> 1001 ->  1002 ->  1003 ->  1004 -> REAR
Queue is FULL
? 4
Queue Size is = 4
Current Queue Elements = 4
? 3
The Top most value is 1001
? 2
The Dequeued Value is 1001
```

```
? 2
The Dequeued Value is 1002
? 5
Content of the Queue is ...
FRONT -> 1003 ->   1004 -> REAR
? 4
Queue Size is = 4
Current Queue Elements = 2
? 2
The Dequeued Value is 1003
? 2
The Dequeued Value is 1004
? 2
Queue Underflow on DEQUEUE ...
? 5
Queue is EMPTY ! ! !
? 6
End of Run of your Program . . .
```

3.5 Linked Lists

Linked list or list is an ordered collection of elements. Each element in the list is referred as a node. Each node contains two fields namely,

- Data field
- Link field.

The data field contains the actual data (i.e., value) of the element to be stored in the list and the link field also referred as the next address field contains the address of the next node in the list. *Fig. 3.6* helps you understand things better.

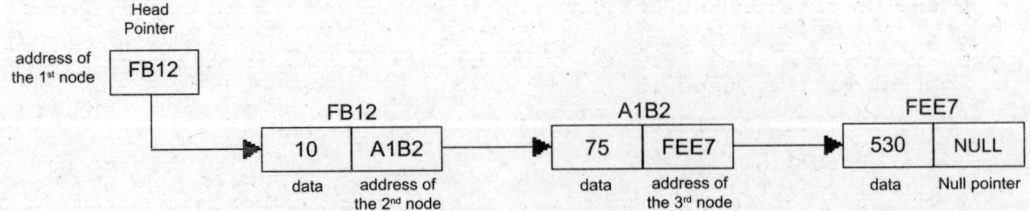

Fig. 3.6 : Representation of a linked list.

The linked list shown in *Fig. 3.6* consists of three nodes, each with a data field and a link field. A linked list contains a pointer, referred as the *head pointer*, which points to the first node in the list that stores the address of the first node of the list (i.e., FB12). The data field contains the actual information which is to be

stored in the list. The data field of the first node stores the value 10. The link field of the first node contains the address of the second node (i.e., A1B2). Similarly, the second node of the list stores the value 75 in the data field and the address of the third node (i.e., FEE7) in the link field. The last node of the list contains only the information part (i.e., 530) in the data field and the address field stores the NULL pointer. This NULL pointer is used to indicate the end of a list. *Note that the nodes in a linked list are not ordered by their physical placement in memory (i.e., they may/may not be placed in contiguous memory locations) but by logical links stored as a part of the data in the node itself.*

The address stored in a linked list are divided into three types namely,

- External address
- Internal address
- Null address.

External address is the address of the first node in the list. This address is stored in the head pointer which points to the first node in the list. The entire linked list can be accessed only with the help of the head pointer. *Internal address* is the address stored in each and every node of the linked list except the last node. The content stored in the link field (also referred as next address field) is the address of the next node. *Null address* is the address stored by the NULL pointer of the last node of the list, which indicates the end of the list.

3.5.1 Types of Linked Lists

There are different types of linked lists. They can be classified as,

- Singly linked list
- Doubly linked list
- Circular linked list.

Singly Linked List

The list that we have seen so far is referred to as the singly linked list, in which each node has a single link to its next node. This list is also referred as a *linear linked list.* The head pointer points to the first node in the list and the null pointer is stored in the link field of the last node in the list, which indicates end of list. *Fig. 3.6* shows a singly linked list. You can traverse (move) in a singly linked list in only one direction (i.e., from head to null in a list). You cannot traverse the list in the reverse direction (i.e., from null to head in a list).

Doubly Linked List

For some applications, especially where it is necessary to traverse lists in both directions, doubly linked lists work much better than singly linked lists. *Doubly linked list is an advanced form of a singly linked list, in which each node contains three fields namely,*

- Previous address field
- Data field
- Next address field.

The previous address field of a node contains address of its previous node. This field is also referred as the ***backward link field***. The data field stores the information part of the node. The next address field contains the address of the next node in the list. This field is also referred as the ***forward link field***. *Fig. 3.7* shows the structure of a doubly linked list.

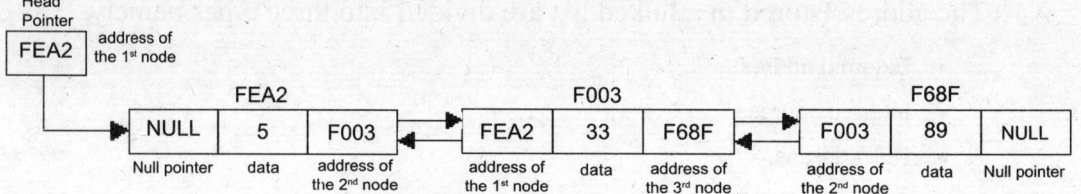

Fig. 3.7 : Doubly linked list.

Circular Linked list

Circular linked list is another form of a linked list in which the last node of the list is connected to the first node in the list. There are different types of circular linked lists. They can be classified as,

- Circular singly linked list
- Circular doubly linked list.

Note that a circular linked list looks like a cyclic list and will not have any end-of-list (i.e., there is no NULL pointer). *Fig. 3.8* shows the structure of a circular linked singly list. *Fig. 3.9* shows the structure of a circular linked doubly list.

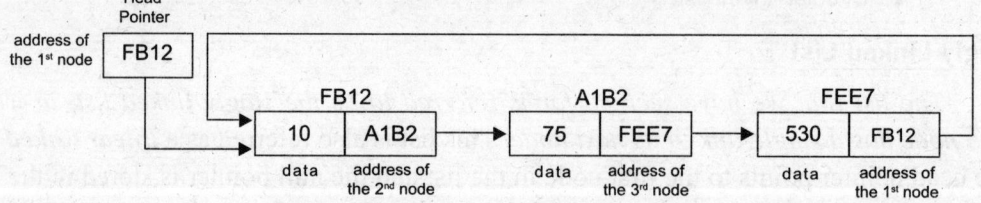

Fig. 3.8 : Circular singly linked list.

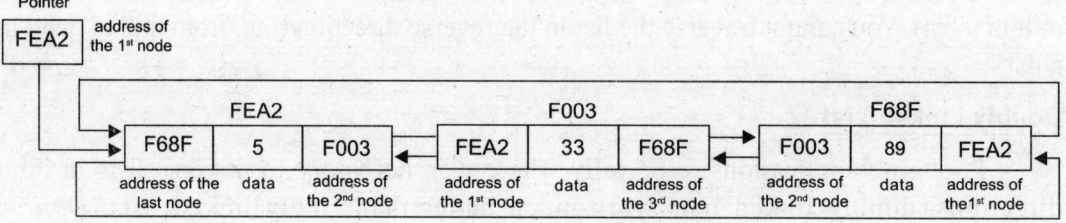

Fig. 3.9 : Circular doubly linked list.

3.6 Linked List based Implementation of Stacks

Another way of representing a stack is by means of a singly linked list. This type of representation has more advantages than representing stacks using arrays. They are,

- It is not necessary to specify the number of elements to be stored in a stack during its declaration (since memory is allocated dynamically at run time when an element is added to the stack).
- Insertions and deletions can be handled easily and efficiently.
- Linked list representation of stacks can grow and shrink in size without wasting the memory space, depending upon the insertion and deletion that occurs in the list.
- Multiple stacks can be represented efficiently using a chain for each stack.

The basic operations that can be performed on a stack are,

- Push operation
- Pop operation
- Peek operation
- View stack size
- View stack contents.

3.6.1 Push Operation

PUSH is an operation used to add a new element in to a stack. The PUSH operation of a stack can also be implemented using singly linked lists. The following procedure helps you to understand things better.

- Create a function PUSH() with two arguments (pointer of stack and VALUE to be inserted).
- Create a new pointer (i.e., NEWPTR) to hold the new element.
- Allocate size of the stack node to the new pointer using GETNODE().
- The size of the stack depends on the heap memory available.
- If GETNODE() returns NULL heap memory is not available.
- Assign the VALUE to the data of new pointer.
- Assign the link of the new pointer to the head (i.e., TOP) of the stack.
- Assign the head of the stack to the new pointer.

Algorithm for Declaration of Structure Stack

Struct STACK
 DATA : Data Field
 ·NEXT : Link Field (Address of next Struct STACK)
End Struct

Algorithm for Allocating Memory for the New Node

GETNODE()
SIZE : INTEGER
NEWNODE : STACK
Step 1 : Set SIZE = get the size of the STACK
Step 2 : Set NEWNODE = Allocate space in memory for SIZE and return the initial address
Step 3 : Return NEWNODE
End GETNODE()

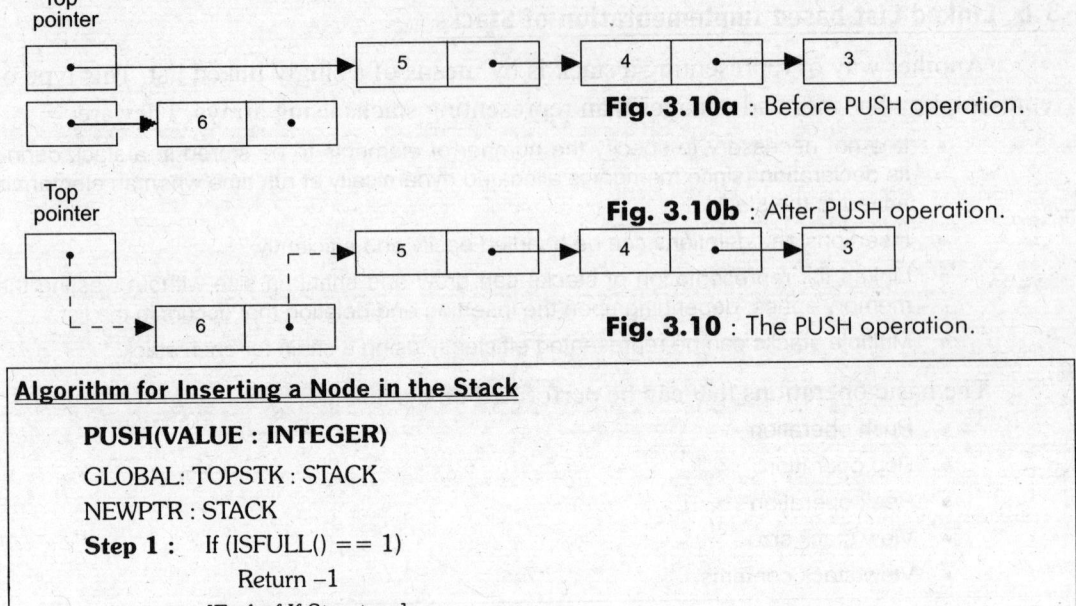

Fig. 3.10a : Before PUSH operation.

Fig. 3.10b : After PUSH operation.

Fig. 3.10 : The PUSH operation.

Algorithm for Inserting a Node in the Stack

    PUSH(VALUE : INTEGER)

    GLOBAL: TOPSTK : STACK

    NEWPTR : STACK

    Step 1 :   If (ISFULL() == 1)

               Return –1

               [End of If Structure]

    Step 2 :   NEWPTR = GETNODE()

    Step 3 :   NEWPTR –> DATA = VALUE

    Step 4 :   NEWPTR –> NEXT = TOPSTK

    Step 5 :   TOPSTK = NEWPTR

    Step 6 :   Return 0

    End PUSH()

Fig. 3.10a shows the stack and the new node before the PUSH operation. In *Fig. 3.10b* the TOP pointer is made to point the NEW pointer of the new node. The element of the new node is made to point the node, which was already pointed by the TOP pointer, and the other nodes remain unchanged.

3.6.2 Pop Operation

POP is an operation used to remove an element from the TOP of the stack. The POP operation of a stack can also be implemented using singly linked lists. When implementing the POP operation, underflow condition of a stack is to be checked. The user should not POP an element from an empty stack. This type of an attempt is illegal and should be avoided. The following procedure can help you to understand things better.

- Create a function POP() with two parameters that stores pointer of the stack and the address of the last element to be deleted.
- Create a temporary pointer (i.e., tempptr) to hold the removed element.
- Assign the TOP pointer to the temporary pointer.
- The TOP pointer is made to point the node after the first node, and the other nodes remain unchanged.
- Free the allocated memory of the temporary pointer by using **RELEASENODE().**

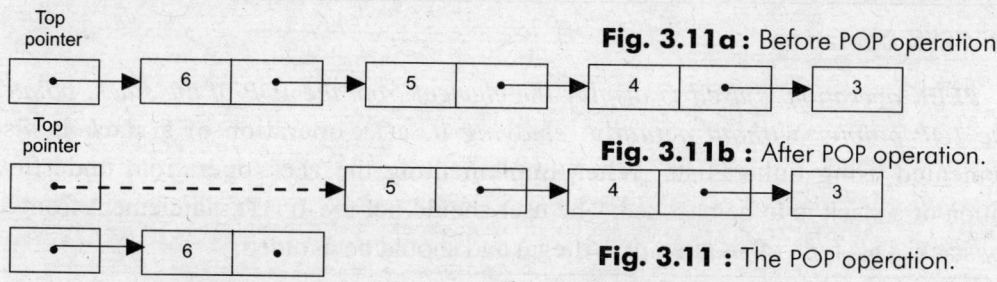

Fig. 3.11a : Before POP operation.

Fig. 3.11b : After POP operation.

Fig. 3.11 : The POP operation.

Algorithm for Releasing the Memory for the Node to be Deleted

RELEASENODE(NEWNODE : STACK)

Step 1 : Deallocate the space for the STACK of NEWNODE

Step 2 : Return

End RELEASENODE()

Algorithm for Removing a Node from the Stack

POP(VALUE : INTEGER)

GLOBAL: TOPSTK : STACK

TEMP : STACK

Step 1 : If (ISEMPTY() == 1)

 Return –1

 [End of If Structure]

Step 2 : TEMP = TOPSTK

Step 3 : TOPSTK = TOPSTK –> NEXT

Step 4 : VALUE = TEMP –> DATA

Step 5 : RELEASENODE(TEMP)

Step 6 : Return 0

End POP()

Fig. 3.11 illustrates the POP operation *Fig. 3.11a* shows the stack before the POP operation. In *Fig. 3.11b* a temporary pointer (i.e., `tempptr`) is created and is made to point the node, which is to be removed (i.e., The node pointer by the TOP pointer). The TOP pointer is made to point the node after the first node, and the other nodes remain unchanged. The memory allocated by the temporary pointer is freed. Thus, the element pointed by the TOP pointer is removed from the stack. *Note that stacks and linked lists are represented identically. The difference between stacks and ordinary linked lists is that insertions and deletions may occur any where in a linked list, but only at the top node in a stack.*

3.6.3 PEEK Operation

PEEK operation is used to display the element from the TOP of the stack, pointed by the TOP pointer without actually removing it. PEEK operation of a stack is also implemented using linked lists. When implementing the PEEK operation, underflow condition of a stack is to be checked. The user should not try to PEEK an element from an empty stack. This type of an attempt is illegal and should be avoided.

Algorithm to Retrieve the Contents from the Top of the Stack

PEEK(VALUE : INTEGER)

GLOBAL: TOPSTK : STACK

TEMP : STACK

Step 1 :   If (ISEMPTY() == 1)

                   Return –1

                   [End of If Structure]

Step 2 :   TEMP = TOPSTK

Step 3 :   VALUE = TEMP –> DATA

Step 4 :   Return 0

End PEEK()

3.6.4 Traversing a Stack

To read the information or to display the information in a stack, you have to traverse (move) a linked list, node by node from the first node, until the end of the list is reached. Traversing a list involves the following steps,

- Assign the address of TOP pointer to a variable.
- Display the information in the data field.
- Traverse the list from one node to another by advancing the pointer upto NULL.

Algorithm for Displaying the Contents of the Stack

VIEW()

GLOBAL: TOPSTK : STACK

TOP : STACK

Step 1 :   If (ISEMPTY() == 1)

                   Print "Stack is EMPTY"

                   Return

                   [End of If Structure]

Step 2 :   TOP = TOPSTK

Step 3 :   Do While(TOP != NULL)

                   Print "Value is", TOP -> DATA

                   TOP = TOP -> NEXT

                   [End of Step 3 While Structure]

End VIEW()

3.6.5 Empty Stack

*If a stack contains no elements, it is referred to as an **empty stack**.* If an element is to be popped from a stack, you have to check whether the stack is empty or not, because elements cannot be popped from an empty stack. In a linked list, if the stack is pointing to NULL the stack is said to be an empty stack. If you want to pop an element, from an empty stack the output, "Stack is underflow" is displayed.

```
Algorithm for IS_EMPTY()

    IS_EMPTY(VALUE : INTEGER)
    GLOBAL: TOPSTK : STACK
    Step 1 :    If (TOPSTK == NULL)
                    Return 1
                Else
                    Return 0
                [End of If Structure]
    End IS_EMPTY()
```

3.6.6 Fully Occupied Stack

If a stack contains elements equal to its size, we say that the stack is full. If an element is to be pushed into a stack, you have to check whether the stack is full or not, since elements cannot be pushed into a stack when it is fully occupied. In case of stacks, represented using linked lists, there is no limit to store the elements in the stack, since the memory required to store the elements of the stack is allocated dynamically using the malloc() function. However, if the memory required for storing the elements of the stack is not available, we cannot insert elements into a stack. If the malloc() function returns NULL in a linked list it is referred as a ***fully occupied stack.*** If you want to insert an element, in a fully occupied stack the output "Stack is overflow" is displayed.

```
Algorithm for IS_FULL()

    IS_FULL(VALUE : INTEGER)
    GLOBAL: TOPSTK : STACK
    TEMP : STACK
    Step 1 :    TEMP = GETNODE()
    Step 2 :    If (TEMP == NULL)
                    Return 1
                Else
                    RELEASENODE(TEMP)
                    Return 0
                [End of If Structure]
    End IS_FULL()
```

3.6.7 Stack Size

In case of stacks represented using linked lists, there is no limit to store the elements, since the memory required to store the elements of the stack is allocated dynamically using the malloc() *function.* To find the size of the stack, (i.e., number of elements that has been stored), to check whether the memory required for storing the elements of the stack is available or not, use the size() function as mentioned in *Program 3.2. Note that elements cannot be stored in a stack , when the memory pointed by the linked list points to NULL.*

Algorithm to Count the Nodes in a Stack

    SIZE(VALUE : INTEGER)
    GLOBAL: TOPSTK : STACK
    TOP : STACK; COUNT : INTEGER
    Step 1 :   TOP = TOPSTK
    Step 2 :   COUNT = 0
    Step 3 :   Do While(TOP != NULL)
                COUNT = COUNT + 1
                TOP = TOP –> NEXT
           [End of **Step 3** While Structure]
    Step 4 :   Return COUNT
    End SIZE()

Program 3.3 :

```
/* Program to implement a stack using singly linked list. */
/* linstack.c */
#include<stdio.h>
#include<conio.h>
#include<stdlib.h>
typedef struct node
  {
    int data;
    struct node *next;
  }stack;
int size();
void view();
int isFull();
int isEmpty();
void displayMenu();
int push(int value);
int pop(int *value);
int peek(int *value);
stack *getNode();
```

```
   void releaseNode(stack *newnode);

stack *topStk = NULL;

void main()
 {
   int data, status, choice;
   displayMenu();
   while(1)
    {
      printf("\n\n ? ");
        scanf("%d", &choice);
      switch(choice)
       {
         case 1 :
                    printf("\n Enter the element : ");
                    fflush(stdin);
                      scanf("%d", &data);
                    status = push(data);
                    if(status == -1)
                      printf("\n Memory is not available ...");
                    break;
         case 2 :
                    status = pop(&data);
                    if(status == -1)
                      printf("\n Stack underflow on POP");
                    else
                      printf("\n The popped value is %d", data);
                    break;
         case 3 :
                    status = peek(&data);
                    if(status == -1)
                      printf("\n Stack is Empty !!!");
                    else
                      printf("\n The Top most value is %d", data);
                    break;
         case 4 :
                    printf("\n Current Stack Elements = %d", size());
                    break;
```

```c
                    case 5 :
                            view();
                            break;
                default:
                            printf("\n End of Run of your Program . . .");
                            exit(0);
            }
        }
    }
    void displayMenu()
    {
        printf("\n Representation of stack using single linked lists ...");
        printf("\n\t 1. Push ");
        printf("\n\t 2. Pop  ");
        printf("\n\t 3. Peek ");
        printf("\n\t 4. Size ");
        printf("\n\t 5. View ");
        printf("\n\t 6. Exit ");
    }
    stack *getNode()
    {
        return((stack *)malloc(sizeof(stack)));
    }
    void releaseNode(stack *newnode)
    {
        free(newnode);
    }
    int push(int value)
    {
        extern stack *topStk;
        stack *newptr;
        if(isFull())
            return -1;
        newptr = getNode();
        newptr -> data = value;
        newptr -> next = topStk;
        topStk = newptr;
        return 0;
    }
```

```c
int pop(int *value)
{
   extern stack *topStk;
   stack *temp;
   if(isEmpty())
      return -1;
   temp = topStk;
   topStk = topStk -> next;
   *value = temp -> data;
   releaseNode(temp);
   return 0;
}
int peek(int *value)
{
   extern stack *topStk;
   stack *temp;
   if(isEmpty())
      return -1;
   temp = topStk;
   *value = temp -> data;
   return 0;
}
int isEmpty()
{
   extern stack *topStk;
   if(topStk == NULL)
      return 1;
   else
      return 0;
}
int isFull()
{
   extern stack *topStk;
   stack *temp;
   temp = getNode();
   if(temp == NULL)
      return 1;
```

```
         else
          {
              releaseNode(temp);
              return 0;
          }
      }
  int size()
   {
      extern stack *topStk;
      stack *top;
      int count = 0;
      for(top = topStk; top != NULL; top = top -> next)
         count++;
      return count;
   }
  void view()
   {
      extern stack *topStk;
      stack *top;
      if(isEmpty())

       {
          printf("\n The stack is Empty !!! ");
          return;
        }
      printf("\n The content of the stack is... TOP");
      for(top = topStk; top != NULL; top = top -> next)
         printf(" --> %d", top -> data);
   }
```

The program displays the following output

```
Representation of stack using single linked lists ...
        1. Push
        2. Pop
        3. Peek
        4. Size
        5. View
        6. Exit
? 2
Stack underflow on POP
```

```
? 3
Stack is Empty
? 4
Current Stack Elements = 0
? 5
The stack is Empty !!!
? 1
Enter the element : 101
? 1
Enter the element : 102
? 1
Enter the element : 103
? 1
Enter the element : 104
? 1
Enter the element : 105
? 5
The content of the stack is... TOP --> 105 --> 104 --> 103 --> 102
--> 101
? 4
Current Stack Elements = 5
? 2
The popped value is 105
? 2
The popped value is 104
? 5
The content of the stack is... TOP --> 103 --> 102 --> 101
? 4
Current Stack Elements = 3
? 2
The popped value is 103
? 2
The popped value is 102
? 5
The content of the stack is... TOP --> 101
? 2
The popped value is 101
? 2
Stack underflow on POP
? 5
The stack is Empty !!!
? 6
End of Run of your Program . . .
```

3.7 Linked List based Implementation of Queue

Another way of representing a queue is by means of a singly linked list. This type of representation has more advantages than representing queues using arrays. They are

- It is not necessary to specify the number of elements to be stored in a queue during its declaration (since memory is allocated dynamically at run time when an element is added to the queue).

- Insertions and deletions can be handled easily and efficiently.

- Linked list representation of queues can grow and shrink in size without wasting the memory space, depending upon the insertion and deletion that occurs in the list.

- Multiple queues can be represented efficiently using a chain for each queue.

The basic operations that can be performed on a queue are,

- Enqueue operation
- Dequeue operation
- Peek operation
- View queue size
- View queue contents.

3.7.1 Enqueue Operation

ENQUEUE is an operation used to add a new element in to a queue. The ENQUEUE operation of a queue can also be implemented using singly linked lists. The following procedure helps you to understand things better.

- Create a function ENQUEUE() with three parameters, FRONT pointer, REAR pointer of the queue and the ELEMENT to be inserted.

- Create a new pointer (i.e., NEWPTR) to hold the new element.

- Allocate size of the queue node to the new pointers dynamically using GETNODE().

- Assign the ELEMENT to the data of new pointer.

- Assign the link of the REAR pointer to the NEWPTR.

- Assign the REAR pointer of the queue to the NEWPTR.

Fig. 3.12a shows the queue and the new node before the ENQUEUE operation. In *Fig. 3.12b,* the link of the REAR pointer is made to point the new node and assign the rear pointer to the new node.

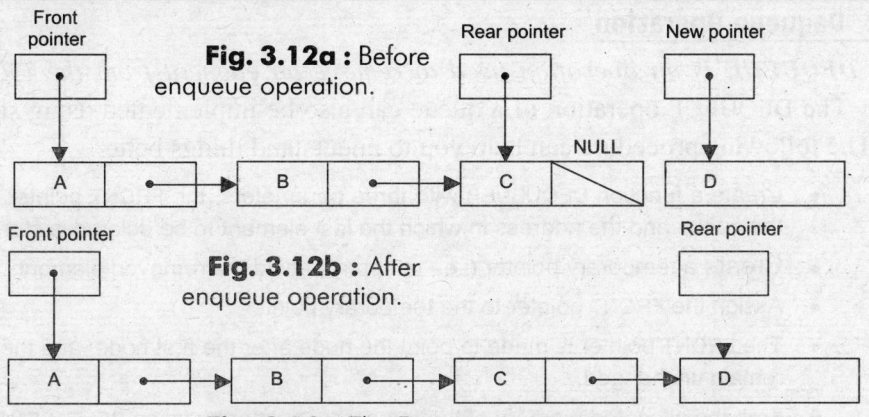

Fig. 3.12a : Before enqueue operation.

Fig. 3.12b : After enqueue operation.

Fig. 3.12 : The Enqueue operation.

Algorithm for Declaration of Structure Queue

Struct QUEUE
    DATA : Data Field
    NEXT : Link Field (Address of next Struct QUEUE)
End Struct

Algorithm for Allocating Memory for the New Node

GETNODE()
SIZE : INTEGER
NEWNODE : QUEUE
Step 1 :   Set SIZE = get the size of the QUEUE
Step 2 :   Set NEWNODE = Allocate the space from memory for the size of SIZE and
            return the initial address
Step 3 :   Return NEWNODE
End GETNODE()

Algorithm for Inserting a Node in the Queue

ENQUEUE(FRONT : QUEUE, REAR : QUEUE, VALUE : DATA)
NEWPTR : QUEUE
Step 1 :   Set NEWPTR = GETNODE()
Step 2 :   If (NEWPTR == NULL)
                Print "Memory is Not Available"
                Return -1
            [End of If Structure]
Step 3 :   Set NEWPTR -> DATA = value
Step 4 :   Set NEWPTR -> NEXT = NULL
Step 5 :   If (FRONT == NULL AND REAR == NULL)
                FRONT = NEWPTR
            Else
                REAR -> NEXT = NEWPTR
            [End of If Structure]
Step 6 :   Set REAR = NEWPTR
Step 7 :   Return 0
End ENQUEUE()

3.7.2 Dequeue Operation

DEQUEUE is an operation used to remove an element from the FRONT of the queue. The DEQUEUE operation of a queue can also be implemented using singly linked lists. The following procedure can help you to understand things better.

- Create a function DEQUEUE() with three parameters, the FRONT pointer, the REAR of the queue and the address in which the last element to be deleted is stored.
- Create a temporary pointer (i.e., tempptr) to hold the removed element.
- Assign the FRONT pointer to the temporary pointer.
- The FRONT pointer is made to point the node after the first node, and the other nodes remain unchanged.
- Free the allocated memory of the temporary pointer by using RELEASENODE().

Algorithm for Releasing the Memory for the Node to be Deleted

RELEASENODE(NEWNODE : QUEUE)

Step 1 : Deallocate the space for the QUEUE of NEWNODE

Step 2 : Return

End RELEASENODE()

Algorithm for Removing a Node from the Queue

DEQUEUE(FRONT : QUEUE, REAR : QUEUE, VALUE : DATA)

TEMP : QUEUE

Step 1 : If (FRONT == NULL AND REAR == NULL)

　　　　　　Print "QUEUE is Underflow"

　　　　　　Return –1

　　　　　　[End of If Structure]

Step 2 : Set TEMP = FRONT

Step 3 : Set VALUE = FRONT -> DATA

Step 4 : If (FRONT == REAR)

　　　　　　Set REAR = NULL

　　　　　　[End of If Structure]

Step 5 : Set FRONT = FRONT -> NEXT

Step 6 : RELEASENODE(TEMP)

Step 7 : Return 0

End DEQUEUE()

Fig. 3.13a shows the queue before the DEQUEUE operation. In *Fig. 3.13b* a temporary pointer (i.e., `tempptr`) is created and is made to point the node, which is to be removed (i.e., The node pointer by the FRONT pointer). The FRONT pointer is made to point the node after the first node, and the other nodes remain unchanged. The memory allocated by the temporary pointer is freed. Thus, the element pointed by the FRONT pointer is removed from the queue.

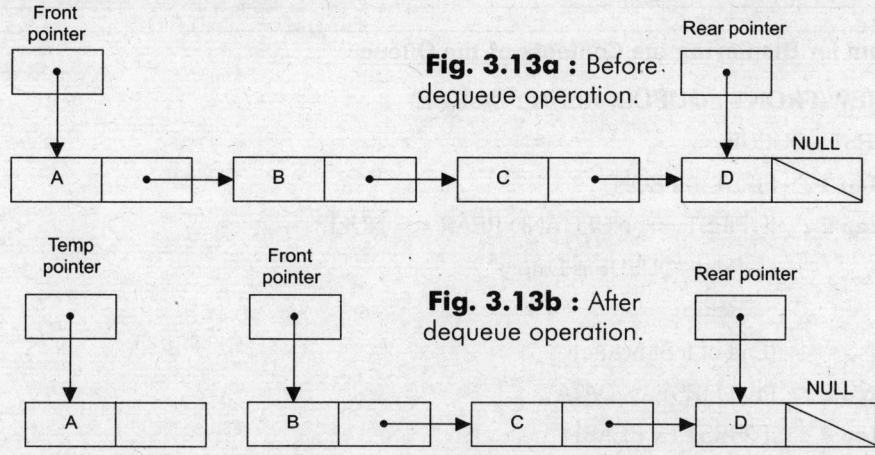

Fig. 3.13 : The Dequeue operation.

Note that queues and linked lists are represented identically. The difference between a queue and an ordinary linked list is that insertions and deletions may occur anywhere in a linked lists, but in a queue, insertions can be made only in the rear end and deletions can be made only in the front end.

3.7.3 PEEK Operation

PEEK operation is used to display the element from the FRONT of the queue, pointed by the FRONT pointer without actually removing it. PEEK operation of a queue is also implemented using linked lists. When implementing the PEEK operation, underflow condition of a queue is to be checked. The user should not try to PEEK an element from an empty queue. This type of an attempt is illegal and should be avoided.

Algorithm to Retrieve the Contents from the Top of the Stack

    PEEK(FRONT : QUEUE, VALUE : DATA)

Step 1 :   If (ISEMPTY(FRONT))

               Print "Queue is Empty"

               Return

            [End of If Structure]

Step 2 :   VALUE = FRONT –> DATA

End PEEK()

3.7.4 Traversing a Queue

To read the information or to display the information in a queue, you have to traverse (move) a linked list, node by node from the FRONT node, until the REAR node is reached. Traversing a list involves the following steps,

- Assign the address of FRONT pointer to a variable.
- Display the information in the data field.
- Traverse the list from one node to another by advancing the pointer upto REAR.

Algorithm for Displaying the Contents of the Queue

VIEW(FRONT : QUEUE, REAR : QUEUE)

FIRST : QUEUE

Step 1 : FIRST = FRONT

Step 2 : If (FIRST == NULL AND REAR == NULL)

Print "QUEUE is Empty"

Return

[End of If Structure]

Step 3 : Print FIRST -> DATA

Step 4 : If (FIRST != REAR)

FIRST = FIRST -> NEXT

Goto **Step 3**

[End of If Structure]

End VIEW()

3.7.5 Empty Queue

*If a queue contains no elements, it is referred as an **empty queue**.* If an element is to be dequeued from a queue, you have to check whether the queue is empty or not, because elements cannot be dequeued from an empty queue. If the FRONT and REAR of the queue points to NULL in a linked list, it is referred as an empty queue. If you want to dequeue an element from an empty queue the output "Queue is underflow" is displayed.

Algorithm for IS_EMPTY()

IS_EMPTY(FRONT : QUEUE)

Step 1 : If (FRONT == NULL)

Return 1

Else

Return 0

[End of If Structure]

End IS_EMPTY()

3.7.6 Fully Occupied Queue

If a queue contains elements equal to its size, we say that the queue is full. If an element is to be enqueued into a queue, you have to check whether the queue is full or not, since elements cannot be enqueued into a queue when it is fully occupied. In case of queues, represented using linked lists, there is no limit to store the elements in the queue, since the memory required to store the elements of the queue is allocated dynamically using the `malloc()` function. However, if the memory required for storing the elements of the queue is not available, we cannot enqueue elements into a queue. If the `malloc()` function returns NULL in a linked list it is referred as an ***fully occupied queue***. If you need to enqueue an element from a fully occupied queue the output "`Queue is overflow`" is displayed.

Algorithm for IS_FULL()

    IS_FULL()

    TEMP : QUEUE

    Step 1 :   TEMP = GETNODE()

    Step 2 :   If (TEMP == NULL)

             Return 1

        Else

             RELEASENODE(TEMP)

             Return 0

        [End of If Structure]

    End IS_FULL()

3.7.7 Queue Size

In case of queues represented using linked lists, there is no limit to store the elements, since the memory required to store the elements of the queue is allocated dynamically using the `malloc()` *function.* To find the size of the queue, (i.e., number of elements that have been stored) to check whether the memory required for storing the elements of the queue is available or not, use the `size()` function as mentioned in ***Program 3.2. Note that elements cannot be stored in a queue, when the memory pointed by the linked list points to NULL.***

Algorithm to Count the Nodes in a Queue

    SIZE(FRONT : QUEUE)

    COUNT : INTEGER

    Step 1 :   COUNT = 0

    Step 2 :   If (FRONT == NULL)

             Return COUNT

        [End of If Structure]

    Step 3 :   Do While(FRONT != NULL)

             COUNT = COUNT + 1

             FRONT = FRONT –> LINK

        [End of **Step 3** While Structure]

    Step 4 :   Return COUNT

    End SIZE()

Program 3.4 :

```
/* Program to implement a Queue using singly linked list. */
/* linqueue.c */
#include<stdio.h>
typedef struct node
 {
    int data;
    struct node *link;
 }queue;
queue* getNode();
void releaseNode(queue *p);
int isFull();
int isEmpty(queue *front);
void enqueue(queue **frontptr, queue **rearptr, int value);
void dequeue(queue **frontptr, queue **rearptr, int *value);
void peek(queue *front, int *value);
void view(queue *front);
int size(queue *front);
void displayMenu(void);
void main()
  {
    queue  *front = NULL, *rear = NULL;
    int choice, item;
    displayMenu();
    while(1)
     {
       printf("\n ? ");
         scanf("%d", &choice);
       switch(choice)
        {
          case 1 :
                  if(isFull())
                     printf("\n Queue Overflow on ENQUEUE");
                  else
                   {
                     printf("\n Enter the element : ");
                     fflush(stdin);
                     scanf("%d", &item);
                     enqueue(&front, &rear, item);
                   }
                  break;
```

```
                case 2 :
                        if(isEmpty(front))
                            printf("\n Queue Underflow on DEQUEUE");
                        else
                         {
                            dequeue(&front, &rear, &item);
                            printf("\nThe dequeued value is %d", item);
                         }
                    break;
            case 3 :
                        if(!isEmpty(front))
                         {
                            peek(front, &item);
                            printf("\n The front value is %d", item);
                         }
                        else
                        printf("\n Queue is Empty");
                        break;
            case 4 :
                        printf("\nCount of Queue Elements=%d",size(front));
                        break;
            case 5 :
                        view(front);
                        break;
            default :
                        printf("\n End of run of your program . . .");
                        exit(0);
        }
    }
}
void displayMenu()
{
    printf("\n Representation of Queue using Linked List ...");
    printf("\n\t 1. Enqueue ");
    printf("\n\t 2. Dequeue ");
    printf("\n\t 3. Peek ");
```

```
    printf("\n\t 4. Size ");

    printf("\n\t 5. View ");

    printf("\n\t 6. Exit ");

}

void releaseNode(queue *p)

{

    free(p);

}

queue* getNode()

{

    int size;

    queue * newnode;

    size = sizeof(queue);

    newnode = (queue *)malloc(size);

    return(newnode);

}

int isEmpty(queue *front)

{

    if(front == NULL)              /* Check for Empty Queue */

        return 1;

    else

        return 0;

}

int isFull()

{

    queue * newnode;

    newnode = getNode();

    if(newnode == NULL)      /* Check for Fully Occupied Queue */

        return 1;

    releaseNode(newnode);

    return 0;

}
```

```c
    void enqueue(queue **frontptr,queue **rearptr, int value)
    {
      queue *newnode;
      if(isFull())
       {
        printf("\n Memory not available");
        return;
       }
      newnode = getNode();
      newnode->data = value;
      newnode->link = NULL;
      if(*frontptr == NULL)
        *frontptr = newnode;
      else
        (*rearptr)->link = newnode;
      *rearptr = newnode;
    }
   void dequeue(queue **frontptr, queue **rearptr, int *value)
    {
      queue *tempnode;
      if(isEmpty(*frontptr))
        return;
      tempnode = *frontptr;
      *frontptr = (*frontptr) -> link;
      if(*frontptr == NULL)
        *rearptr = NULL;
      *value = tempnode -> data;
      releaseNode(tempnode);

    }

   void peek(queue *front, int *value)
   {
      if(isEmpty(front))
       {
        printf("\n The queue is empty !!!");
        return;
       }
      *value = front -> data;
    }
```

```c
   void view(queue *front)
    {
      if(isEmpty(front))
       {
         printf("\n The queue is empty !!!");
         return;
       }
      printf("\n Queue contains ... Front -> ");
      while(front != NULL)
       {
         printf("%d --> ", front ->data);
         front = front -> link;
       }
      printf("Rear \n");
    }
   int size(queue *front)
    {
      int count = 0;
      if(front == NULL)
        return count;
      for(;front != NULL;)
       {
         count++;
         front = front -> link;
       }
      return count;
    }
```

The program displays the following output

```
Representation of Queue using Linked List ...

        1. Enqueue

        2. Dequeue

        3. Peek

        4. Size

        5. View

        6. Exit
? 2
Queue Underflow on DEQUEUE
```

```
? 3
Queue is Empty !!!
? 4
Count of Queue Elements = 0
? 5
The queue is empty !!!
? 1
Enter the element : 1
? 1
Enter the element : 2
? 1
Enter the element : 3
? 1
Enter the element : 4
? 5
Queue contains ... Front -> 11 -> 22 -> 33 -> 44 -> Rear
? 4
Count of Queue Elements = 4
? 3
The front value is 11
? 2
The dequeued value is 11
? 2
The dequeued value is 22
? 4
Count of Queue Elements = 2
? 5
Queue contains ... Front -> 33 -> 44 -> Rear
? 2
The dequeued value is 33
Queue contains ... Front -> 44 -> Rear
? 2
The dequeued value is 44
The queue is empty !!!
? 2
Queue Underflow on DEQUEUE
? 4
Count of Queue Elements = 0
? 5
The queue is empty !!!
? 6
End of Run of your Program . . .
```

3.8 Evaluation of Arithmetic Expressions

An expression consists of two components namely operands and operators. Operators indicate the operation to be carried out on operands. There are two kinds of operators used in evaluating expressions, namely **unary** and **binary** operators. *Unary operators require only one operand to carry out its intended operation whereas binary operators require two operands to carry out its intended operation.* Most operators used are binary in nature. Computers solve arithmetic expressions by restructuring them. So that the order of each calculation is embedded in the expression. Once converted, an expression can then be solved in one pass.

There are three ways of representing expressions in computers. They are,

- Infix notation
- Prefix notation
- Postfix notation.

The prefixes of the notations "in", "pre", and "post" refers to the position of the operands with respect to the operators. All the ways mentioned above, use the operator and operand in different ways in evaluating an expression. *Table 3.1* shows the different ways of representing an expression.

Table 3.1 : Different ways of representing an expression

Notation	Arithmetic expression			Example
INFIX	Operand	OPERATOR	Operand	a+b
PREFIX	OPERATOR	Operand	Operand	+ab
POSTFIX	Operand	Operand	OPERATOR	ab+

The Infix Notation

The normal way of expressing mathematical expressions is called as infix notation. In this form of expressing an arithmetic expression the operator comes in between its operands. For example, an expression to add numbers a and b is written in infix form as

```
( a + b )
```

Note that the operator + is written in between the operands a and b. We are accustomed to this type of infix notations. However, algorithmically, postfix and prefix notations are easier to evaluate than infix notation. Hence, let us discuss the conversion of infix to postfix notation and infix to prefix notations.

Advantages of Infix Notations

- It is the mathematical way of representing the expression.
- It's easier to see visually which operation is done from first to last.

The Prefix Notation

Prefix notation also referred, as *polish notation* is a way of writing algebraic expressions without the use of parenthesis or rules of operator precedence. A Polish mathematician *Jan Lukasiewicz* introduced prefix notation. In this form of expressing an arithmetic expression, the operator is written before its operands. For example, an expression to add numbers a and b is written in prefix form as

```
( + a b )
```

Note that the operator "+" is written before the operands a and b.

The Postfix Notation

Postfix notation also referred, as *suffix form* or *reverse polish notation* (or RPN) is also a way of writing algebraic expressions without the use of parentheses or rules of operator precedence. This was also introduced by Jan Lukasiewicz. In this form of expressing an arithmetic expression the operator is written after its operands.

For example, an expression to add numbers "a" and "b" is written in postfix form as

```
( a b + )
```

Note that the operator + is written after the operands a and b.

Advantages of Postfix Notations

- You need not worry about the rules of precedence.
- You need not worry about the rules for right to left associativity.
- You need not need parenthesis to override the above rules.

To evaluate an infix expression, set of rules such as precedence, associativity of operators have to be followed, which is rather complex to remember. The prefix and postfix notations may look awkward to use as they might look at first. But the reasons for using these notations are that a fairly simple algorithm exists to evaluate arithmetic expressions using a stack.

3.8.1 Conversions of Notations

The main problem in evaluating an expression is to decide the order in which the operations are carried out. The order of evaluation of the expression (a+b) is quite simple. If the expression is complex, we require knowledge of the operators, their precedence to specify the order of evaluation of the expression.

To fix the order of evaluation for an expression, we assign priority for operators. The operators with the highest priority will be evaluated first. Since we give more importance to binary operators, the most important binary operations according to their order of priority are listed in *Table 3.2*. Note that by using parenthesis we can override the default precedence of operators.

Table 3.2 : Binary Opreations in their Order of Priority

Priority	Operation (symbol)
1	Exponentiation (↑)
2	Multiplication (*), Division (/)
3	Addition (+), Subtraction (−)

In *Table 3.2*, the priority is listed in the order of highest to lowest. When unparenthesized operators of the same precedence are scanned, the order of evaluation is assumed from right to left (except for exponentiation whose precedence is left to right).

To every infix expression, there corresponds a postfix expression that has the same effect. The reverse is not true because, there is an ambiguity. The infix expression x+y+z can be represented as either xyz++ or xy+z+ in postfix. The reason for the ambiguity is the lack of parenthesis in the infix expression. If the infix expression were fully bracketed, there would be no ambiguity. Thus (x+y)+z pairs with xy+z+ and x+(y+z) pairs with xyz++.

Rules to be followed during infix to postfix conversion

- Fully parenthesize the expression starting from left to right (During parenthesizing, the operators having higher precedence are first parenthesized).

- Move the operators one by one to their right, such that each operator replaces their corresponding right parenthesis.

- The part of the expression, which has been converted into postfix, is to be treated as single operand.

- Once the expression is converted into postfix form, remove all parenthesis.

Example 3.1

Evaluate the infix expression 3 + 8 * 4 / 2 - (8 - 3) in postfix notation.

SOLUTION :

The order of evaluation of infix expression is as follows,

Infix expression : 3 + 8 * 4 / 2 − (8 − 3)

Postfix notation : 3   8   4   *   2   /   +   8   3   −   −

32

16

19

5

14

∴ The result of the the infix expression in postfix notation is 14.

Example 3.2

        Give the postfix form for the infix expression x + y * z.

SOLUTION :

    The order of evaluation of postfix form is

Parenthesize the operands with operator having the highest priority	=	x + (y * z)
Move the operator inside parenthesis to the right of the operand	=	x + (y z *)
Consider the part of the expression converted to postfix as a single operand	=	x + A [A=(yz*)]
Parenthesize the sub-expression	=	(x + A)
Move the operator inside parenthesis to the right of the operand	=	(x A +)
Substitute the value for A	=	(x(yz*)+)
Remove all parenthesis	=	xyz*+

    ∴ The required postfix form is xyz*+

Example 3.3

        Give the postfix form for the infix expression p + q / r − s.

SOLUTION :

    The order of evaluation of postfix form is

Parenthesize the operands with operator having the highest priority	=	p + (q / r) − s
Move the operator inside parenthesis to the right of the operand	=	p + (q r /) − s
Consider the part of the expression converted to postfix as a single operand	=	p + A − s [A=(qr/)]
Parenthesize the sub-expression with operator having the highest priority	=	(p + A) − s
Move the operator inside parenthesis to the right of the operand	=	(p A +) − s [B=(pA+)]
Consider the part of the expression converted to postfix as a single operand	=	B−s
Parenthesize the sub-expression	=	(B − s)
Move the operator inside parenthesis to the right of the operand	=	(B s −)
Substitute the value for B	=	((pA+) s −)
Substitute the value for A	=	((p(qr/)+) s −)
Remove all parenthesis	=	pqr/+s−

    ∴ The required postfix form is pqr/+s−

Example 3.4

Give the postfix form for the infix expression (m − n) / p * q + r ↑ s * t.

SOLUTION :

The order of evaluation of postfix form is

Move the operator inside parenthesis to the right of the operand	=	(m n −) / p * q + r ↑ s * t
Consider the part of the expression converted to postfix as a single operand	=	A / p * q + r ↑ s * t [A= (mn−)]
Parenthesize the sub-expression with the operator having the highest priority	=	A / p * q + (r ↑ s) * t
Move the operator inside parenthesis to the right of the operand	=	A / p * q + (r s ↑) * t
Consider the part of the expression converted to postfix as a single operand	=	A / p * q + B * t [B= (rs↑)]
Parenthesize the sub-expression with the operator having the highest priority from left to right	=	(A / p) * q + B * t
Move the operator inside parenthesis to the right of the operand	=	(A p /) * q + B * t
Consider the part of the expression converted to postfix as a single operand	=	C * q + B * t [C= (Ap/)]
Parenthesize the sub-expression with the operator having the highest priority from left to right	=	(C * q) + B * t
Move the operator inside parenthesis to the right of the operand	=	(C q *) + B * t
Consider the part of the expression converted to postfix as a single operand	=	D + B * t [D= (Cq*)]
Parenthesize the sub-expression with the operator having the highest priority from left to right	=	D + (B * t)
Move the operator inside parenthesis to the right of the operand	=	D + (B t *)
Consider the part of the expression converted to postfix as a single operand	=	D + E [E= (Bt*)]
Parenthesize the sub-expression with the operator having the highest priority from left to right	=	(D + E)
Move the operator inside parenthesis to the right of the operand	=	(D E +)

Substitute the value for E	=	(D(Bt*)+)
Substitute the value for D	=	((Cq*)(Bt*)+)
Substitute the value for C	=	(((Ap/)q*)(Bt*)+)
Substitute the value for B	=	(((Ap/)q*)((rs↑)t*)+)
Substitute the value for A	=	((((mn-)p/)q*)((rs↑)t*)+)
Remove all parenthesis	=	mn-p/q*rs↑t*+

∴ The required postfix form is mn-p/q*rs↑t*+

Rules to be followed during infix to prefix conversion

- Fully parenthesize the expression starting from left to right (During parenthesizing, the operators having higher precedence are first parenthesized).
- Move the operators one by one to their left, such that each operator replaces their corresponding left parenthesis.
- The part of the expression, which has been converted into prefix, is to be treated as single operand.
- Once the expression is converted into prefix form, remove all parenthesis.

Example 3.5

Evaluate the infix expression 3 + 8 * 4 / 2 - (8 - 3) in prefix notation.

SOLUTION :

The order of evaluation of infix expression is as follows,

Infix expression : $3 + 8 * 4 / 2 - (8 - 3)$

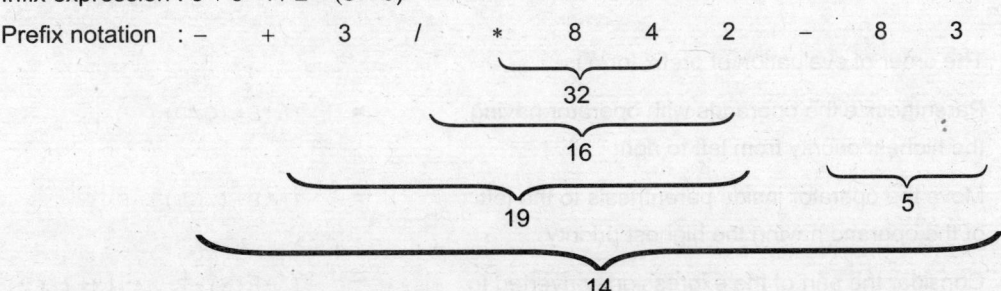

∴ The result of the infix expression in prefix notation is 14

Example 3.6

Give the prefix form for the infix expression p / q * r + s.

SOLUTION :

The order of evaluation of prefix form is

| Parenthesize the operands with operator having the highest priority | = | (p / q) * r + s |

Move the operator inside parenthesis to the left of the operand	=	(/ p q) * r + s
Consider the part of the expression converted to postfix as a single operand	=	A * r + s [A= (/pq)]
Parenthesize the sub-expression with the operator having the highest priority from left to right	=	(A * r) + s
Move the operator inside parenthesis to the left of the operand	=	(* A r) + s
Consider the part of the expression converted to postfix as a single operand	=	B + s [B= (*Ar)]
Parenthesize the sub-expression with the operator having the highest priority from left to right	=	(B + s)
Move the operator inside parenthesis to the left of the operand	=	(+ B s)
Substitute the value for B	=	(+ (*Ar) s)
Substitute the value for A	=	(+ (* (/pq) r) s)
Remove all parenthesis	=	+*/pqrs

∴ The required prefix form is +*/pqrs

Example 3.7

Give the prefix form for the infix expression (A * B + (C / D)) − E.

SOLUTION :

The order of evaluation of prefix form is

Parenthesize the operands with operator having the highest priority from left to right	=	(A*B+(C/D))−E
Move the operator inside parenthesis to the left of the operand having the highest priority	=	(A*B+(/CD))−E
Consider the part of the expression converted to postfix as a single operand	=	(A*B+X)−E [X= (/CD)]
Parenthesize the sub-expression with the operator having the highest priority from left to right	=	((A*B)+X)−E
Move the operator inside parenthesis to the left of the operand	=	((*AB)+X)−E
Consider the part of the expression converted to postfix as a single operand	=	(Y+X)−E [Y= (*AB)]
Move the operator inside parenthesis to the left of the operand	=	(+YX)−E

Consider the part of the expression converted to postfix as a single operand	=	Z−E	$[Z=(+YX)]$
Parenthesize the sub-expression with the operator having the highest priority from left to right	=	(Z−E)	
Move the operator inside parenthesis to the left of the operand	=	(−ZE)	
Substitute the value for z	=	(−(+YX)E)	
Substitute the value for Y	=	(−(+(*AB)X)E)	
Substitute the value for X	=	(−(+(*AB)(/CD))E)	
Remove all parenthesis	=	−+*AB/CDE	

∴ The required prefix form is −+*AB/CDE.

3.8.2 Converting Infix Expression to Postfix Form

Evaluation of an expression using computers can be done in different ways. Some programmers may try to parse the expression and find out the operators that have the highest priority, calculate them and later evaluate the expression with operators of lower priority. But if parentheses and functions are added to the expression it will soon be very complicated to do the calculation. An easier way to calculate it is to convert the expression from infix notation to postfix notation. This may first look very strange but this is much easier when we need to do calculations on a computer.

The following procedure is used to convert infix to postfix expression

- Read the infix string.
- Traverse from left to right of the expression.
- If an operand is encountered, add to the postfix string.
- If an operator is encountered push it on the stack, if any of the following conditions are satisfied.
  - The stack is empty.
  - If the precedence of the operator at the top of the stack is of lower priority than the operator being processed.
- If all the above condition fails, then pop the operator being processed to the postfix string.
- When the infix string is empty, pop the elements of the stack onto the postfix string to get the result.

> **Algorithm for Infix to Postfix Conversion**
>
> **IN_TO_POST(INSTR : STRING, POSTSTR : STRING)**
>
> Where INSTR contains infix expression & POSTSTR stores the resultant expression
>
> **Step 1 :**   Initialize the Stack
>
> **Step 2 :**   While (INSTR != NULL)

Step 3 :   CH = get the character from INSTR

Step 4 :   If (CH == operand) Then

   Append CH into POSTSTR

   Else If (CH == '(') Then

   Push CH into the Stack

   Else If (CH == ')') Then

   Pop the data from the Stack and append the data into POSTSTR until

   we get '(' from the Stack

   Else

   While (precedence(Top of the Stack) >= precedence(CH))

   Pop the data from the Stack and append the data into POSTSTR

   [End of While Structure]

   [End of **Step 4** If Structure]

Step 5 :   Push CH into the Stack.

Step 6 :   [End of **Step 2** While Structure]

Step 7 :   Pop all the data from the Stack and append the data into POSTSTR until

   Stack is Empty

END IN_TO_POST()

Program 3.5 :

```
/* Program to convert infix expression to postfix notation. */
/* in_post.c */
#include<stdio.h>
#include<ctype.h>
#include<string.h>
#define MAX 9

typedef struct node
 {
    char data;
    struct node *next;
 }stack;

int push(char);
int pop(char *);
stack *getNode();
void releaseNode(stack *);
void inToPost(char[], char[]);
int indexPriority(char P[][2], char data);
stack *topStk = NULL;
```

```c
char iP[MAX][2] = {
                    {'(', MAX},{')', 0 },  {'\0', 0}, {'+', 1},
                    {'-', 1}, {'*', 2}, {'/', 2}, {'%', 2},
                    {'^', 3}
                  };
char sP[MAX][2] = {
                    {'(', 0},  {')', -1}, {'\0', 0}, {'+', 1},
                    {'-', 1}, {'*', 2}, {'/', 2}, {'%', 2},
                    {'^', 3}
                  };
void main()
 {
    char inStr[20], postStr[20];
    printf("Enter the infix expression : ");
      scanf("%s", inStr);
    inToPost(inStr, postStr);
    printf("The postfix expression is : %s", postStr);
 }
int push(char value)
 {
    extern stack *topStk;
    stack *newptr;
    newptr = getNode();
    if(newptr == NULL)
      return -1;
    newptr -> data = value;
    newptr -> next = topStk;
    topStk = newptr;
    return 0;
 }
int pop(char *value)
 {
    extern stack *topStk;
    stack *temp;
    if(topstk == NULL)
      return -1;
    temp = topStk;
    topStk = topStk -> next;
    *value = temp -> data;
    releaseNode(temp);
    return 0;
 }
```

```c
stack *getNode()
{
   return ((stack *)malloc(sizeof(stack)) );
}
void releaseNode(stack *newnode)
{
   free(newnode);
}
void inToPost(char inStr[], char postStr[])
{
   char ch, item;
   int i = 0, st = 0, spr, ipr;
   push('\0');
   while((ch = inStr[st++]) != NULL)
    {
      if(tolower(ch) >= 'a' && tolower(ch) <= 'z')
        postStr[i++] = ch;
      else if(ch == '(')
        push(ch);
      else if(ch == ')')
        {
          pop(&item);
          while(item != '(')
            {
              postStr[i++] = item;
              pop(&item);
            }
        }
      else
        {
          pop(&item);
          spr = indexPriority(sP, item);
          ipr = indexPriority(iP, ch);
          while(sP[spr][1] >= iP[ipr][1])
            {
              postStr[i++] = item;
              pop(&item);
              spr = indexPriority(sP, item);
            }
```

```
                push(item) ;
                push(ch);
        }
    }
    while(!pop(&item))
    postStr[i++] = item;
}
int indexPriority(char P[][2], char data)
{
    int ind;
    for(ind = 0; ind < MAX; ind++)
        if(P[ind][0] == data)
            return ind;
}
```

The program displays the following output

RUN1   Enter the infix expression : a-b*c+d/g

       The postfix expression is : abc*--dg/+

RUN2   Enter the infix expression : (A+B)*(C-D)%G

       The postfix expression is : AB+CD-*G/

3.8.3 Evaluating an Expression in Postfix Notation

Evaluating an expression in postfix notation is trivially easy if you use a stack. The postfix expression to be evaluated is scanned from left to right. Variables or constants are pushed onto the stack. When an operator is encountered, the indicated action is performed using the top two elements of the stack, and the result replaces the operands on the stack.

Steps to be noted while evaluating a postfix expression using a stack

- Traverse from left to right of the expression.
- If an operand is encountered, push it onto the stack.
- If you see a binary operator, pop two elements from the stack, evaluate those operands with that operator, and push the result back in the stack.
- If you see a unary operator, pop one element from the stack, evaluate those operands with that operator, and push the result back in the stack.
- When the evaluation of the entire expression is over, the only thing left on the stack should be the final result. If there are zero or more than 1 operands left on the stack, either your program is inconsistent, or the expression was invalid.

Algorithm to Evaluate a Postfix Expression

EVALPOST(POSTEXP : STRING)

Where POSTEXP contains postfix expression

Step 1 :   Initialize the Stack.

Step 2 :   While (POSTEXP != NULL)

Step 3 :   ch = get the character from POSTEXP

Step 4 :   If (ch == Operand) Then

Read the value for CH and Push CH into Stack

Else If (ch == Operator) Then

Pop the two operands from the Stack and perform arithmetic operation

with the operator and Push the resultant value into the Stack

[End of **Step 4** If Structure]

Step 5 :   [End of **Step 2** While Structure]

Step 6 :   Pop the data from the Stack and return the popped data

END EVALPOST()

Program 3.6 :

```c
/* Program to evaluate postfix expression using stacks. */
/* evalpost.c */
#include<stdio.h>
#include<ctype.h>
#include<stdlib.h>
typedef struct node
 {
    int data;
    struct node *next;
 }stack;
stack *topStk = NULL;
int push(int);
int pop(int *);
stack *getNode();
int evalPost(char[]);
int calc(int, int, char);
void printError();
void main()
 {
    char postStr[20];
    printf("Enter the postfix string : ");
      gets(postStr);
    printf("The evaluated value of %s is %d",postStr,evalPost(postStr));
 }
```

```c
stack *getNode()
{
    return((stack *)malloc(sizeof(stack)) );
}
int push(int val)
{
    stack *top;
    top = getNode();
    if(top == NULL)
        return -1;
    top -> data = val;
    top -> next = topStk;
    topStk = top;
    return 0;
}
int pop(int *val)
{
    if(topStk == NULL)
        return -1;
    *val = topStk -> data;
    topStk = topStk -> next;
    return 0;
}
int calc(int a, int b, char c)
{
    switch(c)
    {
        case '+': return a + b;
        case '-': return a - b;
        case '*': return a * b;
        case '/': return a / b;
        case '%': return a % b;
    }
}
int value(char ch)
{
    int val;
    printf("Enter the value of %c : ", ch);
        scanf("%d", &val);
    return val;
}
```

```
    int evalPost(char postExp[])
    {
        int i = 0, op1, op2;
        char ch;
        while((ch = postExp[i++]) != '\0')
        {
            if(ch == ' ')
                continue;
            if((tolower(ch) >= 'a') && (tolower(ch) <= 'z'))
                push(value(ch));
            else if(ch=='+'||ch=='-'||ch=='*'||ch=='/'||ch=='%')
            {
                if(pop(&op2) || pop(&op1))
                    printError();
                push(calc(op1, op2, ch));
            }
        }
        pop(&op1);
        if(!pop(&op2))
            printError();
        return op1;
    }
    void printError()
    {
        printf("The given postfix expression is Wrong");
        exit(0);
    }
```

The program displays the following output

RUN1   Enter the postfix string : ab+cd-*

        Enter the value of a : 10

        Enter the value of b : 20

        Enter the value of c : 30

        Enter the value of d : 40

        The evaluated value of ab+cd-* is -300

RUN2   Enter the postfix string : AB*C/D-

        Enter the value of A : 10

        Enter the value of B : 20

        Enter the value of C : 40

        Enter the value of D : 50

        The evaluated value of AB*C/D- is -45

Important points to be noted while evaluating a postfix expression using a stack

- When you convert infix expression to postfix notation, the operands are always in the same order and the operators are probably in a different order.

- The first element you pop off of the stack in an operation should be evaluated on the right-hand side of the operator. For multiplication and addition, order doesn't matter, but for subtraction and division, your answer will be incorrect if you change your operands around.

Fig. 3.15 shows the evaluation of expression 3 + 8 * 4 / 2 - (8 - 3) in postfix notation evaluated by a stack.

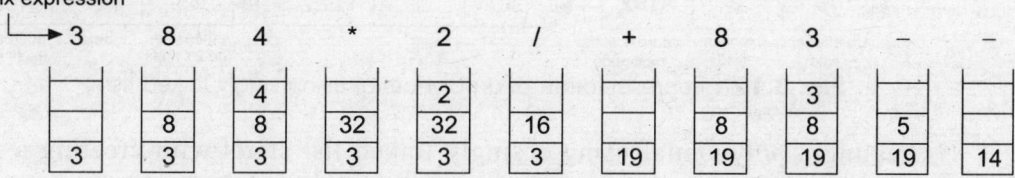

Fig. 3.15 : Expression evaluated by a stack in postfix notation.

3.9 Applications of Linked Lists

Linked lists form the basis of many data structures, so it's worth looking at some applications that are implemented using linked lists. Some important applications using linked lists are

- Polynomial manipulation
- Stacks
- Queues.

3.9.1 Polynomial Manipulation

Linked list is generally used to represent and manipulate polynomials. Polynomials are expressions containing terms with non-zero co-efficients and exponents. For example,

$$P(x) = a_0x^n + a_1x^{n-1} + \ldots + a_{n-1}x + a_n$$

where $P(x)$ is a polynomial in x,

$a_0, a_1, \ldots, a_{n-1}, a_n$ are constants and n is a positive integer.

In the linked list representation of polynomials, each term/element in the list is referred as a node. Each node contains three fields namely,

- Co-efficient field
- Exponent field
- Link field.

The co-efficient and exponent field stores the data of a polynomial. The co-efficient field holds the value of the co-efficient of a polynomial and the exponent field holds the exponent value of a polynomial. The link field contains the address of the next term in the polynomial. The representation of polynomial in a singly linked list is shown in *Fig. 3.16*. The representation of a polynomial using a singly linked list is shown in *Fig. 3.17*.

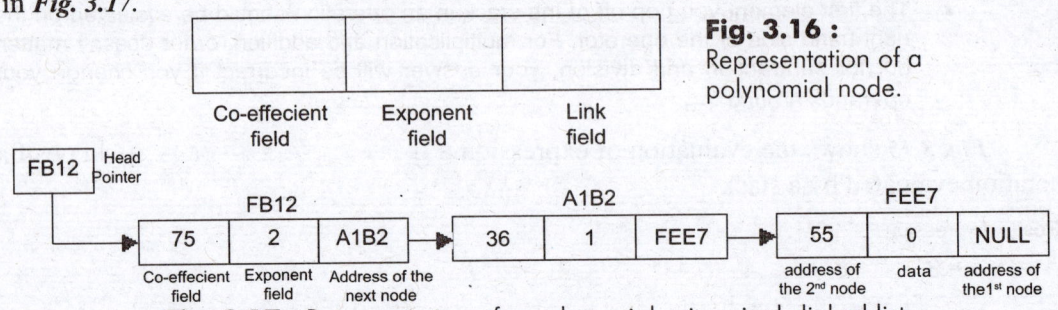

Fig. 3.16 : Representation of a polynomial node.

Fig. 3.17 : Representation of a polynomial using singly linked list.

Creating a polynomial using a singly linked list starts with creating a node. Sufficient memory has to be allocated for creating a node. The information is stored in the memory, allocated by using the `malloc()` function of type node. In *Program 3.7* the function `getNode()` is used for creating a node. After allocating memory for the structure of type `poly`, the information for the terms (i.e., `coef` and `exp`) has to be read from the user. In *Program 3.7* the function `readNode()`, is used for reading details for the node from the user.

Algorithm for Declaration of Structure Poly

    Struct POLY

    COEF   :   FLOAT

    EXP   :   INTEGER

    NEXT   :   Link Field (Address of next Struct POLY)

    End Struct

Algorithm for Allocating Memory for the New Node

    GETNODE()

    SIZE *:* INTEGER; NEWNODE *:* POLY

    Step 1 :   Set SIZE = get the size of the NODE

    Step 2 :   Set NEWNODE = Allocate space in memory for the size of SIZE and return

               the initial address

    Step 3 :   Return NEWNODE

    End GETNODE()

Algorithm for Reading the Content for the New Node (Coefficient, Exponent)

    READNODE(NEWNODE : NODE)

    Step 1 :   Read, NEWNODE –> COEF

    Step 2 :   Read, NEWNODE –> EXP

    Step 3 :   Set NEWNODE –> NEXT = NULL

    Step 4 :   Return

    End READNODE()

Connect the new node with the existing list. If the list is empty, set the head pointer of the list to the new node, otherwise insert the new node in the appropriate position of the list. In *Program 3.7* the function `insertNode()`, is used to connect the new nodes with the list according to the exponent values in decreasing order.

Algorithm for Inserting a Node into the Polynomial List

    INSERTNODE(HEAD : POLY, P : POLY)

Step 1 :   If (P –> COEF == 0) Then

            Return HEAD

            [End of If Structure]

Step 2 :   If (HEAD == NULL) Then

            Return P

Step 3 :   Else If (P –> EXP > HEAD –> EXP) Then

            P –> NEXT = HEAD

            Return P

Step 4 :   Else If (P –> EXP < HEAD –> EXP) Then

            HEAD –> NEXT = INSERTNODE (HEAD –> NEXT, P)

Step 5 :   Else If ((HEAD –> COEF + P –> COEF) != 0) Then

            HEAD –> COEF = HEAD –> COEF + P –> COEF

Step 6 :   Else

            Return HEAD –> NEXT

Step 7 :   [End of **Step 2** If Structure]

Step 8 :   Return HEAD

END INSERTNODE()

The `createPoly()` function is used to read the terms of the polynomial one by one in any order (of exponent / coefficient) as you wish until the exponent value is zero. Using the function `insertNode()`, insert them in their appropriate places and finally return the resultant polynomial.

Algorithm for Creating the Polynomial List

    CREATEPOLY()

HEAD, P : POLY

Step 1 :   Assign HEAD = NULL

Step 2 :   Repeat

Step 3 :   P = GETNODE()

Step 4 :   READNODE(P)

Step 5 :   HEAD = INSERTNODE (HEAD, P)

Step 6 :   [End of **Step 2** Repeat Until (P –> EXP == 0)]

Step 7 :   Return HEAD

END CREATEPOLY()

To read the information or to display the information from a polynomial, you have to traverse (move) a linked list, term by term from the first term, until the end of the list is reached. Traversing a polynomial is carried out by the function `viewPoly()`.

Algorithm for Displaying the Coefficients and Exponents of the Polynomial List

VIEWPOLY(HEAD : POLY)

P : POLY

Step 1 :   Assign P = HEAD

Step 2 :   While (P != NULL)

Step 3 :   Print "Coefficient is ", P –> COEF

Step 4 :   Print "Exponent is ", P –> EXP

Step 5 :   P = P –> NEXT

Step 6 :   [End of **Step 2** While Structure]

END VIEWPOLY()

The function `copyNode()` is used to make an additional copy of an existing node. Using the argument `sign`, you can change the sign of the term coefficient. The sign of the coefficient terms is changed only in case of `polySub()`.

Algorithm for Copying, by Creating a New Node

COPYNODE(P : POLY, SIGN : INTEGER)

NEWNODE : POLY

Step 1 :   NEWNODE = GETNODE()

Step 2 :   NEWNODE –> COEF = SIGN * P –> COEF

Step 3 :   NEWNODE –> EXP = P –> EXP

Step 4 :   NEWNODE –> NEXT = NULL

Step 5 :   Return NEWNODE

END COPYNODE()

The following procedure is used to add two polynomials using singly linked lists.

- Using the function createPoly() read the co-efficient and exponent terms of the first polynomial until exponent term is zero.
- Using the function createPoly() read the co-efficient and exponent terms of the second polynomial until exponent term is zero.
- Using the function polyAdd() add the two polynomials with the following comparisons.
- Copy the first polynomial using the functions copyNode() and insertNode() into the resultant polynomial.
- Traverse each term of the second polynomial and insert them into the resultant polynomial using the functions copyNode() and insertNode() until NULL value of the second polynomial is reached.
- Display the resultant polynomial.

Algorithm for Polynomial Addition

    POLYADD(POLY1 : POLY, POLY2 : POLY)

HEAD : POLY

Step 1 :   Assign HEAD = NULL

Step 2 :   While (POLY1 != NULL)

Step 3 :   HEAD = INSERTNODE(HEAD, COPYNODE(POLY1, 1))

Step 4 :   POLY1 = POLY1 –> NEXT

Step 5 :   [End of **Step 2** While Structure]

Step 6 :   While (POLY2 != NULL)

Step 7 :   HEAD = INSERTNODE(HEAD, COPYNODE(POLY2, 1))

Step 8 :   POLY2 = POLY2 –> NEXT

Step 9 :   [End of **Step 6** While Structure]

Step 10 : Return HEAD

END POLYADD()

The following procedure is used to subtract two polynomials using singly linked lists.

- Using function polySub() subtract the two polynomials with the following comparisons.
- Copy the first polynomial using the functions copyNode() and insertNode() into the resultant polynomial.
- Traverse each term of second polynomial and insert them into the resultant polynomial using the functions copyNode() and insertNode() until NULL value of the second polynomial is reached. (Note: The second polynomial is inserted into the resultant polynomial by changing the sign of the coefficients using the function copyNode()).
- Display the resultant polynomial.

Algorithm for Polynomial Subtraction

    POLYSUB(POLY1 : POLY, POLY2 : POLY)

HEAD : POLY

Step 1 :   Assign HEAD = NULL

Step 2 :   While (POLY1 != NULL)

Step 3 :   HEAD = INSERTNODE(HEAD, COPYNODE(POLY1, 1))

Step 4 :   POLY1 = POLY1 –> NEXT

Step 5 :   [End of **Step 2** While Structure]

Step 6 :   While (POLY2 != NULL)

Step 7 :   HEAD = INSERTNODE(HEAD, COPYNODE(POLY2, –1))

Step 8 :   POLY2 = POLY2 –> NEXT

Step 9 :   [End of **Step 6** While Structure]

Step 10 : Return HEAD

END POLYSUB()

The following procedure is used to multiply two polynomials using singly linked lists.

- Using the function polyMul() multiply two polynomials with the following comparisons.

- Traverse each term of the second polynomial and multiply with each term of the first polynomial using the function mulNode() and insert them into the resultant polynomial using the functions copyNode() and insertNode() until NULL value of the second polynomial is reached.

- Display the resultant polynomial.

Algorithm for Mulitiplying the Multiplicant Polynomial by the Multiplier Node

MULNODE(P1 : POLY, P2 : POLY)

TEMP : POLY

Step 1 :   Assign TEMP = NULL

Step 2 :   If (P1 != NULL) Then

   TEMP = GETNODE()

   TEMP –> COEF = P1 –> COEF * P2 –> COEF

   TEMP –> EXP = P1 –> EXP + P2 –> EXP

   TEMP –> NEXT = MULNODE(P1 –> NEXT, P2)

   [End of If Structure]

Step 3 :   Return TEMP

END MULNODE()

Algorithm for Polynomial Multiplication

POLYMUL(P1 : POLY, P2 : POLY)

Step 1 :   If (P2 == NULL) Then

   Return NULL

   [End of If Structure]

Step 2 :   Return POLYADD(MULNODE(P1,P2), POLYMUL(P1, P2->NEXT))

END POLYMUL()

The following procedure is used to divide two polynomials using singly linked lists.

- The function polyDiv() is used to find the quotient and remainder of the divisor (The function returns the quotient value and assigns the remainder value to the argument in the function).

- If the dividend or divisor is NULL, assign the value of the remainder as NULL and return the value of the quotient as NULL.

- If the first exponent of the dividend is greater than or equal to the first exponent of the divisor,

   ▪ The first coefficient of the dividend is divided by the first coefficient of the divisor, & the first exponent of the dividend is subtracted from the first exponent of the divisor and the corresponding coefficient & exponent is assigned to the resultant polynomial.

- Using the function mulNode() multiply the resultant polynomial with the divisor and subtract the result with the dividend using the function polySub() to get the remainder.

- The function polyDiv() is called recursively with the remainder as the new dividend until the above condition for the division fails.

- If the above condition for the division fails, return the resultant polynomial and assign the dividend value as the remainder.

Algorithm for Polynomial Division

POLYDIV (P1 : POLY, P2 : POLY, REMR : POLY)

RES : POLY

Step 1 :   Assign RES = NULL

Step 2 :   If (P1 == NULL || P2 == NULL) Then

          Set RMDR = NULL

          Return NULL

          [End of If Structure]

Step 3 :   If (P1 –> EXP >= P2 –> EXP) Then

          RES = GETNODE()

          RES –> COEF = P1 –> COEF / P2 –> COEF

          RES –> EXP = P1 –> EXP – P2 –> EXP

          RES –> NEXT=POLYDIV(POLYSUB(P1, MULNODE(P2,RES)), P2, RMDR)

       Else

          RMDR = P1

          [End of If Structure]

Step 4 :   Return RES

END POLYDIV()

Program 3.7 :

```
/* Program to demonstrate polynomial manipulation using singly linked lists. */
/* polymani.c */
#include<stdio.h>
#include<conio.h>
#include<stdlib.h>
#define POSITIVE 1
#define NEGATIVE -1
typedef struct node
 {
   float coef;
   int exp;
   struct node *next;
 }poly;
```

```c
void viewMenu();
void readNode(poly *);
void viewPoly(poly *);
poly *getNode();
poly *createPoly();
poly *copyNode(poly *, int);
poly *insertNode(poly *, poly *);
poly *mulNode(poly *p1, poly *p2);
poly *polyAdd(poly *p1, poly *p2);
poly *polySub(poly *p1, poly *p2);
poly *polyMul(poly *p1, poly *p2);
poly *polyDiv(poly *p1, poly *p2, poly **rmdr);
void main()
 {
    int choice;
    poly *poly1 = NULL, *poly2 = NULL, *res = NULL;
    poly *Quotient, *Remainder = NULL;
    viewMenu();
    while(1)
     {
       printf("\n ? ");
       fflush(stdin);
        scanf("%d", &choice);
       switch(choice)
        {
          case 0 :
                  viewMenu();
                  break;
          case 1 :
                  printf("\n Enter the First Polynomial ...");
                  poly1 = createPoly();
                  printf("\n First Polynomial is : ");
                  viewPoly(poly1);
                  break;
          case 2 :
                  printf("\n Enter the Second Polynomial ...");
                  poly2 = createPoly();
                  printf("\n Second Polynomial is : ");
                  viewPoly(poly2);
                  break;
```

```
        case 3 :
                printf("\n The First Polynomial is ... \n");
                viewPoly(poly1);
                printf("\n The Second Polynomial is ... \n");
                viewPoly(poly2);
                break;
        case 4 :
                printf("\n The Resultant Polynomial after Addition
                                                is ... \n");
                res = polyAdd(poly1, poly2);
                viewPoly(res);
                break;
        case 5 :
                printf("\n The Resultant Polynomial after
                                        Subtraction is ... \n");
                res = polySub(poly1, poly2);
                viewPoly(res);
                break;
        case 6 :
                printf("\n The Resultant Polynomial after
                                Multiplication is ... \n");
                res = polyMul(poly1, poly2);
                viewPoly(res);
                break;
        case 7 :
                printf("\n The Resultant Polynomial after Division
                                                is ... \n");
                Quotient = polyDiv(poly1, poly2, &Remainder);
                printf("\n Quotient : ");
                viewPoly(Quotient);
                printf("\n Remainder : ");
                viewPoly(Remainder);
                break;
        default :
                printf("\n End of Run of your Program . . .");
                exit(0);
    }
  }
}
```

```c
void viewMenu()
{
    printf("\n Polynomial Manipulation using singly linked list ...");
    printf("\n\t 0. View the Main Menu");
    printf("\n\t 1. Create the First Polynomial");
    printf("\n\t 2. Create the Second Polynomial");
    printf("\n\t 3. View the First and Second Polynomial");
    printf("\n\t 4. Polynomial Addition");
    printf("\n\t 5. Polynomial Subtraction");
    printf("\n\t 6. Polynomial Multiplication");
    printf("\n\t 7. Polynomial Division");
    printf("\n\t 8. Exit");
}
poly *createPoly()
{
    poly *p, *head = NULL;
    do
     {
      p = getNode();
       readNode(p);
       head = insertNode(head, p);
     }while(p -> exp != 0);
    return head;
}
poly *getNode()
{
    return (poly *)malloc(sizeof(poly));
}
void readNode(poly* newnode)
{
    int exp;
    float coef;
    printf("\nEnter the Coefficient : ");
       scanf("%f", &coef);
    printf("Enter the Exponent : ");
       scanf("%d", &exp);
    newnode -> coef = coef;
    newnode -> exp = exp;
    newnode -> next = NULL;
}
```

```
    poly *insertNode(poly *head, poly *p)
    {
       if(p -> coef == 0.0f)
         return head;
       if(head == NULL)
         return p;
       else if(p->exp > head->exp)
        {
          p -> next = head;
          return p;
        }
       else if(p->exp < head->exp)
         head -> next = insertNode(head -> next, p);
       else if((head -> coef = head -> coef + p -> coef) == 0.0f)
         return head -> next;
       return head;
    }
  void viewPoly(poly * ply)
    {
       if(ply == NULL)
         printf("NULL \n");
       while(ply != NULL)

        {
           printf("%.2fx^%d", ply -> coef, ply -> exp);
           printf("%s",(ply->next == NULL) ? " = 0\n" : " + ");
           ply = ply -> next;
        }
    }
  poly * polyAdd(poly *poly1, poly *poly2)
    {
       poly *head = NULL;
       while(poly1 != NULL)
        {
           head  = insertNode(head, copyNode(poly1, POSITIVE));
           poly1 = poly1 -> next;
        }
```

```
      while(poly2 != NULL)
       {
          head = insertNode(head, copyNode(poly2, POSITIVE));
          poly2 = poly2 -> next;
       }
      return head;
   }
poly * polySub(poly *poly1, poly *poly2)
   {
      poly *head = NULL;
      while(poly1 != NULL)
       {
          head  = insertNode(head, copyNode(poly1, POSITIVE));
          poly1 = poly1 -> next;
       }
      while(poly2 != NULL)


       {
          head = insertNode(head, copyNode(poly2, NEGATIVE));
          poly2 = poly2 -> next;
       }
      return head;
   }
poly *polyMul(poly *poly1, poly *poly2)
   {
      if(poly2 == NULL)
         return NULL;
      else
         return polyAdd(mulNode(poly1,poly2),polyMul(poly1,poly2->next));
   }
poly * copyNode(poly * p, int sign)
   {
      poly *newnode;
      newnode = getNode();
      newnode -> coef = sign * p -> coef;
      newnode -> exp  = p -> exp;
      newnode -> next = NULL;
       turn newnode;
   }
```

```c
poly *mulNode(poly *p1, poly *p2)
{
    poly *temp = NULL;
    if(p1 != NULL)
    {
        temp = getNode();
        temp -> coef = p1 -> coef * p2 -> coef;
        temp -> exp  = p1 -> exp  + p2 -> exp;
        temp -> next = mulNode(p1->next, p2);
    }
    return temp;
}
/* For Qoutient and Remainder p1 / p2 */
poly *polyDiv(poly *p1, poly *p2, poly **rmdr)
{
    poly *res = NULL;
    if(p1 == NULL || p2 == NULL)
        return (*rmdr = NULL);
    if(p1 -> exp >= p2 -> exp)
    {
        res = getNode();
        res -> coef = p1->coef / p2 -> coef;
        res -> exp  = p1->exp  - p2 -> exp;
        res -> next = polyDiv(polySub(p1,mulNode(p2,res)),p2,rmdr);
    }
    else
        *rmdr = p1;
    return res;
}
```

The program displays the following output

```
Polynomial Manipulation using singly linked list ...
    0. View the Main Menu
    1. Create the First Polynomial
    2. Create the Second Polynomial
    3. View the First and Second Polynomial
    4. Polynomial Addition
    5. Polynomial Subtraction
    6. Polynomial Multiplication
    7. Polynomial Division
    8. Exit
```

```
? 3
The First Polynomial is ...
NULL
The Second Polynomial is ...
NULL
? 4
The Resultant Polynomial after Addition is ...
NULL
? 5
The Resultant Polynomial after Subtraction is ...
NULL
? 6
The Resultant Polynomial after Multiplication is ...
NULL
? 7
The Resultant Polynomial after Division is ...
Solution is Infinity ...
Solution is Infinity ...
Quotient : NULL
Remainder : NULL
? 1
Enter the First Polynomial . . .
Enter the Coefficient : 3
Enter the Exponent : 2
Enter the Coefficient : 2
Enter the Exponent : 1
Enter the Coefficient : 1
Enter the Exponent : 0
First Polynomial is : 3.00x^2 + 2.00x^1 + 1.00x^0 = 0
? 2
Enter the Second Polynomial . . .
Enter the Coefficient : 2
Enter the Exponent : 1
Enter the Coefficient : 3
Enter the Exponent : 1
Enter the Coefficient : 5
Enter the Exponent : 0
Second Polynomial is : 5.00x^1 + 5.00x^0 = 0
```

```
? 3
The First Polynomial is ...
3.00x^2 + 2.00x^1 + 1.00x^0 = 0
The Second Polynomial is ...
5.00x^1 + 5.00x^0 = 0
? 4
The Resultant Polynomial after Addition is ...
3.00x^2 + 7.00x^1 + 6.00x^0 = 0
? 5
The Resultant Polynomial after Subtraction is ...
3.00x^2 + -3.00x^1 + -4.00x^0 = 0
? 6
The Resultant Polynomial after Multiplication is ...
15.00x^3 + 25.00x^2 + 15.00x^1 + 5.00x^0 = 0
? 7
The Resultant Polynomial after Division is ...
Quotient : 0.60x^1 + -0.20x^0= 0
Remainder : 2.00x^0 = 0
? 8
End of Run of your Program . . .
```

Program 3.8 is capable of reading input in any sequence, and at the same time, reading input should end with an exponent value of zero. If you enter the same value of exponent again and again, it will automatically add the co-efficients of same exponent and store it in descending order. You can also change any one of the existing polynomial and manipulate (add, subtract, multiply, divide) the new polynomial with the old polynomial.

Advantages of Linked Lists

- It is not necessary to specify the number of elements in a linked list during its declaration (since memory can be allocated dynamically when a node is added to the list).
- Linked list can grow and shrink in size depending upon the insertion and deletion that occurs in the list.
- Insertions and deletions at any place in a list can be handled easily and efficiently.
- A linked list does not waste any memory space.

Disadvantages of Linked Lists

- Searching a particular element in a list is difficult and time consuming.
- A linked list will use more storage space than an array to store the same number of elements (∴ Each element in a list needs additional memory space for storing the address of the next node).

3.10 Summary

- ✍ Some of the important applications of linked lists are stacks and queues.

- ✍ Stacks can be implemented by arrays or linked lists.

- ✍ A stack is an ordered collection of elements in which insertions and deletions are restricted to one end. The end from which elements are added and/or removed is referred to as top of the stack.

- ✍ Stacks are referred to as Last-In-First-Out (LIFO) lists or **piles** or **push-down lists**.

- ✍ The primitive operations that can be carried out in a stack are push, pop and peek operations.

- ✍ Push is an operation used to add a new element into a stack.

- ✍ Pop is an operation used to remove the top most elements from the stack.

- ✍ Peek is an operation used to display the top most element from the stack.

- ✍ In array representation of stacks, if a stack contains no elements, it is referred to as an empty stack.

- ✍ In array representation of stacks, if a stack contains elements equal to its size, it is referred to as an fully occupied stack.

- ✍ In a linked list, if the stack is pointing to NULL the stack is said to be an empty stack.

- ✍ If the malloc function returns NULL in a linked list it is referred as an fully occupied stack.

- ✍ Some of the important applications of stacks are tower of hanoi, reversing a string, checking for balanced parenthesis and evaluation of arithmetic expressions.

- ✍ Towers of Hanoi is a game in which we are given a tower of n disks, initially stacked in increasing size from top to bottom. Next to this tower, we have two more towers. The objective is to transfer the entire disks from Tower 1 to entire Tower 3 using Tower 2.

- ✍ Given an arithmetic expression, what we exactly mean by balanced is that, an opening parenthesis of various shapes (parenthesis, a square brace, or a flower brace) must match the corresponding last unmatched closing parenthesis symbol of the same shape and all parenthesis symbols must be matched when the arithmetic expression is finished.

- ✍ The different notations for representing arithmetic expressions are infix, prefix and postfix notations.

- ✍ The normal way of expressing mathematical expressions is called as infix notation.

- ✍ Prefix notation also referred, as polish notation is a way of writing algebraic expressions without the use of parenthesis or rules of operator precedence.

✍ Postfix notation also referred, as suffix form or reverse polish notation (or RPN) is also a way of writing algebraic expressions without the use of parentheses or rules of operator precedence.

✍ A queue is an ordered collection of elements in which insertions are made at one end and deletion are made at the other end. The end at which insertions are made is referred to as the rear end, and the end from which deletions are made is referred to as the front end.

✍ The primitive operations that can be carried out in a queue are enqueue and dequeue operations.

✍ Enqueue is an operation used to add a new element into the rear end of a queue.

✍ Dequeue is an operation used to remove an element from the front end of a queue.

✍ Circular queue is another form of a linear queue in which the last node is connected to the first node of the list.

✍ Deque (Double–ended queue) is another form of a queue in which insertions and deletions are made at both the front and rear ends of the queue.

✍ There are two variations of a deque, namely input restricted deque and output restricted deque. The input restricted deque allows insertion at one end (it can be either front or rear) only. The output restricted deque allows deletion at one end (it can be either front or rear) only.

✍ The implementation part of the circular queue is same as that of the linear queue, except that the last position is connected to the first position of the queue. So the front end and rear end moves circularly.

✍ The implementation part of the circular deque is same as that of the linear deque, except that the last position is connected to the first position of the deque. So the front end and rear end moves circularly.

3.11 Short-answer Questions

1. ***Define a stack.***

Stack is an ordered collection of elements in which insertions and deletions are restricted to one end. The end from which elements are added and/or removed is referred to as top of the stack. Stacks are also referred as "piles" and "push-down lists".

2. ***List out the basic opeartions that can be performed on a stack and a queue.***

The basic operations that can be performed on a stack and queue are,
- Push operation
- Pop operation
- View stack contents.

3.   *State the different ways of representing expressions.*

The different ways of representing expressions are,
- Infix notation
- Prefix notation
- Postfix notation.

4.   *State the rules to be followed during infix to postfix conversions.*

The rules to be followed during infix to postfix conversions are,
- Fully parenthesize the expression starting from left to right (During parenthesizing, the operators having higher precedence are first parenthesized).
- Move the operators one by one to their right, such that each operator replaces their corresponding right parenthesis.
- The part of the expression, which has been converted into postfix, is to be treated as single operand.
- Once the expression is converted into postfix form, remove all parenthesis.

5.   *Mention the advantages of representing stacks using linked lists than arrays.*

The advantages of representing stacks using linked lists than arrays.
- It is not necessary to specify the number of elements to be stored in a stack during its declaration (since memory is allocated dynamically at run time when an element is added to the stack).
- Insertions and deletions can be handled easily and efficiently.
- Linked list representation of stacks can grow & shrink in size without wasting the memory space, depending upon the insertion & deletion that occurs in the list.
- Multiple stacks can be represented efficiently using a chain for each stack.

6.   *Define a queue.*

Queue is an ordered collection of elements in which insertions and deletions are restricted to one end. The end from which elements are added and/or removed is referred to as the rear end, and the end from which deletions are made is referred to as the front end.

7.   *State the difference between queues and linked lists.*

The difference between queues and linked lists is that insertions and deletions may occur anywhere in linked list, but in queues insertions can be made only in the rear end and deletions can be made only in the front end.

8.   *List the various type of addresses used in linked lists.*

The addresses used in linked list includes,
- External address
- Internal address
- Null address.

9.   *Define external address.*

External address is the address of the first node in the list. This address is stored in the head pointer which points to the first node in the list.

10.   *Define internal address.*

Internal address is the address stored in each node of the linked list except the last node. The content stored in the link field (also referred as next address field) is the address of the next node.

11.   *Define null address.*

Null address is the address stored by the NULL pointer of the last node of the list, which indicates the end of the list.

12.   *State the use of previous address field and next address field in a linked list.*

The previous address field also referred as the backward link field, of a node contains address of its previous node. The next address field also referred as the forward link field, contains the address of the next node in the list.

13.   *List the basic operations carried out in a linked list.*

The basic operations carried out in a linked list includes,

- Creation of a list
- Insertion of a node
- Deletion of a node
- Modification of a node
- Traversal of the list.

14.   *List the other operations carried out in a linked list.*

The other operations carried out in a linked list includes,

- Searching an element in a list
- Finding the successor element of a node
- Finding the predessor element of a node
- Appending a linked list to another existing list
- Splitting a linked list in to two lists
- Arranging a linked list in ascending or descending order.

15.   *List out the advantages in using a linked list.*

The advantages in using a linked list are

- It is not necessary to specify the number of elements in a linked list during its declaration (since memory can be allocated dynamically when a node is added to the list).
- Linked list can grow and shrink in size depending upon the insertion and deletion that occurs in the list.
- Insertions and deletions at any place in a list can be handled easily and efficiently.
- A linked list does not waste any memory space.

16.   *List out the disadvantages in using a linked list.*

The disadvantages in using a linked list are

- Searching a particular element in a list is difficult and time consuming.
- A linked list will use more storage space than an array to store the same number of elements ($\because$ each element in a list needs additional memory space for storing the address of the next node).

17.   State the difference between arrays and linked list.

Arrays	Linked Lists
a) Size of any array is fixed.	a) Size of a list is variable.
b) It is necessary to specify the number of elements during declaration.	b) It is not necessary to specify the in an array number of elements in during declaration.
c) Insertion and deletions are some what difficult out easily.	c) Insertions and deletions are carried in an array.
d) It occupies less memory than a linked list for the same number of elements.	d) It occupies more memory.

18.   What is meant by static linked lists.

The advantage of using linked lists is to dynamically allocate memory for the nodes when needed. This is achieved using array of structures when the size of the array of structures is fixed, the type of lists is referred to as static linked lists and also is referred to as cursor implementation of linked lists.

19.   Draw the node representation of a singly linked list.

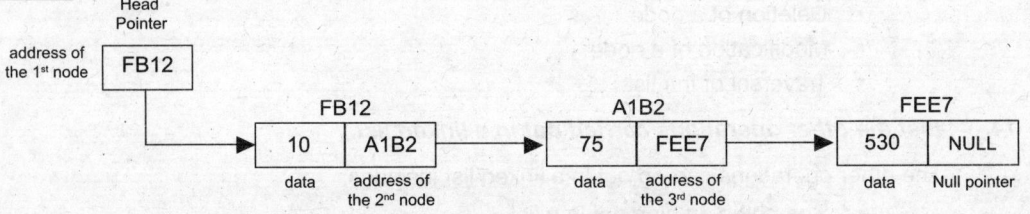

Fig : Representation of a singly linked list.

20.   Draw the node representation of a doubly linked list.

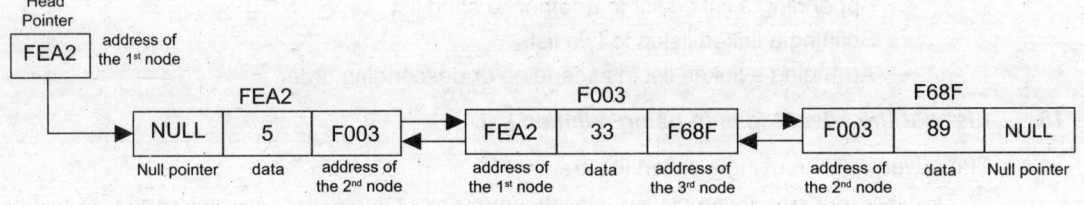

Fig : Representation of a doubly linked list.

21.   Draw the node representation of a circular singly linked list.

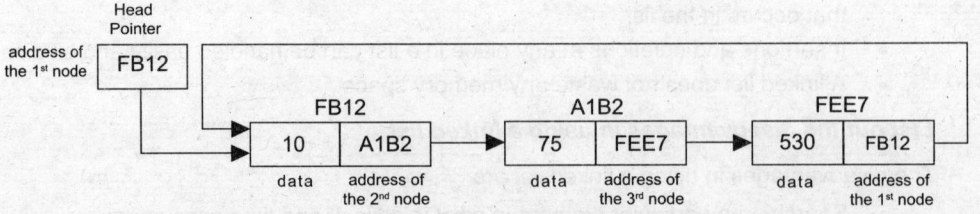

Fig : Representation of a circular singly linked list.

Chapter 4
Non-Linear Data Structures

4.1 Introduction

The data structures that we have seen in *Chapter 3* is linear data structures, since they have a linear relationship between its adjacent elements. The data structures that we are going to see in this chapter are non-linear data structures, since they don't have a linear relationship between its adjacent elements. *In a linear data structure, each node has a link which points to another node, where as in a non-linear data structure, each node may point to several other nodes. Some of the non-linear data structures include, trees and graphs.*

A *tree* is a non-linear, two-dimensional data structure, which represents hierarchical relationships between individual data items. *Fig. 4.1* shows a simple tree structure. Natural trees grow upwards from the ground into the air. But the tree data structure grows downwards from top to bottom. This is the universally accepted way of representing tree data structure. A *tree* is a non-empty collection of nodes and edges that satisfies the following requirements.

- It has a special data item referred as the root of the tree.
- All remaining data items are partitioned into number of subsets, each of which itself is a tree, and are referred as *subtrees*.

Each data item in a tree is referred as a *node*. It is the basic structure of a tree. It specifies the information part of each data item. There are totally 24 nodes in the tree represented in *Fig. 4.1*. An *edge* also referred as a *link* is a connection between two nodes. A *path* in a tree is a sequence of distinct nodes in which successive nodes are connected by edges in the tree. In *Fig. 4.1*, the path between A and X is given by the node pairs (A,C), (C,H), (H,Q) and (Q, X). An important property of a tree is that there should be precisely one path connecting any two nodes. If there is no path between some pair of nodes, or if there is more than one path between some pair of nodes, then we call the structure as a *graph* and not a tree.

The highest level or the first level in a tree is referred as a *root* or *root node*. In *Fig. 4.1,* A is referred as the root node. Each link to the root node refers to a *child* or *child node*. In *Fig. 4.1,* nodes B, C, D and E are referred as child nodes to the parent node A. A link between a parent node and its child node is also referred as a **branch**. All nodes in a tree, except the root node must descend from a parent node via a branch. The child nodes of a given parent node are referred as **siblings**. In **Fig. 4.1,** nodes B, C, D and E are referred as siblings to its parent node A, where as F and G are siblings of its parent node B and so on.

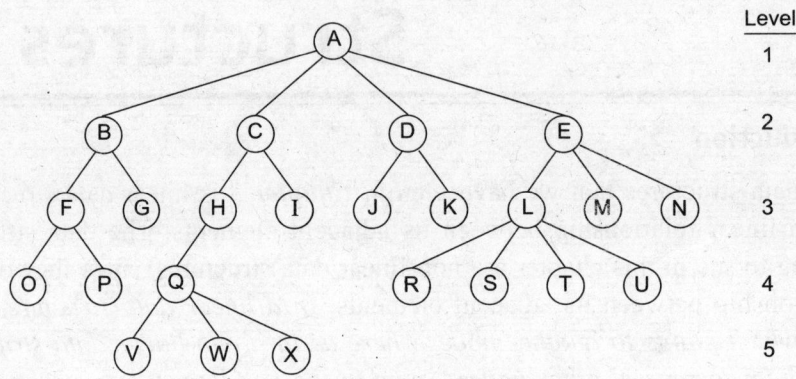

Fig. 4.1 : A tree structure.

In the hierarchy represented by a tree, the child nodes of a parent are one level lower than the parent node. The entire tree structure is leveled in such a way that the root node is always at level 1. Then, its immediate children are at level 2 and their immediate children are at level 3 and so on. In general, if a node is at level n, then its children will be at level (n+1).Thus in **Fig. 4.1,** the root node A is at level 1, its child nodes B, C, D and E are at level 2 and so on. A subset of a tree that is itself a tree is called as a **subtree**. The number of subtrees in a node isn referred as its **degree**. In **Fig. 4.1,** the degree of the root node A is 4, degree of node B is 2, where as the degree of the node L is 4.

The ancestors of a node are all the nodes along the path from the root to that node. The root node of the tree is the ancestor of all the nodes in the tree. In **Fig. 4.1,** the ancestors of node O is A, B, and F. Each node in a tree can be a parent and can have any number of child nodes to it. A node that has no children (i.e., with zero degree) is called as a **leaf** or **terminal node**. All other intermediate nodes that traversing the given tree from its root node to the terminal nodes are referred as **non-terminal nodes**. In **Fig. 4.1,** there are totally 15 terminal nodes (i.e., G, I, K, M, N, O, P, R, S, T, U, V, W and X) and 9 non-terminal nodes (i.e., A, B, C, D, E, F, H, L, Q and X).

The **height** or **depth** of a tree is defined as the maximum level of any node in the tree. In **Fig. 4.1,** root node A has the maximum level, which descends to the bottom of the tree at V, W and X. The height of the tree shown in **Fig. 4.1** is 5. In a given tree, if you remove the root node, then it becomes a forest. In **Fig. 4.1,** if you remove the root node A, then it becomes a forest with four trees.

4.2 Binary Trees

One of the most important tree structures is the binary tree. *A **binary tree** is a tree that has nodes either empty or not more than two child nodes, each of which may be a leaf node.* ***Fig. 4.2*** shows the structure of a binary tree. Each node in a binary tree has two subtrees, the left subtree and the right subtree. All the nodes to the left of a given node in a binary tree are referred as ***left subtrees***. All the nodes to the right of a given node in a binary tree are referred as the ***right subtrees***. In ***Fig. 4.2*** B, D, and F are the left subtrees and C, E and G are the right subtrees.

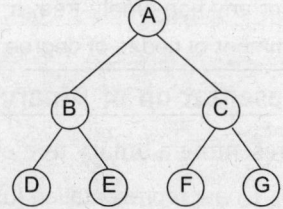

Fig. 4.2 : A binary tree structure.

4.2.1 Types of Binary Trees

Left-skewed binary tree	A binary tree, which has only left child nodes.
Right-skewed binary tree	A a binary tree, which has only right child nodes.
Full binary tree	A binary tree in which all the leaves are on the same level and every non-leaf node has exactly two children.
Complete binary tree	A binary tree in which every non-leaf node has exactly two children not necessarily to be on the same level.

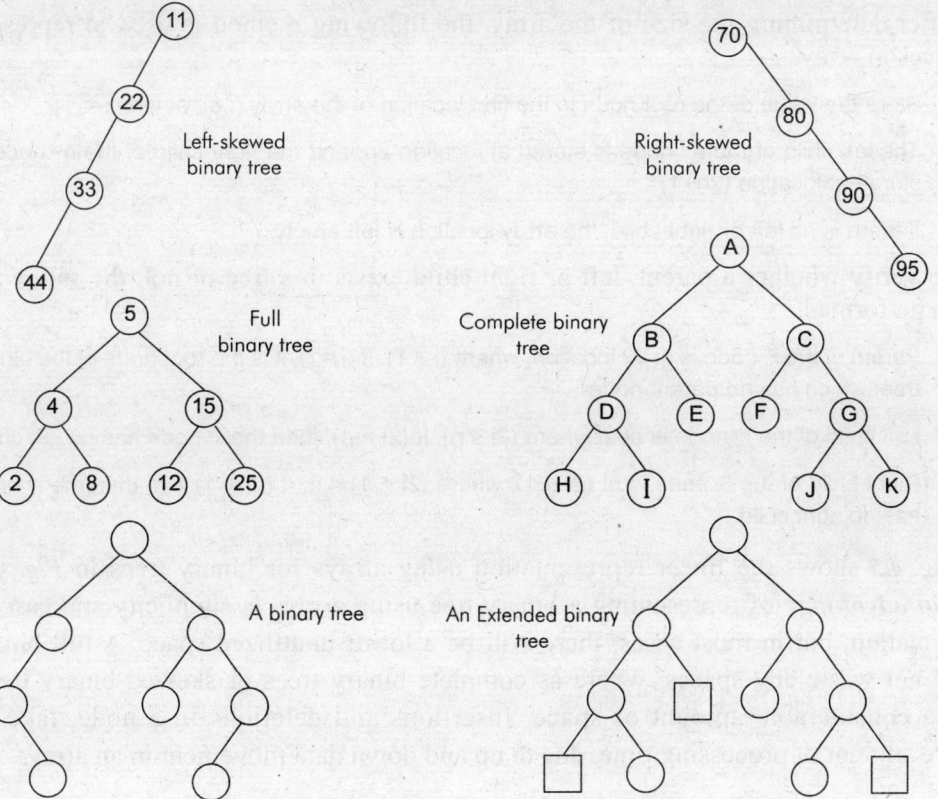

Fig. 4.3 : Examples of binary trees.

Properties of a binary tree

- The maximum number of nodes on level n of a binary tree is 2^{n-1}, where $n \geq 1$.
- The maximum number of nodes in a binary tree of height n is 2^{n-1}, where $n \geq 1$.
- For any non-empty tree, $n_l = n_d + 1$ where n_l is the number of leaf nodes and n_d is the number of nodes of degree 2.

4.3 Representation of Binary Trees

Representing a binary tree can be carried out in two ways. They are,

- Linear representation (using arrays)
- Linked representation (using pointers).

4.3.1 Linear Representation of Binary Trees

An array can be used to store the nodes of a binary tree. The nodes stored in an array are accessed sequentially. Linear representation of binary trees uses a single dimensional array of size $2^{h+1} - 1$ where h is the height of the tree. The binary tree structure shown in *Fig. 4.4 (a,b,c)* has a height of 3 and in *Fig. 4.4(d)* has a height of 4. Therefore, the size of the array required to store the first three binary trees is $2^{3+1} - 1 = 15$ and the size of the array required to store the last binary tree is $2^{4+1} - 1 = 31$.

After determining the size of the array, the following method is used to represent binary trees in arrays.

- Store the value of the root node in the first location of the array (i.e., at index = 1).
- The left child of the nth node is stored at location 2n, and the right child of the nth node is stored at location (2n+1).
- If there is no left or right child, the array location is left empty.

To verify whether a parent, left or right child exists in a tree or not, the following steps are performed.

- Parent of the ith node is at i/2 location, where ($i \neq 1$). If (i =1), it is the root node of the binary tree, which has no parent node.
- Left child of the ith node is at 2i where ($2i \leq n$). If ($2i > n$), then the ith node has no left child.
- Right child of the ith node is at (2i + 1), where ($2i + 1) \leq n$. If ($2i + 1) > n$, then the ith node has no right child.

Fig. 4.5 shows the linear representation using arrays for binary trees in *Fig. 4.4*. The *main advantage* of representing a binary tree using arrays is simplicity and ease of implementation, but in most cases, there will be a lot of unutilized space. A full binary tree will not waste any spaces, where as complete binary trees or skewed binary trees, utilises a considerable amount of space. Insertions and deletions in a node, take an **exc**essive amount of processing time, due to up and down data movement in an array.

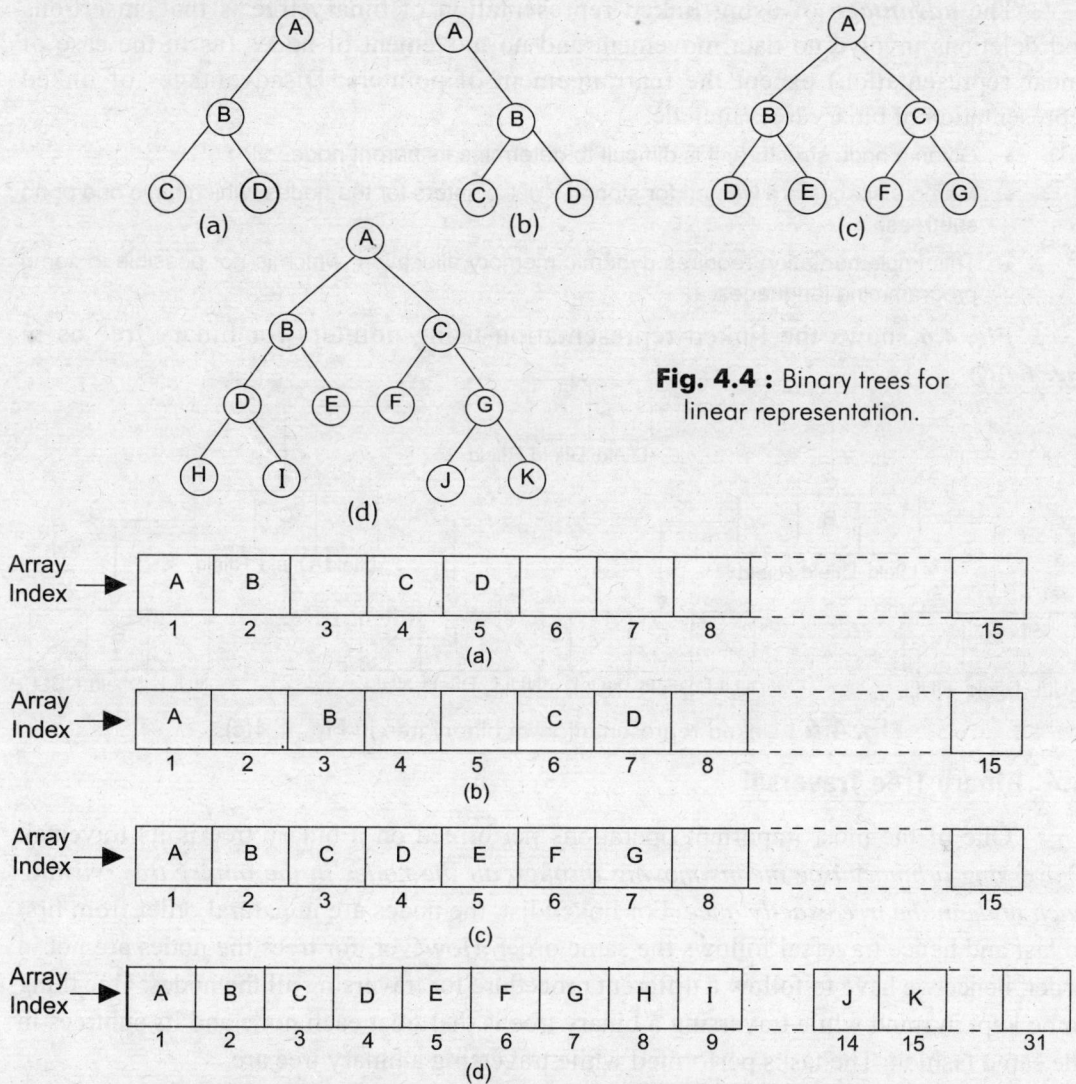

Fig. 4.4 : Binary trees for linear representation.

Fig. 4.5 : Linear representation of binary trees in **Fig. 4.4**.

4.3.2 Linked Representation of Binary Trees

The disadvantages of linear representation of binary trees, such as wasting memory space and difficulty in inserting and deleting a node in a tree are overcome by representing binary trees in linked representation. As we know that each node in a binary tree has two child nodes, each node in a linked representation requires 3 fields, the left child field (denoted as *Lfield*), the right child field (denoted as *Rfield*) and the data field (denoted as *Dfield*). Further, the root field contains the location of the root node. Hence, it requires 3 arrays and a variable to implement linked representation of binary trees. If any subtree is empty, then the corresponding pointers (i.e., Lfield and Rfield) will store a NULL value. If the tree itself is empty, then the root field will store a NULL value.

The **advantage** of using linked representation of binary tree is that, insertions and deletions involve no data movement and no movement of nodes (as in the case of linear representation) except the rearrangement of pointers. Disadvantages of linked representation of binary trees include,

- Given a node structure, it is difficult to determine its parent node.
- Memory spaces are wasted for storing NULL pointers for the nodes, which have one or no subtrees.
- This implementation requires dynamic memory allocation, which is not possible in some programming languages.

Fig. 4.6 shows the linked representation using pointers for binary tree as in **Fig. 4.4(c)**.

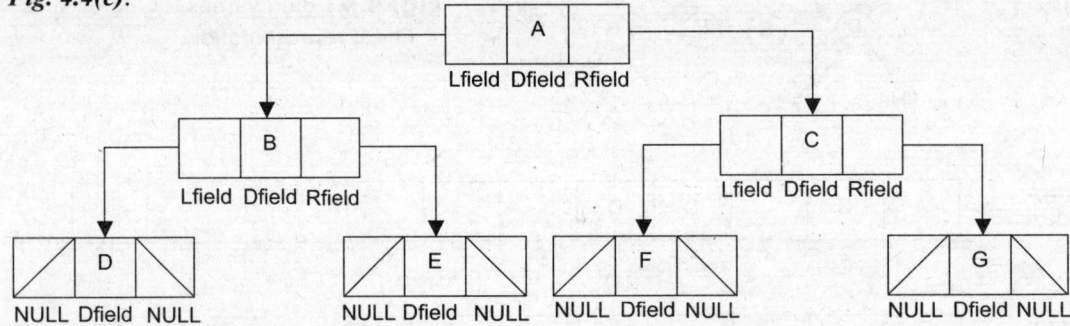

Fig. 4.6 : Linked representation of binary tree in **Fig. 4.4(c)**.

4.4 Binary Tree Traversal

One of the most important operations performed on a binary tree is its traversal. *Traversing a binary tree means moving through all the nodes in the binary tree, visiting each node in the tree exactly once.* For linked list, the nodes are in natural order from first to last and hence traversal follows the same order. However, for trees the nodes are not in order, hence we have to follow a different procedure for traversing all the nodes. One thing to be kept in mind while traversing a binary tree is that treat each node and its subtrees in the same fashion. The tasks performed while traversing a binary tree are,

- Visiting a node (denoted by the letter V)
- Traversing the left subtree (denoted by the letter L)
- Traversing the right subtree (denote by the letter R).

Depending upon the three tasks there are six possible combinations of traversals. They are VLR, LVR, LRV, VRL, RVL, and RLV. These six traversals can be reduced to three combinations by adopting the convention, that we always traverse the left subtree before the right subtree. Depending on the factor, the possible combinations are,

- V L R (designated as Preorder traversal)
- L V R (designated as Inorder traversal)
- L R V (designated as Postorder traversal).

4.4.1 Preorder Traversal

The steps for traversing a binary tree in preorder traversal are,

- Process the root node.
- Traverse the left subtree.
- Traverse the right subtree.

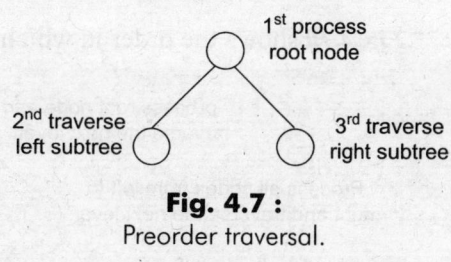

Fig. 4.7 :
Preorder traversal.

The value of in each node is processed as the node is visited. Preorder traversal leads to prefix expressions. *Fig. 4.7* shows the order in which each node is visited in preorder traversal.

4.4.2 Inorder Traversal

The steps for traversing a binary tree in preorder traversal also referred as *symmetric order traversal* are,

- Traverse the left subtree.
- Process the root node.
- Traverse the right subtree.

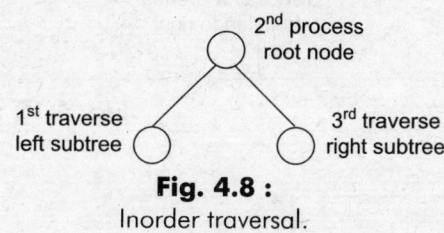

Fig. 4.8 :
Inorder traversal.

Inorder traversal leads to infix expressions.
Fig. 4.8 shows the order in which each node is visited in inorder traversal. The inorder traversal of a binary tree points the values always in ascending order, and hence this process is referred as the binary tree sort.

4.4.3 Postorder Traversal

The steps for traversing a binary tree in postorder traversal also referred as *end order traversal* are,

- Traverse the left subtree.
- Traverse the right subtree.
- Process the root node.

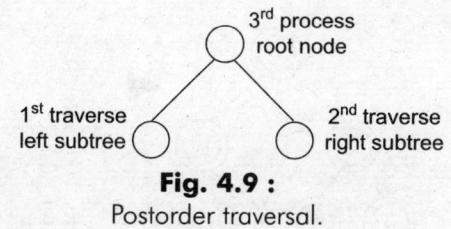

Fig. 4.9 :
Postorder traversal.

Postorder traversal leads to postfix expressions. *Fig. 4.9* shows the order in which each node is visited in postorder traversal.

4.4.4 Level order Traversal

In level order traversal, we traverse the nodes according to their levels. The steps for traversing a binary tree in level order traversal are,

- Process the root node at level 1.
- Traverse the next level (i.e, level 2), below the root node.

- Process the nodes from left to right in that level.
- Similarly traverse the next level and process the nodes from left to right and continue till the end of levels.

Fig. 4.10 shows the order in which each node is visited in level order traversal.

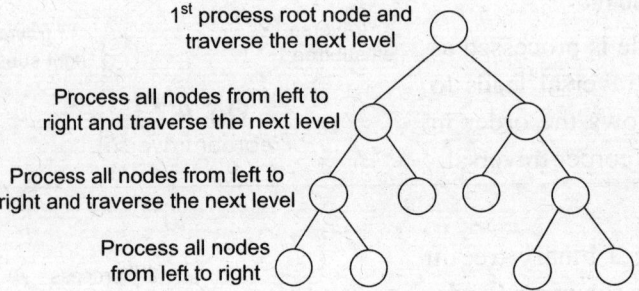

1st process root node and
traverse the next level

Process all nodes from left to
right and traverse the next level

Process all nodes from left to
right and traverse the next level

Process all nodes
from left to right

Fig. 4.10 : Level order traversal.

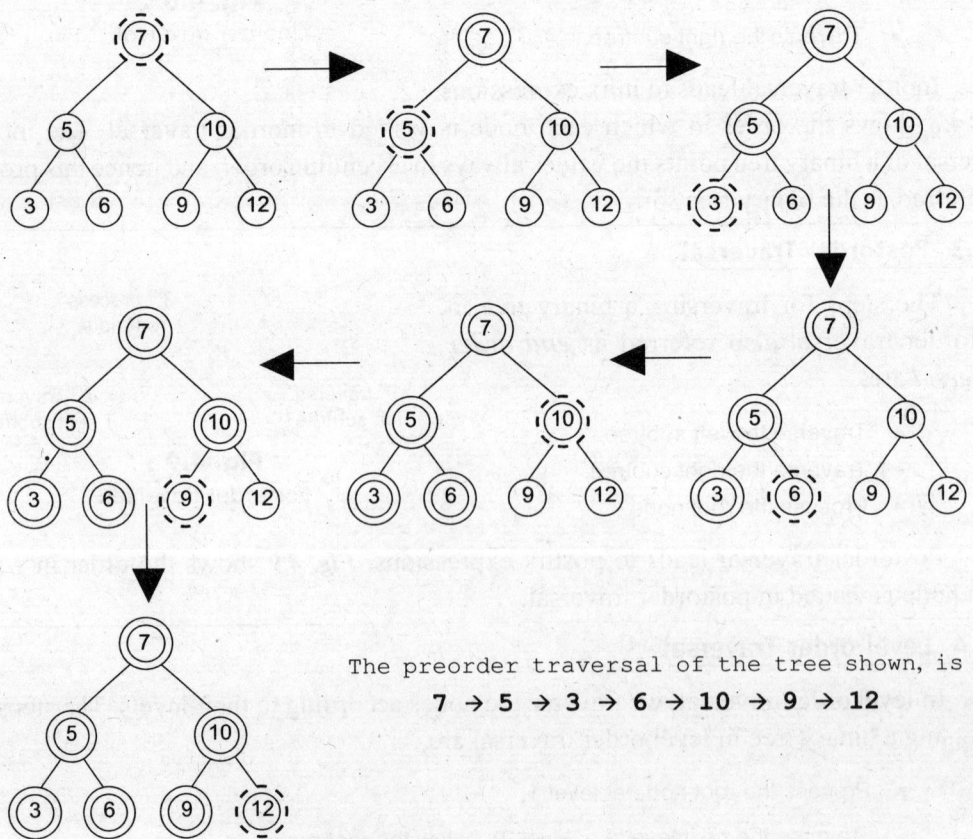

The preorder traversal of the tree shown, is

7 → 5 → 3 → 6 → 10 → 9 → 12

Fig. 4.11 : Preorder traversal.

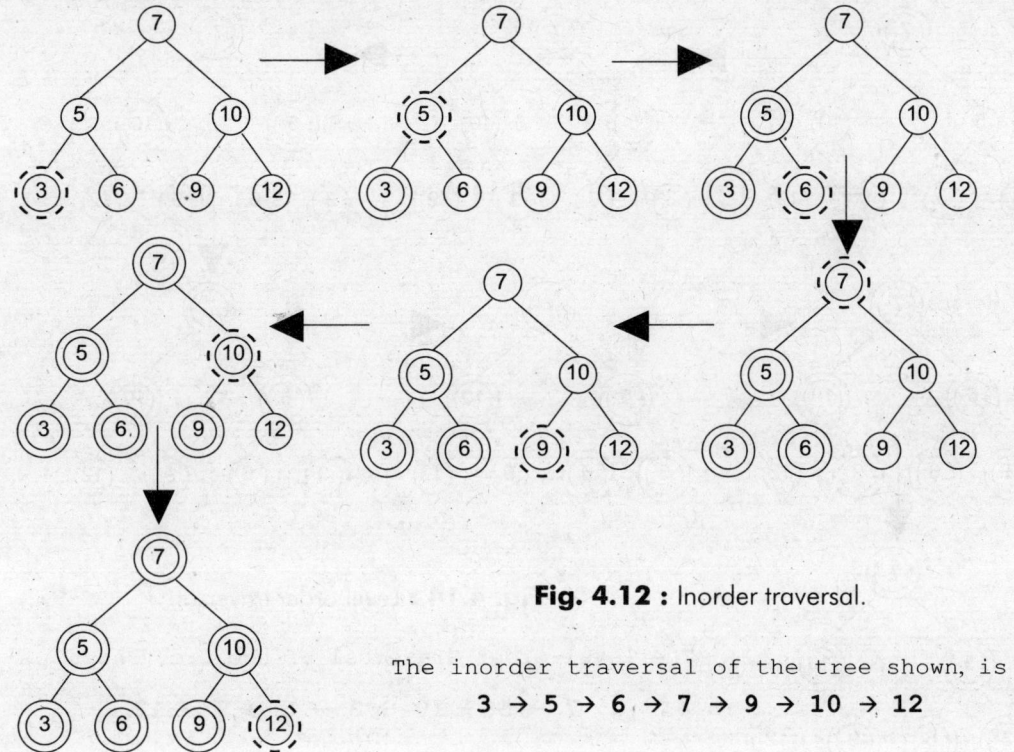

Fig. 4.12 : Inorder traversal.

The inorder traversal of the tree shown, is

3 → 5 → 6 → 7 → 9 → 10 → 12

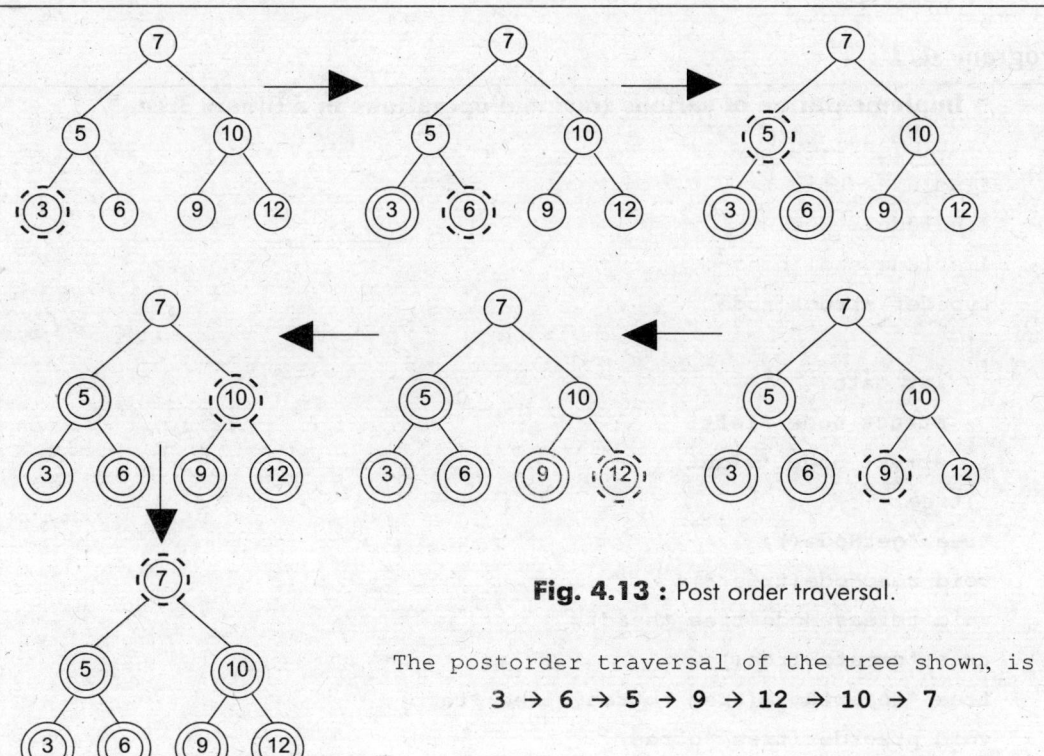

Fig. 4.13 : Post order traversal.

The postorder traversal of the tree shown, is

3 → 6 → 5 → 9 → 12 → 10 → 7

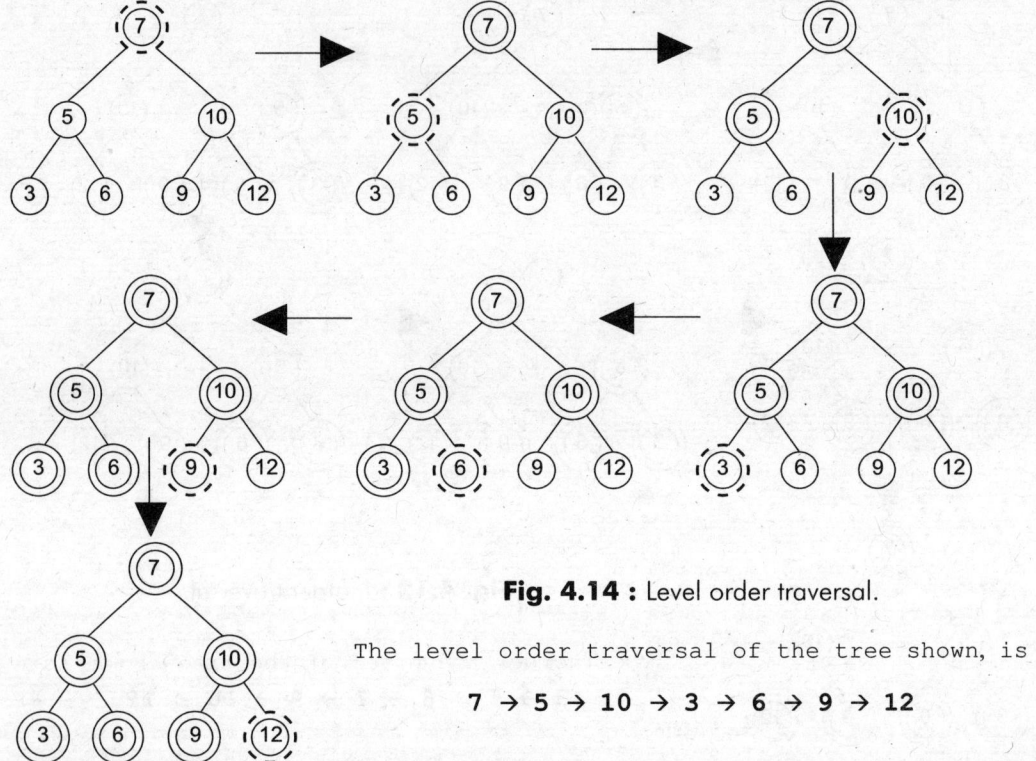

Fig. 4.14 : Level order traversal.

The level order traversal of the tree shown, is

$$7 \rightarrow 5 \rightarrow 10 \rightarrow 3 \rightarrow 6 \rightarrow 9 \rightarrow 12$$

Program 4.1 :

```
/* Implementation of various traversal operations in a Binary Tree. */
/* traverse.c */
#include<stdio.h>
#include<alloc.h>
#include<stdlib.h>
typedef struct node
 {
   int data;
   struct node *left;
   struct node *right;
 }tree;
tree *getNode();
void readNode(tree *);
void releaseNode(tree *head);
tree *createBTree();
tree *insertNode(tree *btree, tree *temp);
void preorder(tree *btree);
```

```c
void inorder(tree *btree);
void postorder(tree *btree);
void levelorder(tree *btree);
tree *queue[30];
int front, rear;
int isEmpty();
void enqueue(tree* btree);
int dequeue(tree **dqdata);
void main()
 {
    int choice;
    tree *btree;
    printf("\nCreate a New Binary Tree");
    btree = createBTree();
    printf("\n TREE TRAVERSAL ");
    printf("\n\t 1. Preorder Traversal ");
    printf("\n\t 2. Inorder Traversal ");
    printf("\n\t 3. Postorder Traversal ");
    printf("\n\t 4. Levelorder Traversal ");
    printf("\n ? ");
      scanf("%d", &choice);
    if(btree == NULL)
     {
       printf("Binary Tree is Empty");
       return;
     }
    switch(choice)
     {
       case 1 :
               printf("Preorder Binary tree traversal is ... \n");
               preorder(btree);
               break;
       case 2 :
               printf("Inorder Binary tree traversal is ... \n");
               inorder(btree);
               break;
       case 3 :
               printf("Postorder Binary tree traversal is ... \n");
               postorder(btree);
               break;
```

```c
        case 4 :
                printf("Levelorder Binary tree traversal is ... \n");
                levelorder(btree);
                break;
        }
    releaseNode(btree);
}
tree* getNode()
{
    int size;
    tree * newnode;
    size = sizeof(tree);
    newnode = (tree *)malloc(size);
    return(newnode);
}
void readNode(tree* newnode)
{
    printf("\nEnter the Data : ");
        scanf("%d", &newnode->data);
    newnode->left  = NULL;
    newnode->right = NULL;
}
void releaseNode(tree* head)
{
    free(head);
}
tree *createBTree()
{
    char ch;
    tree *btree = NULL, *temp;
    do
     {
        temp = getNode();
        readNode(temp);
        btree = insertNode(btree, temp);
        fflush(stdin);
        printf("Do u wish to Add Data in the Tree (y/n) ? ");
            scanf("%c", &ch);
     }while( (ch == 'y') || (ch == 'Y') );
    return btree;
}
tree *insertNode(tree *btree, tree *temp)
{
    if(btree == NULL)
        return temp;
    else if(temp->data < btree->data)
        btree->left = insertNode(btree->left,temp);
```

```
    else if(temp->data > btree->data)
      btree->right = insertNode(btree->right,temp);
    else if(temp->data == btree->data)
     {
       printf("\n Data is already Existing ... ");
       return btree;
     }
    return btree;
  }
  void inorder(tree *btree)
  {
    if(btree != NULL)
     {
       inorder(btree->left);
       printf(" %d ", btree->data);
       inorder(btree->right);
     }
  }
  void preorder(tree *btree)
  {
    if(btree != NULL)
     {
       printf(" %d ", btree->data);
       preorder(btree->left);
       preorder(btree->right);
     }
  }
  void postorder(tree *btree)
  {
    if(btree != NULL)
     {
       postorder(btree->left);
       postorder(btree->right);
       printf(" %d ", btree->data);
     }
  }
  void levelorder(tree *btree)
  {
    enqueue(btree);
    while(!isEmpty())
     {
       dequeue(&btree);
       if(btree != NULL)
        {
          printf(" %d ", btree->data);
          enqueue(btree->left);
          enqueue(btree->right);
        }
     }
  }
```

```
int isEmpty()
{
   if(front == -1 && rear == -1)
      return 1;
   else
      return 0;
}
void enqueue(tree* btree)
{
   if(isEmpty())
      front = rear = 0;
   else
      rear = rear + 1;
   queue[rear] = btree;
   return 0;
}
int dequeue(tree **dqdata)
{
   if(isEmpty())
      return -1;
   *dqdata = queue[front];
   if(front == rear)
      front = rear = -1;
   else
      front = front + 1;
   return 0;

}
```

The program displays the following output

RUN 1

```
Enter data to Create a New Binary Tree ...

Enter the Data : 7
Do u wish to Add Data in the Tree (y/n) ? y

Enter the Data : 5
Do u wish to Add Data in the Tree (y/n) ? y

Enter the Data : 10
Do u wish to Add Data in the Tree (y/n) ? y

Enter the Data : 3
Do u wish to Add Data in the Tree (y/n) ? y

Enter the Data : 6
Do u wish to Add Data in the Tree (y/n) ? y

Enter the Data : 9
Do u wish to Add Data in the Tree (y/n) ? y

Enter the Data : 12
Do u wish to Add Data in the Tree (y/n) ? n
```

```
     TREE TRAVERSAL
               1. Preorder Traversal
               2. Inorder Traversal
               3. Postorder Traversal
               4. Levelorder Traversal
     ? 1
```

Preorder Binary tree traversal is ...
7   5   3   6   10   9   12

RUN2

```
     Enter data to Create a New Binary Tree ...

     Enter the Data : 7
     Do u wish to Add Data in the Tree (y/n) ? y

     Enter the Data : 5
     Do u wish to Add Data in the Tree (y/n) ? y

     Enter the Data : 10
     Do u wish to Add Data in the Tree (y/n) ? y

     Enter the Data : 3
     Do u wish to Add Data in the Tree (y/n) ? y

     Enter the Data : 6
     Do u wish to Add Data in the Tree (y/n) ? y

     Enter the Data : 9
     Do u wish to Add Data in the Tree (y/n) ? y

     Enter the Data : 12
     Do u wish to Add Data in the Tree (y/n) ? n

     TREE TRAVERSAL
               1. Preorder Traversal
               2. Inorder Traversal
               3. Postorder Traversal
               4. Levelorder Traversal
     ? 2
```

Inorder Binary tree traversal is ...
3   5   6   7   9   10   12

RUN3

```
     Enter data to Create a New Binary Tree ...

     Enter the Data : 7
     Do u wish to Add Data in the Tree (y/n) ? y

     Enter the Data : 5
     Do u wish to Add Data in the Tree (y/n) ? y

     Enter the Data : 10
     Do u wish to Add Data in the Tree (y/n) ? y

     Enter the Data : 3
     Do u wish to Add Data in the Tree (y/n) ? y
```

```
Enter the Data : 6
Do u wish to Add Data in the Tree (y/n) ? y

Enter the Data : 9
Do u wish to Add Data in the Tree (y/n) ? y

Enter the Data : 12
Do u wish to Add Data in the Tree (y/n) ? n
  TREE TRAVERSAL
              1. Preorder Traversal
              2. Inorder Traversal
              3. Postorder Traversal
              4. Levelorder Traversal
  ? 3


Postorder Binary tree traversal is ...
3    6    5    9    12    10    7

Enter data to Create a New Binary Tree ...
Enter the Data : 7
Do u wish to Add Data in the Tree (y/n) ? y
Enter the Data : 5
Do u wish to Add Data in the Tree (y/n) ? y

Enter the Data : 10
Do u wish to Add Data in the Tree (y/n) ? y

Enter the Data : 3
Do u wish to Add Data in the Tree (y/n) ? y

Enter the Data : 6
Do u wish to Add Data in the Tree (y/n) ? y

Enter the Data : 9
Do u wish to Add Data in the Tree (y/n) ? y

Enter the Data : 12
Do u wish to Add Data in the Tree (y/n) ? n
  TREE TRAVERSAL
              1. Preorder Traversal
              2. Inorder Traversal
              3. Postorder Traversal
              4. Levelorder Traversal
  ? 4

Level Binary tree traversal is ...
7    5    10    3    6    9    12
```

4.5 Binary Search Trees

A *binary search tree* is a special binary tree, that satisfy the following characteristics.

- Every node has a value and no two nodes should have the same value (i.e., the values in the binary search tree are distinct).
- The values in any left subtree is less than the value of its parent node.
- The values in any right subtree is greater than the value of its parent node.
- The left and right subtrees of each node are again binary search trees.

Fig. 4.15 shows the structure of a binary search tree.

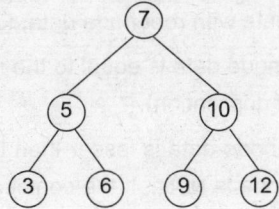

Fig. 4.15 : A binary search tree.

4.5.1 Basic Operations in a Binary Search Tree

The basic operations that can be performed on a binary search tree are,

- Creation of a binary search tree.
- Insertion of a node
- Deletion of a node
- Searching a node
- Modification of a node
- View the contents of the binary search tree.

Creation of a Binary Search Tree

Every node of a binary search tree have two parts, data part and the link part. Creation of binary search tree involves three processes. They are,

- Creating a node.
- Read details for the node from the user.
- Insert the node in the binary search tree.

In *Program 4.2,* getNode() function is used to create a node. After allocating memory for the structure of type node, the information for the node has to be read from the user. In *Program 4.2,* readNode() function is used for reading details for the node from the user. In *Program 4.2,* insertNode() function is used for inserting a new node in the binary search tree. The functions getNode(), readNode() and insertNode() are called repeatedly, to insert any number of nodes in the binary search trees.

Insertion of a node

One of the most primitive operations that can be done in a binary search tree is the insertion of a node. *Memory is to be allocated for the new node (in a similar way that is done while creating a tree) before reading the data.* The new node will contain empty data field and empty link field. The data field of the new node is then stored with the information read from the user. The link field of the new node is assigned to NULL.

- Check whether the root node of the binary search tree is NULL.

- If the condition is true, it means, that the binary search tree is empty, hence consider the new node as the root node. Otherwise, follow the next steps.

- Compare the new node data with root node data, for the following three conditions.

- If the content of the new node data is equal to the root node data, insertion operation is terminated (because of duplication).

- If the content of the new node data is lesser than the root node data, check whether the left child of the root node is NULL. If the condition is true, insert the new data and terminate the process. Otherwise, consider the left child of the root node as the root node and check for the three conditions again.

- If the content of the new node data is greater than the root node data, check whether the right child of the root node is NULL. If the condition is true, insert the new data and terminate the process. Otherwise, consider the right child of the root node as the root node and check for the three conditions again.

In *Program 4.2,* `insertNode()` function is used for inserting a new node in the binary search tree. *Fig. 4.16* explains insertion operation in a binary search in detail.

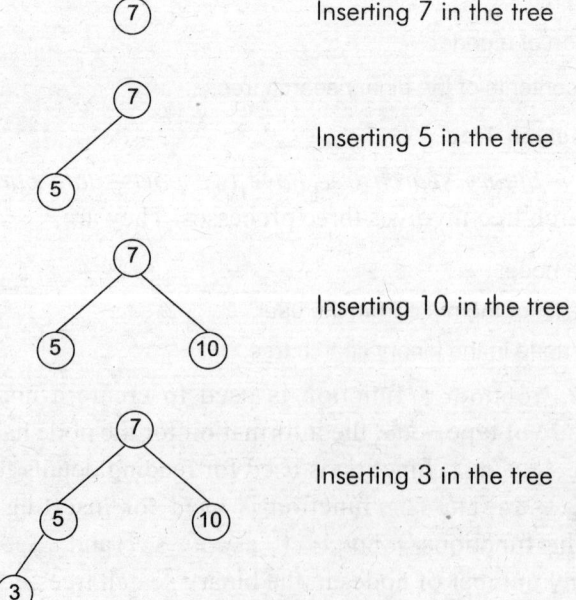

Inserting 7 in the tree

Inserting 5 in the tree

Inserting 10 in the tree

Inserting 3 in the tree

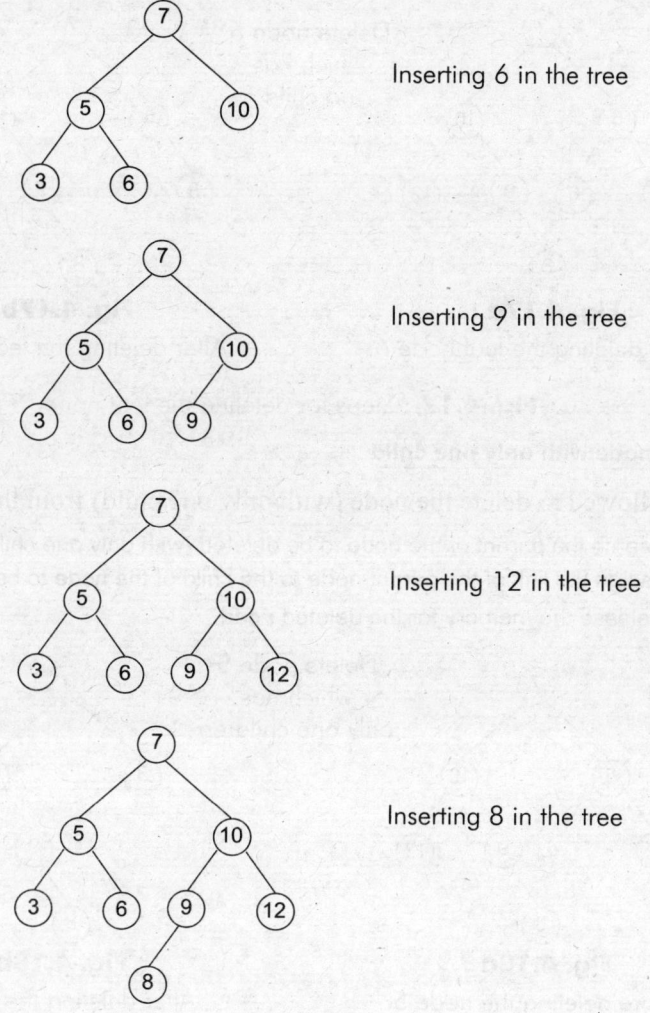

Inserting 6 in the tree

Inserting 9 in the tree

Inserting 12 in the tree

Inserting 8 in the tree

Fig. 4.16 : Steps for insertion operatiom in a binary search tree.

Deletion of a Node

Another primitive operation that can be done in a binary search tree is the deletion of a node. *Memory is to be released for the node to be deleted.* A node can be deleted from the tree from three different places namely,

- Deleting the leaf node
- Deleting the node with only one child
- Deleting the node with two children.

Deleting the leaf node

Steps followed to delete a leaf node from the binary search tree.

- Search the parent of the leaf node, and make the link to the leaf node as NULL.
- Release the memory for the deleted node.

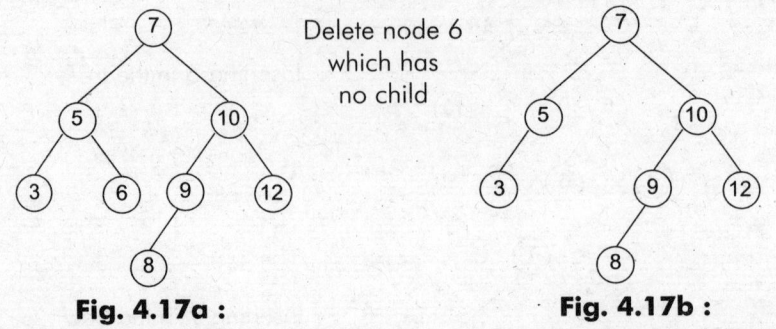

Fig. 4.17a :
Before deleting the leaf node 6.

Fig. 4.17b :
After deleting the leaf node 6.

Fig. 4.17 : Steps for deleting the leaf node.

Deleting the node with only one child

Steps followed to delete the node (with only one child) from the binary search tree.

- Search the parent of the node to be deleted (with only one child).
- Assign the link of the parent node to the child of the node to be deleted.
- Release the memory for the deleted node.

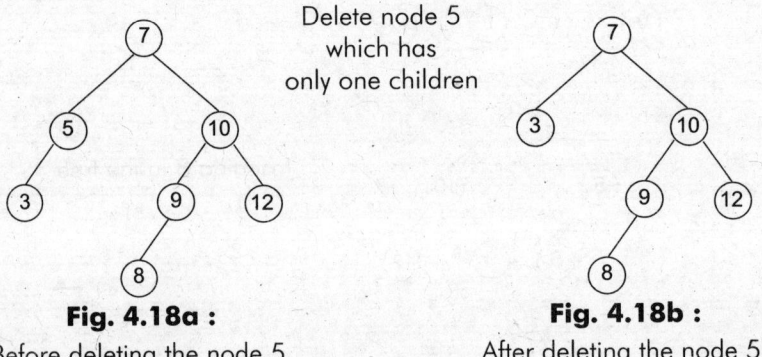

Fig. 4.18a :
Before deleting the node 5.

Fig. 4.18b :
After deleting the node 5.

Fig. 4.18 : Steps for deleting a node that has only one children.

Deleting the node with two children

Steps followed to delete the node (with two children) from the binary search tree.

- Search the parent of the node to be deleted (with two children).
- Copy the content of the inorder successor to the node to be deleted (with two children).
- Delete the inorder successor node. If the inorder successor node has no child, follow the steps given in the *deleting the leaf node*. If the inorder successor node has only one child, follow the steps given in the *deleting the node with one child*.
- Release the memory for the inorder successor node.
- Replace the contents of the node to be deleted with the copy of the content of the inorder successor node.

In *Program 4.2,* deleteNode() function is used for deleting the node from binary search tree.

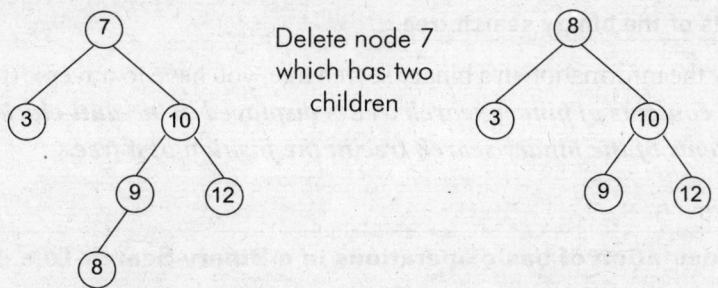

Fig. 4.19a :   **Fig. 4.19b :**

Before deleting the node 7.   After deleting the node 7.

Fig. 4.19 : Steps of deleting a node that has two children.

Searching a node

One of the most important operations that can be done in a binary search tree is searching a node.

- Check whether the root node of the binary search tree is NULL. If yes, binary search tree is empty and hence terminate the search operation.

- If the binary search tree is not empty, compare the new node data with root node data, for the following three conditions.

- If the content of the new node data is equal to the root node data, search element is found and hence terminate the process.

- If the content of the new node data is lesser than the root node data, check whether the left child of the root node is NULL. If the condition is true, search element is not found and terminate the process. Otherwise, consider the left child of the root node as the root node and check for the three conditions again.

- If the content of the new node data is greater than the root node data, check whether the right child of the root node is NULL. If the condition is true, search element is not found, and terminate the process. Otherwise, consider the right child of the root node as the root node and check for the three conditions again.

In *Program 4.2* searchNode() function is used for searching a node in the binary search tree.

Modification of a Node

Steps followed to modify a node in a binary search tree.

- First search the node in the binary search tree using the steps given in the *searching of a node* and then modify the content of the node.

In *Program 4.2,* modifyNode() function is used for modifying a node in the binary search tree.

View the contents of the binary search tree

To display the information in a binary search tree, you have to traverse (move) a binary search tree. *The contents of binary search tree is displayed in 90° anti-clockwise position, to view the contents of the binary search tree in the fashion of a tree.*

Program 4.2 :

```c
/* Implementation of basic operations in a Binary Search Tree. */
/* bst.c */
#include<stdio.h>
#include<alloc.h>
#include<stdlib.h>
typedef struct node
 {
   int data;
   struct node *left, *right;
 }tree;
tree *getNode();
void displayMenu();
void readNode(tree *);
void releaseNode(tree* head);
tree *createBTree();
tree *insertNode(tree *btree, tree *temp);
tree *deleteNode(int digit, tree *btree);
tree *searchNode(tree *btree, int key);
void view(tree *btree, int level);
tree *findParNode(tree *btree, int item, tree **par);
tree *delNoChild(tree *btree, tree *par, tree *loc);
tree *delOneChild(tree *btree, tree *par, tree *loc);
tree *delTwoChild(tree *btree, tree *par, tree *loc);
void main()
 {
   int choice, key;
   tree *btree = NULL, *temp, *par, *loc;
   displayMenu();
   while(1)
    {
      printf("\n ? ");
        scanf("%d", &choice);
```

```
    switch(choice)
     {
       case 0 :
                displayMenu();
                break;
       case 1 :
                btree = NULL;
                printf("\nCreate a New Binary Tree");
                btree = createBTree();
                break;
       case 2 :
                printf("\nInsert the Node in the Tree");
                temp = getNode();
                readNode(temp);
                btree = insertNode(btree,temp);
                break;
       case 3 :
                if(btree == NULL)
                   printf("Binary Tree is Empty ...");
                else
                  {
                    printf("\nDelete the Node from the Tree");
                    printf("\nEnter the Element for Deleting the Node : ");
                        scanf("%d", &key);
                    btree = deleteNode(key, btree);
                  }
                break;
       case 4 :
                if(btree == NULL)
                   printf("Binary Tree is Empty ...");
                else
                  {
                    printf("\nSearch the Node in the Tree");
                    printf("\nEnter the Searching Element : ");
                      scanf("%d", &key);
                    temp = searchNode(btree, key);
                    if(temp == NULL)
                      printf("Search Element %d is Not Found",key);
                    else
                      printf("Search Element %d is Found",temp->data);
                  }
                break;
```

```
                    case 5 :
                              if(btree == NULL)
                                 printf("Binary Tree is Empty ...");
                              else
                               {
                                 printf("Binary Search tree is ...\n");
                                 view(btree, 1);
                               }
                              break;
                    default :
                              printf("\n End of Run of your Program . . .");
                              releaseNode(btree);
                              exit(0);
                   }
            }
      }

void displayMenu()
  {
     printf("\n Basic operations in a Binary Search Tree ");
     printf("\n\t 0. Show Menu ");
     printf("\n\t 1. Create a Binary Tree ");
     printf("\n\t 2. Insert a Node ");
     printf("\n\t 3. Delete a Node ");
     printf("\n\t 4. Search a Node ");
     printf("\n\t 5. View the Binary Tree ");
     printf("\n\t 6. Exit ");
  }

tree* getNode()
  {
     int size;
     tree *newnode;
     size = sizeof(tree);
     newnode = (tree *)malloc(size);
     return(newnode);
  }

void readNode(tree* newnode)
  {
     printf("\nEnter the Data : ");
       scanf("%d", &newnode->data);
     newnode->left  = NULL;  newnode->right = NULL;
  }
```

```c
    void releaseNode(tree* head)
    {
      free(head);
    }
tree *createBTree()
  {
    char ch;
    tree *btree = NULL, *temp;
    do
     {
        temp = getNode();
        readNode(temp);
        btree = insertNode(btree, temp);
        fflush(stdin);
        printf("Do u wish to Add Data in the Tree (y/n) ? ");
          scanf("%c", &ch);
     }while( (ch == 'y') || (ch == 'Y') );
    return btree;
  }
  tree *findParNode(tree *btree, int data, tree **par)
  {
    if(btree == NULL)
      return NULL;
    else if(data < btree->data)
      {
       *par = btree;
       return findParNode(btree->left,data, par);
      }
    else if(data > btree->data)
      {
       *par = btree;
       return findParNode(btree->right, data, par);
      }
    else if(data == btree->data)
      return btree;
  }
```

```
tree *insertNode(tree *btree, tree *temp)
{
   tree *par = NULL, *loc = NULL;
   loc = findParNode(btree, temp->data, &par);
   if(loc != NULL)
    {
     printf("\n Data is already Existing ... ");
     return btree;
    }
   if(btree == NULL)
     return temp;
   else if(temp->data < par->data)
     par->left = temp;
   else
     par->right = temp;
   return btree;
}
tree *deleteNode(int key, tree *btree)
{
   tree *par = NULL, *loc = NULL;
   loc = findParNode(btree, key, &par);
   if(loc == NULL)
    {
      printf("\nItem Not Present ");
      return btree;
    }
   if(loc->left == NULL && loc->right == NULL)
     btree = delNoChild(btree, par, loc);
   if((loc->left != NULL && loc->right == NULL) ||
     (loc->left == NULL && loc->right != NULL))
   btree = delOneChild(btree, par, loc);
   if(loc->left != NULL && loc->right != NULL)
     btree = delTwoChild(btree, par, loc);
   releaseNode(loc);
   return btree;
}
```

```
    tree *delNoChild(tree *btree, tree *par, tree *loc)
    {
       if(par == NULL)
         return NULL;
       else if(loc == par->left)
         par->left = NULL;
       else if(loc == par->right)
         par->right = NULL;
       return btree;
    }
    tree *delOneChild(tree *btree, tree *par, tree *loc)
    {
       tree *temp;
       if(loc->left != NULL)
         temp = loc->left;
       else
         temp = loc->right;
       if(par == NULL)
         btree = temp;
       else if(loc == par->left)
         par->left = temp;
       else if(loc == par->right)
         par->right = temp;
       return btree;
    }
    tree *delTwoChild(tree *btree, tree *par, tree *loc)
    {
       tree *suc, *parSuc;
       parSuc = loc;
       for(suc = loc -> right; suc->left != NULL;suc = suc->left)
         parSuc = suc;
       if(suc->left==NULL && suc->right==NULL)
         delNoChild(btree, parSuc, suc);
       else
         delOneChild(btree, parSuc, suc);
       if(par == NULL)
         btree = suc;
       else if(loc == par->left)
         par->left = suc;
       else if(loc == par->right)
         par->right = suc;
```

```
         suc->left = loc->left;
         suc->right = loc->right;
         return btree;
    }

 tree *searchNode(tree *btree, int key)
   {
      if(btree == NULL)
         return NULL;
      else if(key < btree->data)
         return searchNode(btree->left,key);
      else if(key > btree->data)
         return searchNode(btree->right,key);
      else if(key == btree->data)
         return btree;
   }

 void view(tree *btree, int level)
   {
      int k;
      if(btree == NULL)
         return;
      view(btree -> right, level+1);
      printf("\n");
      for(k = 0; k < level; k++)
         printf("      ");
      printf("%d", btree -> data);
      view(btree -> left, level+1);

   }
```

The program displays the following output

```
Basic operations in a Binary Search Tree
         0. Show Menu
         1. Create a Binary Tree
         2. Insert a Node
         3. Delete a Node
         4. Search a Node
         5. View the Binary Tree
         6. Exit
```

```
    ? 3
    Binary Tree is Empty ...
    ? 4
    Binary Tree is Empty ...
    ? 5
    Binary Tree is Empty ...

    ? 1
    Create a New Binary Tree
    Enter the Data : 20
    Do u wish to Add Data in the Tree (y/n) ? y
    Enter the Data : 10
    Do u wish to Add Data in the Tree (y/n) ? y
    Enter the Data : 30
    Do u wish to Add Data in the Tree (y/n) ? n
    ? 5
    Binary search tree is ...
            30
        20
            10
    ? 2
    Insert the Node in the Tree
    Enter the Data : 25
    ? 5
    Binary search tree is ...
            30
                25
        20
            10
    ? 2
    Insert the Node in the Tree
    Enter the Data : 15
    ? 5
    Binary search tree is ...
            30
                25
        20
                15
            10
```

```
? 2

Insert the Node in the Tree

Enter the Data : 5

? 5

Binary search tree is ...
            30
                    25
        20
                    15
            10
                5
? 2

Insert the Node in the Tree

Enter the Data : 40
? 5

Binary search tree is ...
                    40
            30
                    25
        20
                    15
            10
                5
? 4

Search the Node in the Tree

Enter the Searching Element : 50

Search Element 50 is Not Found
? 4

Search the Node in the Tree

Enter the Searching Element : 5

Search Element 5 is Found
? 4

Search the Node in the Tree

Enter the Searching Element : 40

Search Element 40 is Found
```

```
? 3

Delete the Node from the Tree

Enter the Element for Deleting the Node : 5

? 5

Binary search tree is ...
                40
          30
                25
       20
                15
          10


? 3

Delete the Node from the Tree

Enter the Element for Deleting the Node : 30

? 5

Inorder Binary tree traversal is ...
          40
                25
       20
                15
          10

? 3

Delete the Node from the Tree

Enter the Element for Deleting the Node : 10

? 5

Binary search tree is ...
          40
                25
       20
          15

? 3

Delete the Node from the Tree

Enter the Element for Deleting the Node : 20

? 5

Binary search tree is ...
          40
       25
          15
```

```
? 3

Delete the Node from the Tree

Enter the Element for Deleting the Node : 25

? 5

Binary search tree is ...

     40

          15

? 3

Delete the Node from the Tree

Enter the Element for Deleting the Node : 40

? 5

Binary search tree is ...

     15

? 3

Delete the Node from the Tree

Enter the Element for Deleting the Node : 15

? 5

Binary Tree is Empty ...

? 6

End of Run of your Program . . .
```

4.6 Applications of Trees

1. Trees represent hierarchical relationships between individual data items. We can use trees to store information that naturally forms a hierarchy. For example, in Windows, go to command line and type tree. You can see the folder structure organized and stored as a tree structure.

2. Binary trees are used to represent arithmetic expressions, This feature is used in calculators and in compilers for parsing the expressions and executing them.

3. Binary search tree arranges its node elements in a sorted manner, so the information can be easily inserted, deleted and searched.

4. Spell check and auto correct suggestions during typing are enabled using tree structure.

5. Used in router algorithms to identify where the next path/route of the packet to be sent.

4.7 The Set Representation, Union and Find Operation

The Set is a collection of data items that are accessed by operations such as Union, Intersection and Set Difference. A Set is an unbound collection of distinct elements. Consider the following collection of elements of set S shown in *Fig. 4.21.*

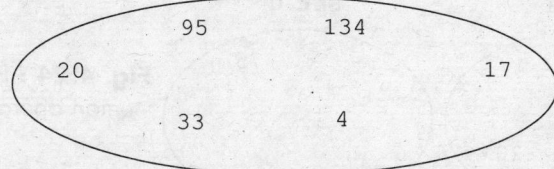

Fig. 4.21 : An example of set.

The following are some of the different sets formed from the above collection of elements.

```
S = {95, 20, 17, 33, 4, 134}
S = {33, 95, 17, 4, 20, 134}
S = {134, 20, 95, 4, 17, 33}
S = {134, 17, 20, 4, 95, 33}
S = {4, 20, 134, 33, 17, 95}
S = {17, 134, 33, 95, 4, 20}
S = {20, 4, 95, 17, 134, 20}
```

Two sets are said to be disjoint set if they do not share any elements. Consider *Fig. 4.21* which shows the two disjoint sets.

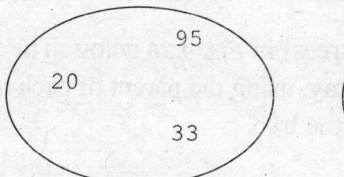 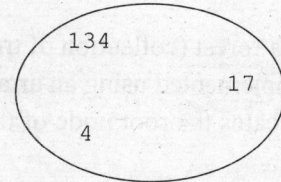

Fig. 4.22 : An example of disjoint set.

The above two sets are disjoint, since they do not share any elements between each other.

The different operations that are carried in a set / disjoint set are as follows,

- find(x)   -   Find the unique representative of the set that contains "x".
- union(x,y)   -   Combine the sets containing "x" and "y" in to a single set (note that before union operation "x" and "y" must be in different sets).
- make(x)   -   Create a single set, containing the element "x".

Set A   **Set B**   **Set C**

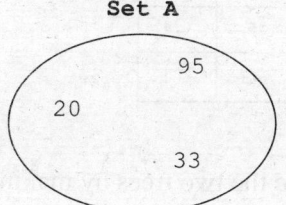 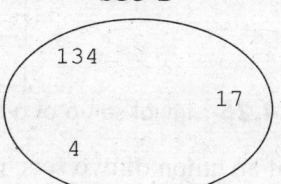 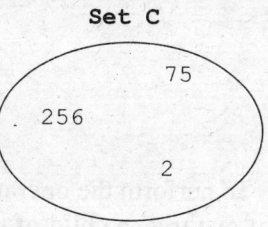

Fig. 4.23 : Sets to explain find operation

Fig. 4.23 shows three sets A, B and C. The find operation of find(256) returns the Set C, whcih contains the element 256. find(256) = Set C.

The union operation is shown in *Fig. 4.25* for the elements 95 and 134 for the three sets A, B and C is shown in *Fig. 4.24*.

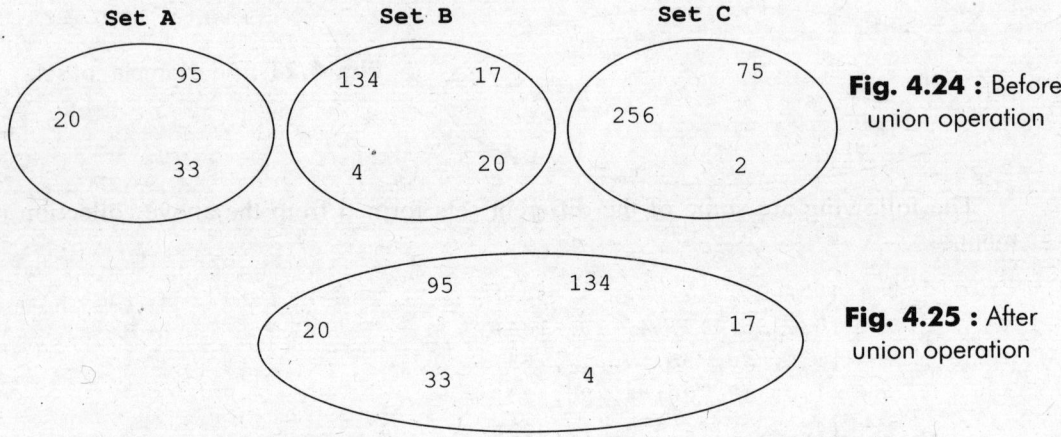

Fig. 4.24 : Before union operation

Fig. 4.25 : After union operation

4.7.1 Smart Union Algorithms

The two different smart union algorithms are as follows,

- Union-by-size
- Union-by-height.

Consider the initial setup of a forest (collection of trees) in *Fig. 4.26* below. The trees (not necessarily binary trees) are implemented using an array, using the parent of each node. The value of the array with –1 indicates the root node of the tree.

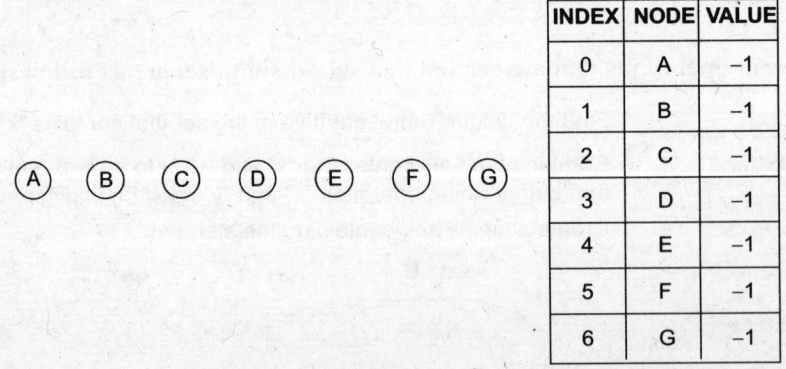

INDEX	NODE	VALUE
0	A	–1
1	B	–1
2	C	–1
3	D	–1
4	E	–1
5	F	–1
6	G	–1

Fig. 4.26 : Initial setup of a forest.

To perform the operation of an union of two sets, merge the two trees by making the root of one tree, a child of the root of the other tree. *Fig. 4.27* shows the forest after union of D and E. *Fig. 4.28* shows the forest after union of F and G. *Fig. 4.29* shows the forest after union of D and F.

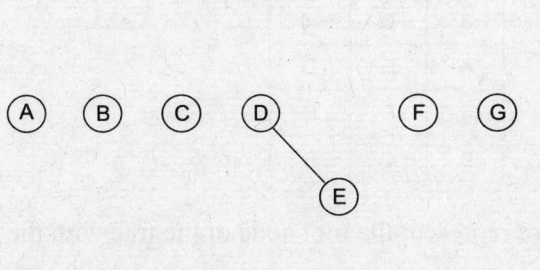

INDEX	NODE	VALUE
0	A	−1
1	B	−1
2	C	−1
3	D	−1
4	E	D
5	F	−1
6	G	−1

Fig. 4.27 : Setup of a forest after union(D,E)

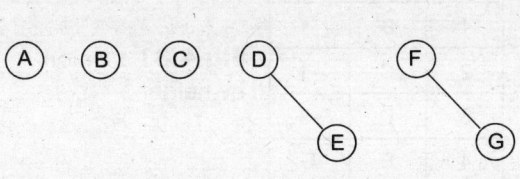

INDEX	NODE	VALUE
0	A	−1
1	B	−1
2	C	−1
3	D	−1
4	E	D
5	F	−1
6	G	F

Fig. 4.28 : Setup of a forest after union(F,G)

INDEX	NODE	VALUE
0	A	−1
1	B	−1
2	C	−1
3	D	−1
4	E	D
5	F	D
6	G	F

Fig. 4.29 : Setup of a forest after union(D,G)

In case of union-by-size algorithm, we represent the root node of the tree with the size of the tree, hence for node D we represent the value as -4 since it contains 4 trees as shown in *Fig. 4.30*. In union-by-size algorithm, we always make the smaller tree a subtree of the larger tree. When depth increases, the tree is smaller than the other side. Thus, after union, it is at least twice as large as the earlier stage.

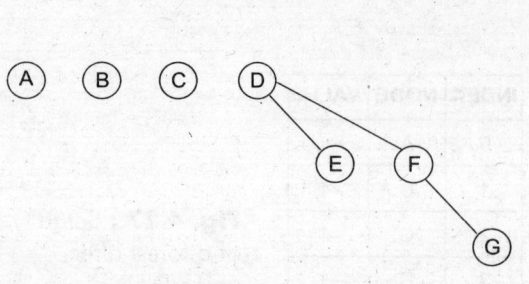

INDEX	NODE	VALUE
0	A	–1
1	B	–1
2	C	–1
3	D	–4
4	E	D
5	F	D
6	G	F

Fig. 4.30 : Union-by-size

In case of union-by-height algorithm, we represent the root node of the tree with the height of the tree, hence for node D we represent the value as –3 since the height is 3 as shown in *Fig. 4.31*. In case of union-by-height algorithm, we always make the shorter tree a subtree of the higher tree.

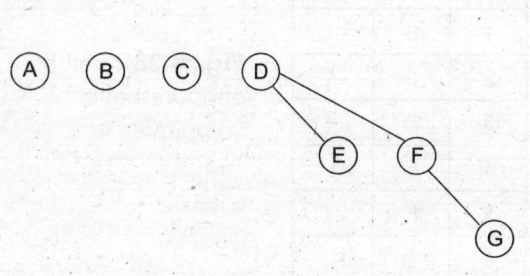

INDEX	NODE	VALUE
0	A	–1
1	B	–1
2	C	–1
3	D	–3
4	E	D
5	F	D
6	G	F

Fig 4.31 : Union-by-height

4.7.2 Path Compression

Consider we have an algorithm, that does all the basic computations (i.e., insert, delete, modify, view and traversal) in a relatively faster time, but sometimes the search operation can take relatively more time because of the shape in a given tree. To decrease the time taken for the search operation, after we do a search on a node N, remember the path of the node N to the root. Then traverse the path a second time and make all the nodes in the path to point to the root of the tree. This shortens the length of the path from the node accessed to the root node and makes the tree shorter. This technique is referred to as *path compression*. *Fig. 4.32* shows the tree before path compression in which node G is to be searched. *Fig. 4.33* shows the tree after path compression, where all the nodes of the tree and connected to the root node of the tree "A".

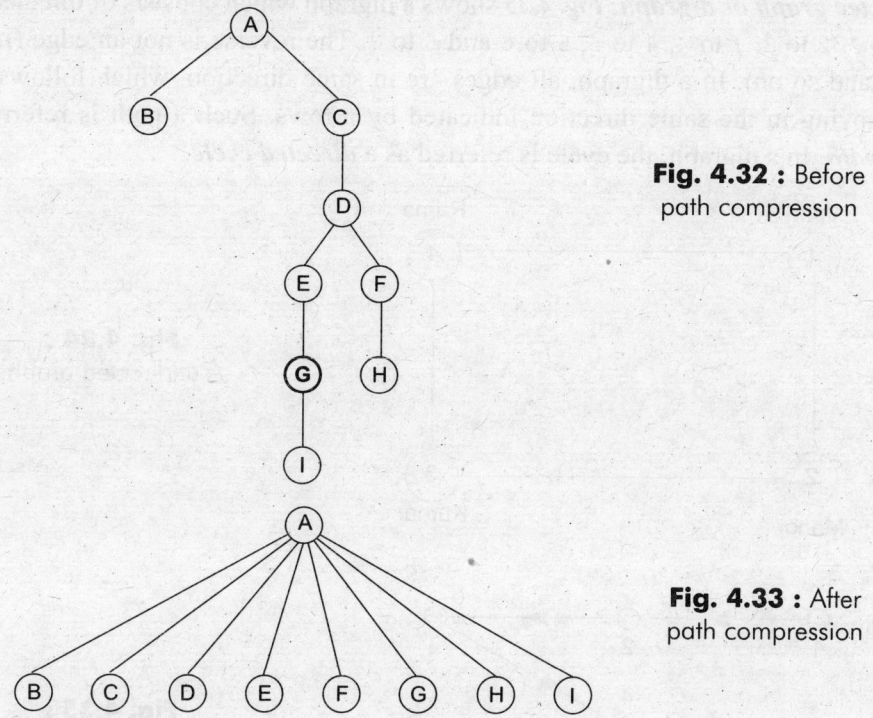

Fig. 4.32 : Before path compression

Fig. 4.33 : After path compression

4.8 Graphs

*A **graph** is a non-linear data structure that represents less relationship between its adjacent elements. There is no hierarchical relationship between the adjacent elements in case of graphs.* A **graph** consists of a set of non-empty vertices (referred as nodes in case of trees), together with a set of edges, and each edge joins two different vertices. A **path** is a sequence of distinct vertices each adjacent to the next, except possibly the first vertex and last vertex is different. In *Fig. 4.34* 1, 2, 3 is path and 3, 4, 1 is also a path. A **cycle** is a path containing atleast three vertices such that the starting and the ending vertices are the same. In *Fig. 4.34,* the path formed by vertices 1, 2, 3, 4 is a cycle.

4.8.1 Definitions in Graphs

Graphs are generally classified as,

* Undirected graph
* Directed graph.

*If an edge between any two nodes is not directionally oriented, a graph is referred as an **undirected graph** or **unqualified graph***. *Fig. 4.34* shows an undirected graph in which 12 or 21, 23 or 32, 34 or 43 and 14 or 41 are edges that are not oriented directionally. *If an edge between any two nodes is directionally oriented, a graph is referred* as a ***directed graph*** or ***digraph***. ***Fig. 4.35*** shows a digraph which consists of directed edges from 1 to 2, 2 to 3, 3 to 4, 4 to 5, 5 to 6 and 6 to 7. The reverse is not an edge (i.e., 2 to 1, 3 to 2 and so on). In a digraph, all edges are in same direction, which follows a path always moving in the same direction indicated by arrows. Such a path is referred as a ***directed path***. In a digraph, the cycle is referred as a ***directed cycle***.

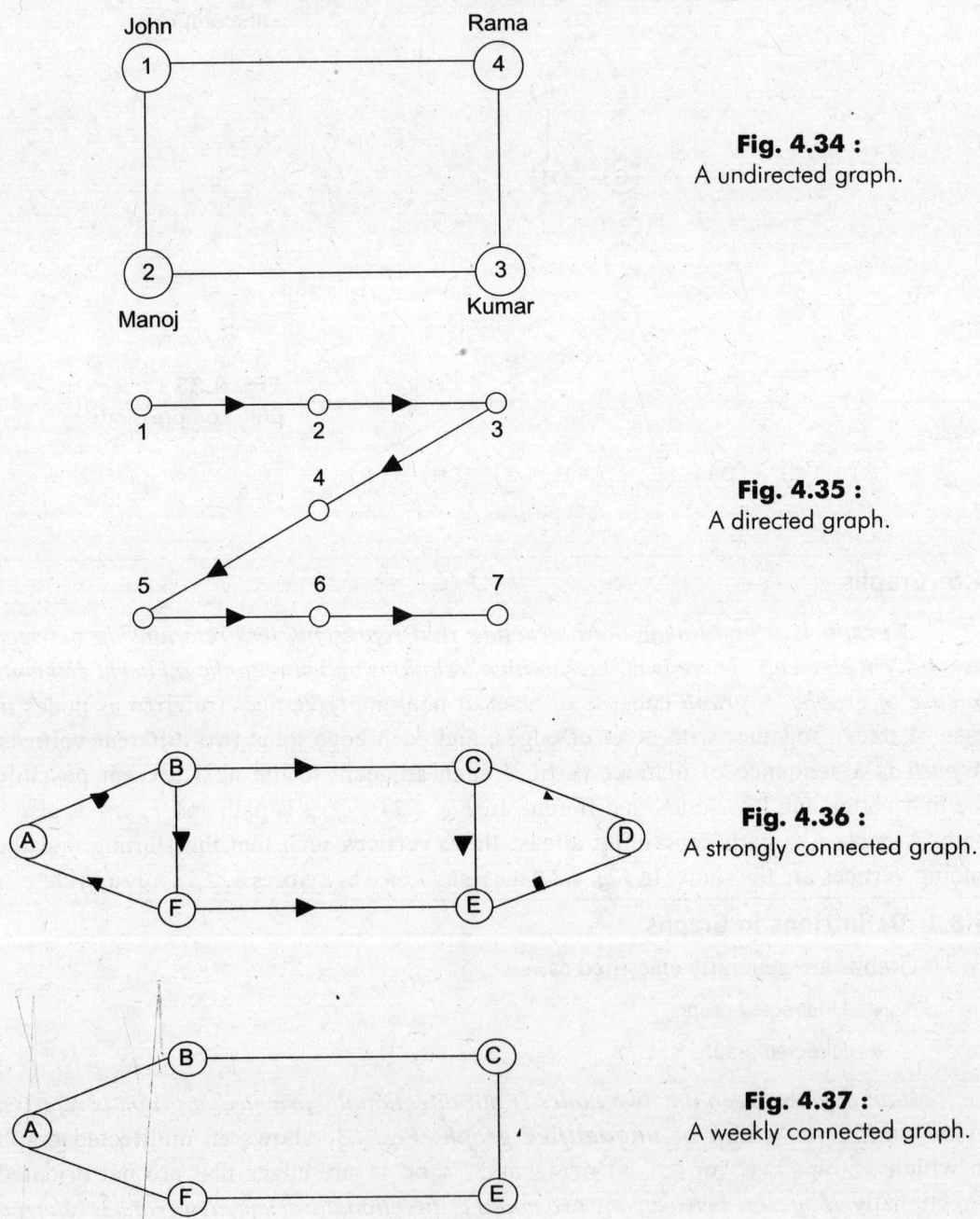

Fig. 4.34 :
A undirected graph.

Fig. 4.35 :
A directed graph.

Fig. 4.36 :
A strongly connected graph.

Fig. 4.37 :
A weekly connected graph.

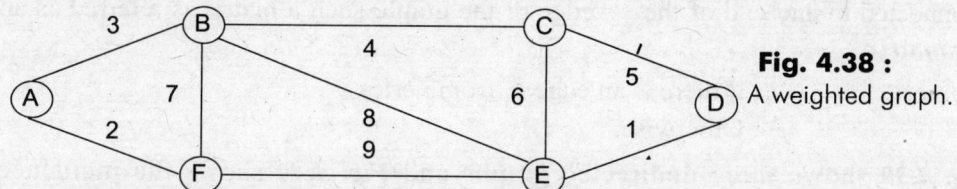

Fig. 4.38 :
A weighted graph.

A graph is called as a **null graph**, if it has no vertices. A graph is called as a **sub graph** of another graph, if all the vertices and edges, are also available in the another graph. A graph is called as a **connected graph**, if there exists a path from any one vertex to all other vertex. In a pair of distinct vertices if there is a directed path from every vertex to all other vertex then the directed graph is said to be **strongly connected graph**. It is also referred as a **complete graph**. A complete digraph with n vertices will have n(n-1) edges. A complete undirected graph with n vertices will have n(n-1)/2 edges. **Fig. 4.36** shows a strongly connected digraph. A directed graph is said to be **weakly connected graph**,if any vertex does not have a directed path to any other vertices. **Fig. 4.37** shows the weakly connected digraph.

The **indegree** of a vertex in a digraph is the number of edges entering into the vertex. The **outdegree** of a vertex in a digraph is the number of edges exiting from the vertex. The **degree** of a vertex is calculated by the sum of indgree and outdegree in a digraph. In **Fig. 4.37,** the indegree of the vertex c is 2, whereas the outdegree of the vertex c is 0. A vertex whose indegree is 0 is referred as **source vertex** and the vertex whose outdegree is 0 is referred as a **sink vertex**. A vertex, which has no edge is referred to as an **isolated vertex**. A vertex whose indegree is 1 and outdegree is 0 is referred to as **pendant vertex**. A graph is said to be a **weighted graph** if every edge in the graph is assigned some weight or value. The weight of an edge is a positive value that may be representing the distance between the vertices or the weights of the edges along the path. **Fig. 4.38** shows the weighted graph.

4.9 Graph Representation and its Operations

Graphs are generally represented in the following scheme as,

- Incidence Matrix
- Adjacency Matrix
- Adjacency List.

4.9.1 Incidence Matrix

A graph containing m vertices and n edges can be represented by a matrix with m rows and n columns. The matrix is formed by storing 1 in its ith row and j^{th} column if there exists a ith vertex, connected to one end of the jth edge, and a 0, if there is no i^{th}

vertex, connected to any end of the jth edge of the graph, such a matrix is referred as an *incidence matrix*.

      IncMat[i][j] = 1, if there is an edge E_j from vertex V_i.
                    = 0, Otherwise.

    Fig. 4.39 shows three undirected graphs and *Fig. 4.40* shows the incidence matrices for the undirected graphs shown in *Fig. 4.39*.

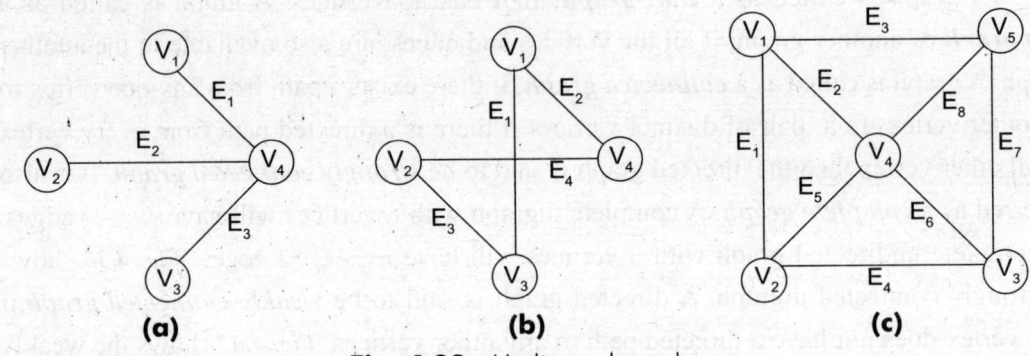

Fig. 4.39 : Undirected graphs.

	E_1	E_2	E_3
V_1	1	0	0
V_2	0	1	0
V_3	0	0	1
V_4	1	1	1

(a)

	E_1	E_2	E_3	E_4
V_1	1	1	0	0
V_2	0	0	1	1
V_3	1	0	1	0
V_4	0	1	0	1

(b)

	E_1	E_2	E_3	E_4	E_5	E_6	E_7	E_8
V_1	1	1	1	0	0	0	0	0
V_2	1	0	0	1	1	0	0	0
V_3	0	0	0	1	0	1	1	0
V_4	0	1	0	0	1	1	0	1
V_5	0	0	1	0	0	0	1	1

(c)

Fig. 4.40 :
Incidence matrices for the undirected graphs shown in Fig. 4.39.

Program 4.3 :

```
/* Program to create and view the incidence Matrix. */
#include <stdio.h>
#define VSIZE 20
int nVertex, nEdges, incMat[VSIZE][VSIZE];
void insertEdge(int vStart, int vEnd);
void createGraph();
void viewGraph();
```

```
void main()
{
    createGraph();
    printf("\n Incidence Matrix is . . . \n");
    viewGraph();
}

void createGraph()
{
    int r, c;
    printf("\n Enter the no. of Vertices : ");
        scanf("%d", &nVertex);
    for(r = 0; r < nVertex; r++)
        for(c = 0; c < nVertex; c++)
            if(r != c)
                insertEdge(r+1, c+1);
}

void insertEdge(int vStart, int vEnd)
{
    int i, ie;
    if(vStart > nVertex || vStart < 1 || vEnd > nVertex || vEnd < 1)
        return;
    printf("Enter the Weight of the Edge from V%d to V%d : ",
            vStart, vEnd);
        scanf("%d", &ie);
    if(ie <= 0)
        return;
    for(i = 0;  i < nEdges; i++)
        if(incMat[vEnd-1][i] == 1 && incMat[vStart-1][i] == 1)
            return;
    incMat[vEnd-1][nEdges] = incMat[vStart-1][nEdges] = 1;
    nEdges++;
}
```

```
    void viewGraph()
    {
       int e, v, c;
       for(e = 0; e < nEdges; e++)
          printf("      E%d",e+1);
       for(v = 0; v < nVertex; v++)
       {
          printf("\nV%-2d    ",v+1);
          for(c = 0; c < nEdges; c++)
             printf("%-2d     ", incMat[v][c]);
       }
       printf("\n Total No. of Vertices : %d", nVertex);
       printf("\n Total No. of Edges : %d", nEdges);
    }
```

The program displays the following output

RUN 1   Enter the no. of Vertices : 4

            Enter the Weight of the Edge from V1 to V2 :0

            Enter the Weight of the Edge from V1 to V3 :0

            Enter the Weight of the Edge from V1 to V4 :1

            Enter the Weight of the Edge from V2 to V1 :0

            Enter the Weight of the Edge from V2 to V3 :0

            Enter the Weight of the Edge from V2 to V4 :1

            Enter the Weight of the Edge from V3 to V1 :0

            Enter the Weight of the Edge from V3 to V2 :0

            Enter the Weight of the Edge from V3 to V4 :1

            Enter the Weight of the Edge from V4 to V1 :1

            Enter the Weight of the Edge from V4 to V2 :1

            Enter the Weight of the Edge from V4 to V3 :1

     Incidence Matrix is

          E1   E2   E3

     V1   1   0   0

     V2   0   1   0

     V3   0   0   1

     V4   1   1   1

     Total No. of Vertices : 4

     Total No. of Edges : 3
```
```

```
RUN2      Enter the no. of Vertices : 5
          Enter the Weight of the Edge from V1 to V2 : 1
          Enter the Weight of the Edge from V1 to V3 : 0
          Enter the Weight of the Edge from V1 to V4 : 1
          Enter the Weight of the Edge from V1 to V5 : 1
          Enter the Weight of the Edge from V2 to V1 : 1
          Enter the Weight of the Edge from V2 to V3 : 1
          Enter the Weight of the Edge from V2 to V4 : 1
          Enter the Weight of the Edge from V2 to V5 : 0
          Enter the Weight of the Edge from V3 to V1 : 0
          Enter the Weight of the Edge from V3 to V2 : 1
          Enter the Weight of the Edge from V3 to V4 : 1
          Enter the Weight of the Edge from V3 to V5 : 1
          Enter the Weight of the Edge from V4 to V1 : 1
          Enter the Weight of the Edge from V4 to V2 : 1
          Enter the Weight of the Edge from V4 to V3 : 1
          Enter the Weight of the Edge from V4 to V5 : 1
          Enter the Weight of the Edge from V5 to V1 : 1
          Enter the Weight of the Edge from V5 to V2 : 0
          Enter the Weight of the Edge from V5 to V3 : 1
          Enter the Weight of the Edge from V5 to V4 : 1
      Incidence Matrix is . . .
```

	E1	E2	E3	E4	E5	E6	E7	E8
V1	1	1	1	0	0	0	0	0
V2	1	0	0	1	1	0	0	0
V3	0	0	0	1	0	1	1	0
V4	0	1	0	0	1	1	0	1
V5	0	0	1	0	0	0	1	1

```
      Total No. of Vertices : 5
      Total No. of Edges : 8
```

4.9.2 Adjacency Matrix

A graph containing n vertices can be represented by a matrix with n rows and n columns. The matrix is formed by storing the edge weight in its ith row and j^{th} column of the matrix, if there exists an edge between i^{th} and j^{th} vertex of the graph, and a 0, if there is no edge between i^{th} and j^{th} vertex of the graph. Such matrix is referred as an *adjacency matrix*. *Note that for an unweighted graph, the edge weight is 1.*

AdjMat[i][j] = Weight of the edge, if there is a path from vertex v_i to v_j.
            = 0, Otherwise.

Fig. 4.41 shows the adjacency matrices for the undirected graphs shown in *Fig. 4.39*. Even though adjacency matrix is an easy way of representing graphs, a graph with considerable number of nodes could use a large amount of memory. Most of the adjacency matrices will be sparse matrices (a matrix with lot of zeros in it) only.

	V_1	V_2	V_3	V_4
V_1	0	0	0	1
V_2	0	0	0	1
V_3	0	0	0	1
V_4	1	1	1	0

(a)

	V_1	V_2	V_3	V_4
V_1	0	0	1	1
V_2	0	0	1	1
V_3	1	1	0	0
V_4	1	1	0	0

(b)

(c)

	V_1	V_2	V_3	V_4	V_5
V_1	0	1	0	1	1
V_2	1	0	1	1	0
V_3	0	1	0	1	1
V_4	1	1	1	0	1
V_5	1	0	1	1	0

Fig. 4.41 :
Adjacency matrices for the undirected graphs shown in Fig. 4.40.

Program 4.4 :

```c
/* Program to create, insert, delete and view the adjacency matrix. */
/* adjmat.c */
#include <stdio.h>
#define VSIZE 20
int checkWt();
int checkDir();
void insertVertex();
void deleteVertex(int vDel);
void insertEdge(int vStart, int vEnd);
void deleteEdge(int vStart, int vEnd);
void createGraph();
void viewGraph();
void display_menu();
int nVertex, adjMat[VSIZE][VSIZE];
void main()
{
    char choice = 'y';
    int ch, vs, ve, vd;
```

```
        display_menu();
     while( (choice == 'y') || (choice == 'Y') )
      {
         printf("\n? ");
         fflush(stdin);
           scanf("%d", &ch);
         switch(ch)
          {
          case 0 :
                      display_menu();
                      break;
          case 1 :
                      createGraph();
                      break;
          case 2 :
                      insertVertex();
                      break;
          case 3 :
                      printf("\n Enter the Starting & Ending Vertex to
                      insert an edge : ");
                         scanf("%d %d", &vs, &ve);
                      insertEdge(vs, ve);
                      break;
          case 4 :
                      printf("\n Enter the Vertex to delete : ");
                         scanf("%d", &vd);
                      deleteVertex(vd);
                      break;
          case 5 :
                      printf("\n Enter the Starting & Ending Vertex to
                      delete an edge : ");
                         scanf("%d %d", &vs, &ve);
                      deleteEdge(vs, ve);
                      break;
          case 6 :
                      viewGraph();
                      break;
          default:
                      printf("End of run of your program . . .");
                      exit(0);
          }
      }
 }
```

```c
void insertVertex()
{
    int rc;
    nVertex++;
    for(rc = 0; rc < nVertex; rc++)
        adjMat[rc][nVertex-1] = adjMat[nVertex-1][rc] = 0;
}

void insertEdge(int vStart, int vEnd)
{
    int ie;
    if(vStart > nVertex || vStart < 1 || vEnd > nVertex || vEnd < 1)
        return;
    printf("Enter Weight of the Edge from V%d to V%d : ", vStart, vEnd);
        scanf("%d",&adjMat[vStart-1][vEnd-1]);
}

void deleteVertex(int vDel)
{
    int r, c;
    if(vDel > nVertex || vDel < 1)
        return;
    for(r = vDel-1; r < nVertex; r++)
        for(c = 0; c < nVertex; c++)
            adjMat[r][c] = adjMat[r+1][c];
    for(c = vDel-1; c < nVertex; c++)
        for(r = 0; r < nVertex; r++)
            adjMat[r][c] = adjMat[r][c+1];
    nVertex--;
}

void deleteEdge(int vStart, int vEnd)
{
    if(vStart > nVertex || vStart < 1 || vEnd > nVertex || vEnd < 1)
        return;
    if(!checkDir()) adjMat[vEnd-1][vStart-1] = 0;
        adjMat[vStart-1][vEnd-1] = 0;
}
```

```
    int checkDir()
    {
        int r, c;
        for(r = 0; r < nVertex; r++)
          for(c = 0; c < nVertex; c++)
              if(adjMat[r][c] != adjMat[c][r])
                  return 1;
        return 0;
    }
    int checkWt()
    {
        int r, c;
        for(r = 0; r < nVertex; r++)
          for(c = 0; c < nVertex; c++)
            if(adjMat[r][c] > 1)
                return 1;
        return 0;
    }
    void createGraph()
    {
        int r, c;
        printf("\n Enter the no. of Vertices : ");
          scanf("%d", &nVertex);
        for(r = 0; r < nVertex; r++)
          for(c = 0; c < nVertex; c++)
            {
              adjMat[r][c] = 0;
              if(r != c)
                 insertEdge(r+1, c+1);
            }
    }
    void viewGraph()
    {
        int v, r, c, edges, inDeg[VSIZE], outDeg[VSIZE];
        for(v = 0; v < nVertex; v++)
          printf("     V%d",v+1);
        for(r = 0; r < nVertex; r++)
          {
              printf("\nV%-2d    ", r+1);
```

```
            for(c = 0; c < nVertex; c++)
                printf("%-2d      ", adjMat[r][c]);
        }
    for(v = 0; v < nVertex; v++)
        inDeg[v] = outDeg[v] = 0;
    edges = 0;
    for(r = 0; r < nVertex; r++)
        for(c = 0; c < nVertex; c++)
            if(adjMat[r][c] != 0)
            {
                edges++;
                outDeg[r]++;
                inDeg[c]++;
            }
    if(!checkDir())
        edges = edges / 2;
    printf("\n %s Graph",(checkDir())? "Directed" : "Undirected");
    printf("\n %s Graph",(checkWt())? "Weighted" : "Unweighted");
    printf("\nTotal No. of Vertices : %d", nVertex);
    printf("\nTotal No. of Edges : %d", edges);
    printf("\nVertex Indegree Outdegree");
    for(r = 0; r < nVertex; r++)
        {
            printf("\n V%-2d     %-9d %-9d", r+1, inDeg[r], outDeg[r]);
            if(inDeg[r] == 0 && outDeg[r] != 0)
                printf(": %s", "SOURCE");
            if(inDeg[r] != 0 && outDeg[r] == 0)
                printf(": %s", "SINK");
            if(inDeg[r] == 1 && outDeg[r] == 0)
                printf(": %s", "PENDANT");
            if(inDeg[r] == 0 && outDeg[r] == 0)
                printf(": %s", "ISOLATED");
        }
}

void display_menu()
{
    printf("\n\nBasic Operations in an Adjacency Matrix . . . ");
    printf("\n\t 0. Display Menu");
    printf("\n\t 1. Creation of Graph");
    printf("\n\t 2. Insert a Vertex ");
```

```
      printf("\n\t 3. Insert an Edge");

      printf("\n\t 4. Delete a Vertex");

      printf("\n\t 5. Delete an Edge");

      printf("\n\t 6. View the Graph");

      printf("\n\t 7. Exit");
  }
```

The program displays the following output

```
Basic Operations in an Adjacency Matrix . . .

            0. Display Menu

            1. Creation of Graph

            2. Insert a Vertex

            3. Insert an Edge

            4. Delete a Vertex

            5. Delete an Edge

            6. View the Graph

            7. Exit

  ? 1

Enter the no. of Vertices : 5

Enter the Weight of the Edge from V1 to V2 :1

Enter the Weight of the Edge from V1 to V3 :0

Enter the Weight of the Edge from V1 to V4 :1

Enter the Weight of the Edge from V1 to V5 :1

Enter the Weight of the Edge from V2 to V1 :1

Enter the Weight of the Edge from V2 to V3 :1

Enter the Weight of the Edge from V2 to V4 :1

Enter the Weight of the Edge from V2 to V5 :0

Enter the Weight of the Edge from V3 to V1 :0

Enter the Weight of the Edge from V3 to V2 :1

Enter the Weight of the Edge from V3 to V4 :1

Enter the Weight of the Edge from V3 to V5 :1

Enter the Weight of the Edge from V4 to V1 :1

Enter the Weight of the Edge from V4 to V2 :1

Enter the Weight of the Edge from V4 to V3 :1

Enter the Weight of the Edge from V4 to V5 :1

Enter the Weight of the Edge from V5 to V1 :1

Enter the Weight of the Edge from V5 to V2 :0

Enter the Weight of the Edge from V5 to V3 :1

Enter the Weight of the Edge from V5 to V4 :1
```

```
? 6
        V1      V2      V3      V4      V5
V1      0       1       0       1       1
V2      1       0       1       1       0
V3      0       1       0       1       1
V4      1       1       1       0       1
V5      1       0       1       1       0
```

Undirected Graph

Unweighted Graph

Total No. of Vertices : 5

Total No. of Edges : 8

Vertex Indegree Outdegree

Vertex	Indegree	Outdegree
V1	3	3
V2	3	3
V3	3	3
V4	4	4
V5	3	3

```
? 2
? 6
        V1      V2      V3      V4      V5      V6
V1      0       1       0       1       1       0
V2      1       0       1       1       0       0
V3      0       1       0       1       1       0
V4      1       1       1       0       1       0
V5      1       0       1       1       0       0
V6      0       0       0       0       0       0
```

Undirected Graph

Unweighted Graph

Total No. of Vertices : 6

Total No. of Edges : 8

Vertex Indegree Outdegree

Vertex	Indegree	Outdegree	
V1	3	3	
V2	3	3	
V3	3	3	
V4	4	4	
V5	3	3	
V6	0	0	: ISOLATED

```
? 3
Enter the Starting & Ending Vertex to insert edge : 1 6
Enter the Weight of the Edge from V1 to V6 : 1
? 6
        V1      V2      V3      V4      V5      V6
V1      0       1       0       1       1       1
V2      1       0       1       1       0       0
V3      0       1       0       1       1       0
V4      1       1       1       0       1       0
V5      1       0       1       1       0       0
V6      0       0       0       0       0       0

Directed Graph
Unweighted Graph
Total No. of Vertices : 6
Total No. of Edges : 17

Vertex Indegree Outdegree
V1      3               4
V2      3               3
V3      3               3
V4      4               4
V5      3               3
V6      1               0               : SINK: PENDANT
? 4
Enter the Vertex to delete : 5
? 6
        V1      V2      V3      V4      V5
V1      0       1       0       1       1
V2      1       0       1       1       0
V3      0       1       0       1       0
V4      1       1       1       0       0
V5      0       0       0       0       0

Directed Graph
Unweighted Graph
Total No. of Vertices : 5
Total No. of Edges : 11
```

```
Vertex Indegree Outdegree
  V1        2          3
  V2        3          3
  V3        2          2
  V4        3          3
  V5        1          0           : SINK: PENDANT
? 5

Enter the Starting & Ending Vertex to delete edge : 4   3
? 6
        V1      V2      V3      V4      V5
 V1      0       1       0       1       1
 V2      1       0       1       1       0
 V3      0       1       0       1       0
 V4      1       1       0       0       0
 V5      0       0       0       0       0
Directed Graph
Unweighted Graph
Total No. of Vertices : 5
Total No. of Edges : 10
Vertex Indegree Outdegree
  V1        2          3
  V2        3          3
  V3        1          2
  V4        3          2
  V5        1          0           : SINK: PENDANT
? 7
End of run of your program . . .
```

4.9.3 Adjacency List

A graph containing m vertices and n edges can be represented using a linked list, referred to as *adjacency list*. The number of vertices in a graph forms a singly linked list. Each vertex has a seperate linked list, with nodes equal to the number of edges connected from the corresponding vertex. *Fig. 4.42* and *Fig. 4.43* show the undirected and directed graph. *Fig. 4.44* shows the adjacency lists for the graph shown in *Fig. 4.42* and *Fig. 4.43*. Even though adjacency list is a difficult way of representing graphs, a graph with large number of nodes will use small amount of memory.

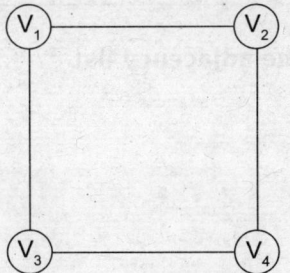

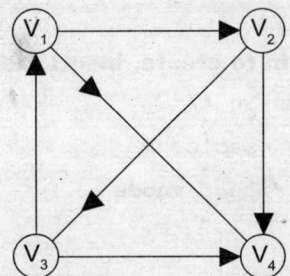

Fig. 4.42 : An undirected graph. **Fig. 4.43 :** A directed graph.

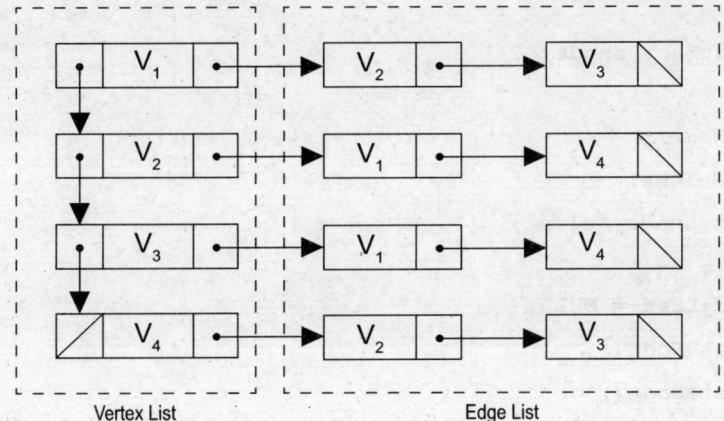

Fig. 4.44 : Adjacency list for the undirected graph shown in **Fig. 4.42.**

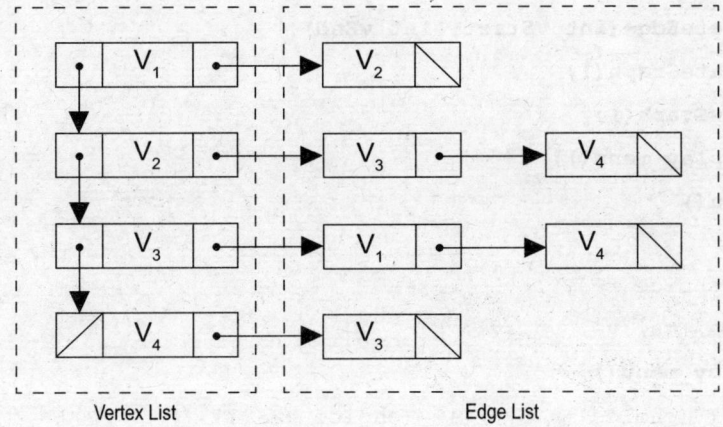

Fig. 4.45 : Adjacency list for the directed graph shown in **Fig. 4.43.**

Program 4.5 :

```
/* Program to create, insert, delete and view the adjacency list. */
/* adjlist.c */
#include <stdio.h>
typedef struct vnode
 {
    int vName;
    struct vnode *vlink;
    struct enode *elink;
 }vnode;
typedef struct enode
 {
    int vName;
    int eWeight;
    struct enode *elink;
 }enode;
vnode *adjList = NULL;
vnode* getvNode();
enode* geteNode();
void insertVertex();
void insertEdge(int vStart, int vEnd);
void deleteVertex(int vDel);
void deleteEdge(int vStart, int vEnd);
void createGraph();
void viewGraph();
void display_menu();
void main()
 {
    char choice = 'y';
    int ch, vs, ve, vd, vi;
    display_menu();
    while( (choice == 'y') || (choice == 'Y') )
     {
        printf("\n? ");
        fflush(stdin);
          scanf("%d", &ch);
```

```
        switch(ch)
        {
          case 0 :
                    display_menu();
                    break;
          case 1 :
                    createGraph();
                    break;
          case 2 :
                    insertVertex();
                    break;
          case 3 :
                    printf("\n Enter the Starting & Ending Vertex to
                            insert Edge : ");
                      scanf("%d %d", &vs, &ve);
                    insertEdge(vs, ve);
                    break;
          case 4 :
                    printf("\n Enter the Vertex to delete : ");
                      scanf("%d", &vd);
                    deleteVertex(vd);
                    break;
          case 5 :
                    printf("\n Enter the Starting & Ending Vertex to
                            delete Edge : ");
                      scanf("%d %d", &vs, &ve);
                    deleteEdge(vs, ve);
                    break;
          case 6 :
                    viewGraph();
                    break;
          default :
                    printf("End of run of your program . . .");
                    exit(0);
        }
    }
}
```

```c
vnode* getvNode()
{
    int size;
    vnode *newvnode;
    size = sizeof(vnode);
    newvnode = (vnode *)malloc(size);
    newvnode->vlink = NULL;
    newvnode->elink = NULL;
    return(newvnode);
}
enode* geteNode()
{
    int size;
    enode *newenode;
    size = sizeof(enode);
    newenode = (enode *) malloc(size);
    newenode->elink = NULL;
    return (newenode);
}
void insertVertex()
{
    vnode *tv, *nv;
    nv = getvNode();
    nv->vName = 1;
    if(adjList == NULL)
    {
        adjList = nv;
        return;
    }
    for(tv = adjList; tv->vlink != NULL; tv=tv->vlink);
    tv->vlink = nv;
    nv->vName = tv->vName + 1;
}
```

```
    void insertEdge(int vStart, int vEnd)
     {
        vnode *pv;

        enode *te, *pe;

        for(pv = adjList; pv != NULL && pv->vName != vStart; pv = pv->vlink);

        if(pv == NULL)

            return;

        te = geteNode();

        printf("Enter Edge Weight from V%d to V%d : ", vStart, vEnd);

            scanf("%d",&te->eWeight);

        te->vName = vEnd;

        if(pv->elink == NULL)

          {

            pv->elink = te;

            return;

          }

        for(pe = pv->elink; pe->elink != NULL; pe = pe->elink);

        pe->elink = te;

     }

    void deleteVertex(int vDel)
     {
        vnode *pv, *tv;

        enode *pe, *te;

        if(adjList == NULL)

            return;

        if(adjList->vName == vDel)

          {

            tv = adjList;

            adjList = adjList->vlink;

            free(tv);

            return;

          }

        for(pv = adjList; pv->vlink != NULL && pv -> vlink -> vName != vDel;

            pv = pv->vlink);

        if(pv ->vlink == NULL)

            return;
```

```c
    else
     {
       tv = pv -> vlink;
       pv->vlink = pv->vlink->vlink;
       free(tv);
     }
    for(pv = adjList; pv != NULL; pv = pv->vlink)
     {
       if(pv->elink == NULL)
          continue;
       if(pv->elink->vName == vDel)
        {
          te = pv->elink;
          pv->elink = pv->elink->elink;
          free(te);
          continue;
        }
    for(pe = pv->elink; pe->elink != NULL && pe -> elink -> vName != vDel;
        pe = pe -> elink);
    if( pe->elink!= NULL)
     {
       te = pe->elink;
       pe->elink = pe->elink->elink;
       free(te);
     }
   }
 }

void deleteEdge(int vStart, int vEnd)
 {
   vnode *pv, *tv;
   enode *pe, *te;
   if(adjList == NULL)
      return;
   for(pv = adjList; pv != NULL && pv->vName!=vStart;pv=pv->vlink);
```

```
    if(pv == NULL)
        return;
    if(pv->elink == NULL)
        return;
    if(pv->elink->vName == vEnd)
     {
        te = pv->elink;
        pv->elink = pv->elink->elink;
        free(te);
        return;
     }
    for(pe = pv->elink; pe->elink != NULL && pe -> elink -> vName != vEnd;
        pe = pe->elink);
    if(pe->elink!= NULL)
     {
        te = pe->elink;
        pe->elink = pe->elink->elink;
        free(te);
     }
  }
void createGraph()
 {
    int r, c, v, nVertex;
    printf("\n Enter the no. of Vertices : ");
      scanf("%d", &nVertex);
    adjList = NULL;
    for(v = 0; v < nVertex; v++)
        insertVertex();
    for(r = 0; r < nVertex; r++)
       for(c = 0; c < nVertex; c++)
          if(r != c)
             insertEdge(r+1, c+1);
 }
void viewGraph()
 {
    int edges = 0, wStatus = 0, nVertex = 0;
    vnode *pv;  enode *pc;
```

```c
    for(pv = adjList; pv!= NULL; pv = pv->vlink)
     {
        printf("\nV%-2d to ", pv->vName);
        nVertex++;
        for(pe = pv->elink; pe != NULL; pe = pe->elink)
         {
            edges++;
            printf("V%-2d(%d) -> ", pe->vName, pe->eWeight);
            if(pe->eWeight > 1)
               wStatus = 1;
         }
        printf("NULL");
     }
    if(adjList != NULL)
     {
        printf("\n %s Graph", (wStatus) ? "Weighted" : "Unweighted");
        printf("\n Number of Vertices : %5d", nVertex);
        printf("\n Number of Edges    : %5d", edges);
     }
}
void display_menu()
 {
    printf("\n\n Basic Operations in a Adjacency List . . .");
    printf("\n\t 0. Display Menu");
    printf("\n\t 1. Creation of Graph");
    printf("\n\t 2. Insert a Vertex ");
    printf("\n\t 3. Insert an Edge");
    printf("\n\t 4. Delete a Vertex");
    printf("\n\t 5. Delete an Edge");
    printf("\n\t 6. View the Graph");
    printf("\n\t 7. Exit");
 }
```

The program displays the following output

```
Basic Operations in an Adjacency List . . .
          0. Display Menu
          1. Creation of Graph
          2. Insert a Vertex
          3. Insert an Edge
```

```
            4. Delete a Vertex
            5. Delete an Edge
            6. View the Graph
            7. Exit
? 1
Enter the no. of Vertices : 4
Enter Edge Weight from V1 to V2 : 1
Enter Edge Weight from V1 to V3 : 1
Enter Edge Weight from V1 to V4 : 0
Enter Edge Weight from V2 to V1 : 1
Enter Edge Weight from V2 to V3 : 0
Enter Edge Weight from V2 to V4 : 1
Enter Edge Weight from V3 to V1 : 1
Enter Edge Weight from V3 to V2 : 0
Enter Edge Weight from V3 to V4 : 1
Enter Edge Weight from V4 to V1 : 0
Enter Edge Weight from V4 to V2 : 1
Enter Edge Weight from V4 to V3 : 1
? 6
V1  to  V2 (1) -> V3 (1) -> V4 (0) -> NULL
V2  to  V1 (1) -> V3 (0) -> V4 (1) -> NULL
V3  to  V1 (1) -> V2 (0) -> V4 (1) -> NULL
V4  to  V1 (0) -> V2 (1) -> V3 (1) -> NULL
Unweighted Graph
Number of Vertices :      4
Number of Edges    :      12
? 2
? 6
V1  to  V2 (1) -> V3 (1) -> V4 (0) -> NULL
V2  to  V1 (1) -> V3 (0) -> V4 (1) -> NULL
V3  to  V1 (1) -> V2 (0) -> V4 (1) -> NULL
V4  to  V1 (0) -> V2 (1) -> V3 (1) -> NULL
V5  to  NULL
Unweighted Graph
Number of Vertices :      5
Number of Edges    :      12
```

```
? 3

Enter the Starting & Ending Vertex to insert Edge : 5 2

Enter Edge Weight from V5 to V2 : 1

? 6

V1  to  V2 (1) -> V3 (1) -> V4 (0) -> NULL

V2  to  V1 (1) -> V3 (0) -> V4 (1) -> NULL

V3  to  V1 (1) -> V2 (0) -> V4 (1) -> NULL

V4  to  V1 (0) -> V2 (1) -> V3 (1) -> NULL

V5  to  V2 (1) -> NULL

Unweighted Graph

Number of Vertices :      5

Number of Edges    :      13

? 4

Enter the Vertex to delete : 3

? 6

V1  to  V2 (1) -> V4 (0) -> NULL

V2  to  V1 (1) -> V4 (1) -> NULL

V4  to  V1 (0) -> V2 (1) -> NULL

V5  to  V2 (1) -> NULL

Unweighted Graph

Number of Vertices :      4

Number of Edges    :      7

? 5

Enter the Starting & Ending Vertex to delete Edge : 4 2

? 6

V1  to  V2 (1) -> V4 (0) -> NULL

V2  to  V1 (1) -> V4 (1) -> NULL

V4  to  V1 (0) -> NULL

V5  to  V2 (1) -> NULL

Unweighted Graph

Number of Vertices :      4

Number of Edges    :      6

? 7

End of run of your program . . .
```

4.10 Graph Traversals

Traversing a graph means visiting all the nodes in the graph. In many practical applications, traversing a graph is important, such that each vertex is visited once systematically by traversing through minimum number of paths. The two important graph traversal methods are

- Depth-first traversal (or) Depth-first search (DFS)
- Breadth-first traversal (or) Breadth-first search (BFS).

4.10.1 Depth–First Traversal

The logic of Depth First Search (DFS) is similar to the preorder traversal of a tree. Visit the first node initially and then find the unvisited node which is adjacent to the first node. Then the unvisited node is visited and a depth first search is initiated from the adjacent node (considering it as the first node). If all the adjacent nodes have been visited, backtrack to the last node visited, and find another adjacent node and again initiate the depth-first search from the adjacent node. This traversal continues until all nodes have been visited once. The following algorithm describes DFS in a graph.

Algorithm for Depth First Traversal
Step 1 : Consider that the DFS is beginning from the starting vertex A. Process the vertex A and mark it as visited.
Step 2 : Using the adjacency matrix of the graph find the vertex along the path which begins vertex A, that has not been visited yet. Process the vertex and consider this as the new vertex and mark the vertex as visited.
Step 3 : Repeat **Step 2** using the new search vertex. If no vertices (i.e., if a dead end if reached) back track to the previous node and continue the search from there.
Step 4 : When backtracking to the previous search node in **Step 3** is impossible, the search from the originally chosen search node is complete.
Step 5 : If the graph still contains unvisited nodes, choose any vertex that has not been visited and repeat **Step 1** to **Step 4**.

The above algorithm executes in a recursive manner, which can also be implemented non-recursively using stacks. To understand the traversal clearly, consider the graph shown in *Fig. 4.39(c)* and its adjacency matrix shown in *Fig. 4.41(c)*. *Program 4.6* helps you to understand things better.

Program 4.6 :

```
/* Program to demonstrate Depth First Search. */
/* dfs.c */
#include<stdio.h>
int a[10][10], visited[10], n;
void searchFrom(int k);
```

```
    void main()
    {
        int i, j;
        printf("Enter the no. of Nodes : ");
          scanf("%d", &n);
        printf("Enter the adjacency matrix . . . \n");
        for(i = 1; i <= n; i++)
          for(j = 1; j <= n; j++)
            if(i != j)
              {
                  printf("Enter the value of %d, %d element : ", i, j);
                    scanf("%d", &a[i][j]);
              }
        printf("Nodes are visited in this order ");
        for(i = 1; i <= n; i++)
          if(visited[i] == 0)
            searchFrom(i);
    }
    void searchFrom(int k)
    {
        int i;
        printf("-> %d ", k);
        visited[k] = 1;
        for(i = 1; i <= n; i++)
          if(visited[i] == 0)
            if(a[k][i] != 0)
                searchFrom(i);
    }
```

The program displays the following output

```
Enter the no. of Nodes : 5
Enter the adjacency matrix . . .
Enter the value of 1, 2 element : 1
Enter the value of 1, 3 element : 0
Enter the value of 1, 4 element : 1
Enter the value of 1, 5 element : 0
Enter the value of 2, 1 element : 0
Enter the value of 2, 3 element : 1
```

```
        Enter the value of 2, 4 element : 1
        Enter the value of 2, 5 element : 0
        Enter the value of 3, 1 element : 0
        Enter the value of 3, 2 element : 0
        Enter the value of 3, 4 element : 0
        Enter the value of 3, 5 element : 1
        Enter the value of 4, 1 element : 0
        Enter the value of 4, 2 element : 0
        Enter the value of 4, 3 element : 1
        Enter the value of 4, 5 element : 1
        Enter the value of 5, 1 element : 1
        Enter the value of 5, 2 element : 0
        Enter the value of 5, 3 element : 0
        Enter the value of 5, 4 element : 0
        Nodes are visited in this order -> 1 -> 2 -> 3 -> 5 -> 4
```

4.10.2 Breadth–First Traversal

The breadth-first traversal *technique begins at a given vertex and then proceeds to all the vertices connected to that vertex*. Visit the first node initially, and then find all the unvisited nodes which is adjacent to the first node. Then unvisited nodes are visited and a breadth first search is initiated from the adjacent node (considering it as the first node). This traversal continues until all nodes have been visited once. The following algorithm describes BFS in a graph.

Algorithm for Breadth First Traversal

Step 1 : Consider any vertex in the graph. Process the vertex and mark it as visited.

Step 2 : Using the adjacency matrix of the graph proceed to the next vertex which has an edge connection wise the vertex considered in **Step 1** .

Step 3 : Backtrack to the vertex considered in **Step 1** descend along an edge towards an unvisited vertex and mark the new vertex as visited.

Step 4 : Repeat **Step 3** until all vertices adjacent to the node in **Step 1** have been marked as visited.

Step 5 : Repeat **Step 1** to **Step 4** starting from vertex visited in **Step 2**, then start again from vertices visited in **Step 3** in the order visited.

The above algorithm executes in a recursive manner, which can also be implemented non-recursively using queues. To understand the traversal clearly, consider the graph shown in in *Fig. 4.35(c)* and its adjacency matrix shown in *Fig. 4.37(c)*. *Program 4.7* helps you to understand things better.

Program 4.7 :

```c
/* Program to demonstrate Breadth-First Search. */
/* bfs.c */
#include<stdio.h>
int visited[10], a[10][10], n;
void searchFrom(int k);
void main()
 {
    int i,j;
    printf("Enter the No. of Nodes : ");
      scanf("%d", &n);
    printf("Enter the adjacency matrix ... \n");
    for(i = 1; i <= n; i++)
      for(j = 1; j <= n; j++)
        if(i != j)
          {
              printf("Enter the value of %d, %d element : ", i, j);
               scanf("%d", &a[i][j]);
          }
    printf("Nodes are visited in this order ");
    for(i = 1; i <= n; i++)
      if(visited[i] == 0)
         searchFrom(i);
 }
void searchFrom(int k)
 {
    int i;
    printf("--> %d ", k);
    visited[k] = 1;
    for(i = 1; i <= n; i++)
     {
       if(visited[i] == 0)
        {
```

```
        if(a[k][i] != 0)
          {

             searchFrom(i);

          }

       }

     break;

   }

 }
```

The program displays the following output

```
Enter the No. of Nodes : 5
Enter the adjacency matrix ...
Enter the value of 1, 2 element : 1
Enter the value of 1, 3 element : 0
Enter the value of 1, 4 element : 1
Enter the value of 1, 5 element : 0
Enter the value of 2, 1 element : 0
Enter the value of 2, 3 element : 1
Enter the value of 2, 4 element : 1
Enter the value of 2, 5 element : 0
Enter the value of 3, 1 element : 0
Enter the value of 3, 2 element : 0
Enter the value of 3, 4 element : 0
Enter the value of 3, 5 element : 1
Enter the value of 4, 1 element : 0
Enter the value of 4, 2 element : 0
Enter the value of 4, 3 element : 1
Enter the value of 4, 5 element : 1
Enter the value of 5, 1 element : 1
Enter the value of 5, 2 element : 0
Enter the value of 5, 3 element : 0
Enter the value of 5, 4 element : 0
```
Nodes are visited in this order --> 1 --> 2 --> 3 --> 4 --> 5

4.11 Summary

✍ A tree is a non-linear, two-dimensional data structure, which represents hierarchical relationship between individual data items.

✍ A path in a tree is a sequence of distinct nodes in which successive nodes are connected by edges in the tree.

✍ A node that has no children is called as a terminal node. It is also referred as a leaf node. These nodes have zero degree.

✍ All intermediate nodes that traverse the given tree from its root node to the terminal nodes are referred as non-terminal nodes.

✍ A binary tree is a tree, which has nodes either empty or not more than two child nodes, each of which may be a leaf node.

✍ A full binary tree is a tree in which all the leaves are on the same level and every non-leaf node has exactly two children.

✍ A complete binary tree is a tree in which every non-leaf node has exactly two children not necessarily to be on the same level.

✍ Postorder traversal leads to postfix expressions. Preorder traversal leads to prefix expressions. Inorder traversal leads to infix expressions.

✍ **Left-skewed binary tree is a** binary tree, which has only left child nodes.

✍ **Right-skewed binary tree is a** binary tree, which has only right child nodes.

✍ An extended binary tree is a transformation of any binary tree in to a complete binary tree, which replaces every null subtree of original binary tree with special nodes. The nodes from the original tree are referred as internal nodes, while the special nodes are referred as external nodes.

✍ A **general tree** (i.e., a tree with nodes having any number of children) can be converted into an equivalent binary tree using the leftmost child right siblings representation.

✍ Traversing a binary tree, means moving through all the nodes in the binary tree, visiting each node in the tree only once.

✍ The different binary tree traversal techniques are preorder traversal, inorder traversal, and postorder traversal.

✍ In linked representation of binary trees, we can see that all leaf nodes and some non-leaf nodes have NULL values. Instead of storing NULL values in the left and right pointer fields, we can store some useful information, in left and right pointer fields. These links are considered as threads. A binary tree, which implements these threads are referred to as threaded binary tree.

✍   If you use only the left pointer fields (for storing inorder predecessor) as threads, then the binary tree is referred to as left in-threaded binary tree.

✍   If you use only the right pointer fields (for storing inorder successor) as threads, then the binary tree is referred to as right in-threaded binary tree.

✍   If you use both left and right pointer fields as threads, then the binary tree is referred to as fully in-threaded binary tree.

✍   An appropriate data structure that supports the operations for inserting a new element and deleting the largest element is referred to as a **priority queue.**

✍   A **binary heap** is an array that is viewed as a complete binary tree. A complete binary tree is completely filled on all levels except possibly the lowest, and the lowest level is filled from the left.

✍   A graph is a non-linear data structure that represents less relationship between its adjacent elements. There is no hierarchical relationship between the adjacent elements in case of graphs.

✍   If an edge between any two nodes in a graph is not directionally oriented, a graph is called as undirected graph. It is also referred as unqualified graph.

✍   If an edge between any two nodes in a graph is directionally oriented, a graph is called as directed graph. It is also referred as a digraph.

✍   A path in a graph is defined as a sequence of distinct vertices each adjacent to the next, except possibly the first vertex and last vertex is different.

✍   A cycle is a path containing atleast three vertices such that the starting and the ending vertices are the same.

✍   A directed graph is said to be a strongly connected graph if, for every pair of distinct vertices there is a directed path from every vertex to every other vertex. It is also referred as a complete graph.

✍   A directed graph is said to be weakly connected graph, if any vertex doesn't have a directed path to any other vertices.

✍   A graph is said to be a weighted graph if every edge in the graph is assigned some weight or value. The weight of an edge is a positive value that may be representing the distance between the vertices or the weights of the edges along the path.

✍   Adjacency matrix is a representation used to represent a graph with zeros and ones. A graph containing n vertices can be represented by a matrix with n rows and n columns.

✍   Traversing a graph means visiting all the nodes in the graph. The two important graph traversal methods are depth-first traversal and breadth-first traversal.

4.12 Short-answer Questions

1.   ***Define a tree.***

A tree is a non-linear, two-dimensional data structure, which represents hierarchical relationship between individual data items.

2.   ***Define a path in a tree.***

A path in a tree is a sequence of distinct nodes in which successive nodes are connected by edges in the tree.

3.   ***Define terminal nodes in a tree.***

A node that has no children is called as a terminal node. It is also referred as a leaf node. These nodes have degree has zero.

4.   ***Define a binary tree.***

A binary tree is a tree, which has nodes either empty or not more than two child nodes, each of which may be a leaf node.

5.   ***Define a full binary tree.***

A full binary tree is a tree in which all the leaves are on the same level and every non-leaf node has exactly two children.

6.   ***Define a complete binary tree.***

A complete binary tree is a tree in which every non-leaf node has exactly two children not necessarily to be on the same level.

7.   ***State the properties of a binary tree.***

The properties of a binary tree includes,

- The maximum number of nodes on level n of a binary tree is 2^{n-1}, where $n \geq 1$.
- The maximum number of nodes in a binary tree of height n is 2^{n-1}, where $n \geq 1$.
- For any non-empty tree, $n_l = n_d + 1$ where n_l is the number of leaf nodes and n_d is the number of nodes of degree 2.

8.   ***What are the different binary tree traversal techniques ?***

The different binary tree traversal techniques are

- Preorder traversal.
- Inorder traversal.
- Postorder traversal.
- Levelorder traversal.

9.   ***What are the tasks performed while traversing a binary tree ?***

The tasks performed while traversing a binary tree are,

- Visiting a node.
- Traverse the left subtree.
- Traverse the right subtree.

10.   ***What are the tasks performed during preorder traversal.***

The tasks performed during preorder traversal,

- Process the root node.
- Traverse the left subtree.
- Traverse the right subtree.

11.   What are the tasks performed during inorder traversal.

The tasks performed during inorder traversal,

- Traverse the left subtree.
- Process the root node.
- Traverse the right subtree.

12.   What are the tasks performed during postorder traversal.

The tasks performed during postorder traversal,

- Traverse the left subtree.
- Traverse the right subtree.
- Process the root node.

13.   What are the tasks performed during levelorder traversal.

The tasks performed during levelorder traversal,

- Process the root node at level 1.
- Traverse the next level (i.e, level 2), below the root node.
- Process the nodes from left to right in that level.
- Similarly traverse the next level and process the nodes from left to right and continue till the end of levels.

14.   State the merits and demerits of linked representation of a binary tree.

The merits of linked representation of binary trees include

- Insertions and deletions in a node, involves no data movement except the re-arrangement of pointers, hence less processing time.

The demerits of linked representation of binary trees include

- Given a node structure, it is difficult to determine its parent node.
- Memory spaces are wasted for storing null pointers for the nodes, which have one or no subtrees.
- It requires dynamic memory allocation, which is not possible in some programming languages.

15.   Define a binary search tree.

A binary search tree is a special binary tree, which is either empty or if it is empty it should satisfy the following characteristics.

- Every node has a value and no two nodes should have the same value (i.e., the values in the binary search tree are distinct.
- The values in any left subtree is less than the value of its parent node.
- The values in any right subtree is greater than the value of its parent node.
- The left and right subtrees of each node are again binary search trees.

16.   Define a graph.

A graph is a non-linear data structure that represents less relationship between its adjacent elements. There is no hierarchical relationship between the adjacent elements in case of graphs.

17. *Define undirected graph.*

If an edge between any two nodes in a graph is not directionally oriented, a graph is called as undirected graph. It is also referred as unqualified graph.

18. *Define directed graph.*

If an edge between any two nodes in a graph is directionally oriented, a graph is called as directed graph. It is also referred as a digraph.

19. *Define a path in a graph.*

A path in a graph is defined as a sequence of distinct vertices each adjacent to the next, except possibly the first vertex and last vertex is different.

20. *Define a cycle in a graph.*

A cycle is a path containing atleast three vertices such that the starting and the ending vertices are the same.

21. *Define a strongly connected graph.*

A directed graph is said to be a strongly connected if, for every pair of distinct vertices their is a directed path from every vertex to every other vertex. It is also referred as a complete graph.

22. *Define a weakly connected graph.*

A directed graph is said to be weakly connected graph, if any vertex doesn't have a directed path to any other vertices.

23. *Define a weighted graph.*

A graph is said to be a weighted graph if every edge in the graph is assigned some weight or value. The weight of an edge is a positive value that may be representing the distance between the vertices or the weights of the edges along the path.

24. *Define incidence matrix.*

incidence matrix is a representation used to represent a graph with zeros and ones. A graph containing m vertices and n edges can be represented by a matrix with m rows and n columns. The matrix is formed by storing 1 in its ith row and j^{th} column corresponding to the matrix, if there exists a ith vertex, connected to one end of the jth edge, and a 0, if there is no i^{th} vertex, connected to any end of the jth edge of the graph.

25. *Define adjacency matrix.*

Adjacency matrix is a representation used to represent a graph with zeros and ones. A graph containing n vertices can be represented by a matrix with n rows and n columns. The matrix is formed by storing 1 in its ith row and j^{th} column of the matrix, if there exists an edge between i^{th} and j^{th} vertex of the graph, and a 0, if there is no edge between i^{th} and j^{th} vertex of the graph.

26. *Define adjacency list.*

A graph containing m vertices and n edges can be represented using a linked list, referred to as adjacency list.

27.   What is meant by traversing a graph ? State the different ways of traversing a graph.

Traversing a graph means visiting all the nodes in the graph. In many practical applications, traversing a graph is important, such that each vertex is visited once systematically by traversing through minimum number of paths. The two important graph traversal methods are

- Depth-first traversal (or) Depth-first search (DFS)
- Breadth-first traversal (or) Breadth-first search (BFS).

28.   Determine the incidence matrix for the undirected graph shown below.

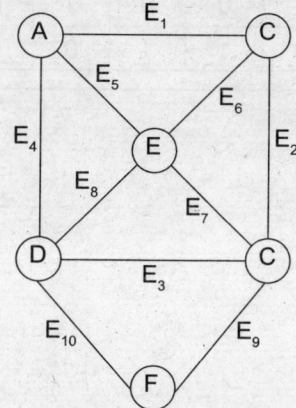

The incidence matrix for the undirected graph shown above is as follows,

	E_1	E_2	E_3	E_4	E_5	E_6	E_7	E_8	E_9	E_{10}
A	1	0	0	1	1	0	0	0	0	0
B	1	1	0	0	0	1	0	0	0	0
C	0	1	1	0	0	0	1	0	1	0
D	0	0	1	1	0	0	0	1	0	1
E	0	0	0	0	1	1	1	1	0	0
F	0	0	0	0	0	0	0	0	1	1

29.   Determine the incidence matrix for the directed graph shown below.

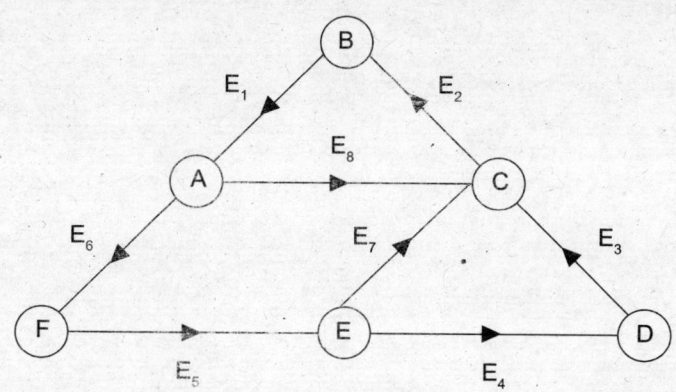

The incidence matrix for the directed graph shown above is as follows,

	E_1	E_2	E_3	E_4	E_5	E_6	E_7	E_8
A	0	0	0	0	0	0	0	1
B	1	0	0	0	0	0	0	0
C	0	1	0	0	0	0	0	0
D	0	0	1	0	0	0	0	0
E	0	0	0	1	0	0	1	0
F	0	0	0	0	1	1	0	0

Chapter 5

Searching and Sorting Algorithms

5.1 Introduction

In this chapter, we will discuss two most important techniques used in programming, sorting and searching. *Sorting a group or sequence of elements or data items means rearranging them in either ascending or descending order depending upon the relationship among the data items present in the group. Searching is a programming technique that determines whether an element or a data item is present in a list or not.*

Sorting is, without doubt, the most fundamental algorithmic problem. Many different approaches lead to useful sorting algorithms and these ideas can be used to solve many other problems. Finding better algorithms to sort a given set of data is an ongoing problem in the field of computer science. Simple sorting techniques sort data in ascending and descending order. The same algorithm can be used for sorting in increasing or decreasing order but all we need is to change $\leq$ to $\geq$ in the comparison algorithm as we desire. In this chapter we will see some important sorting techniques and the data is sorted only in ascending order. The factors to be considered while choosing a sorting technique are,

- Programming time of the sorting technique.
- Execution time of the sorting technique.
- Number of comparisons required for sorting the list.
- Main or auxiliary memory space needed for the sorting technique.

One reason why sorting is so important is that once a set of items is sorted, many other problems become easy. Speeding up searching is perhaps the most important application of sorting. Once the elements are placed in sorted order in an array, the kth largest element can be found in constant time by simply looking in the kth position of the array.

All data processing requires accessing records efficiently and quickly. Search techniques are most efficient only when the data items are already sorted according to some specified keys. If the list is not sorted, searching a record takes more time when the list is large. Therefore sorting becomes a necessary technique. A sorting method is said to be stable when it has minimum number of swaps i.e., if the two data items of matching value are guaranteed not to be rearranged with respect to each other when the algorithm progresses.

Some of the important sorting techniques are

- Bubble sort
- Insertion sort
- Quick sort
- Merge sort.

Some of the important searching techniques are

- Linear search
- Binary search.

5.2 Linear Search

Linear search also referred as ***sequential search***, is the simplest searching technique. *The search begins at one end of the list and searches for the required element one by one until the element is found or till the end of the list is reached.* The search is said to be successful if the search element is found and unsuccessful if the search element is not found in the list. The data items in the list need not be in a sorted order for linear search technique.

Algorithm for Linear Search

LINEAR_SEARCH (ARR, N, FIND)

where ARR is an array of N elements in which the element FIND is to be searched.

Step 1 :   Repeat For I = 0, 1, 2, . . . , N–1

Step 2 :   If (ARR[I] = FIND) Then

        RETURN I

        [End of If structure]

Step 3 :   [End of **Step 1** For loop]

Step 4 :   RETURN –1

END LINEAR_SEARCH()

Program **5.1** :

```
/* Program to search an element using linear search. */
/* linear.c */
#include<stdio.h>
int linear_search(int [], int, int);
void main()
 {
   int i, n, pos, find, list[100];
   printf("Enter the limit : ");
     scanf("%d", &n);
   printf("Enter the elements : ");
```

```
     for(i = 0 ; i < n ;  i++)
         scanf("%d", &list[i]);
      printf("Enter the element to be searched : ");
        scanf("%d", &find);
     if((pos = linear_search(list, n, find)) == -1)
        printf("The element is not present");
      else
        printf("The element is in the %d position", pos+1);

   }

  int linear_search(int list[], int n, int find)

  {

     int i;

     for(i = 0 ; i < n ; i++)

       if(list[i] == find)

          return i;

     return -1;

  }
```

The program displays the following output

RUN 1
```
Enter the limit : 5

Enter the elements : 78   88   89   98   65

Enter the element to be searched : 65

The element is in the 5 position
```
RUN 2
```
Enter the limit : 5

Enter the elements : 90   8   3   2   1

Enter the element to be searched : 5

The element is not present
```

The given list may be in any particular order and it is just as likely that the element to be found may be present in the first position or in the last position. The best case is 1, which occurs when the element to be searched occurs in the first position of the list. The worst case is n, which occurs when the element to be searched occurs in the last position of the list. The still worst case is n+1, which occurs when the element to be searched is not present in the list. On an average, the number of comparisons carried by the linear search algorithm, is half the elements present in the list and is given by (n+1)/2. The linear searching method searches well for small-unsorted arrays, but becomes inefficient when the size of the list is large.

5.3 Binary Search

The *binary search algorithm* is one of the most efficient searching techniques, which requires the list to be sorted in ascending order. *To search for an element in the list, the binary search algorithm splits the list and locates the middle element of the list. It is then compared with the search element.* If the search element is less than the middle element, the first part of the list is searched else the second part of the list is searched. The algorithm again reduces the list into two halves, locates the middle element, and compares with the search element. If the search element is less than the middle element, the first part of the list is searched. This process continues until the search element is equal to the middle element or the list consists of only one element that is not equal to the search element.

Algorithm for Binary Search

BINARY_SEARCH (ARR, FIND, BEG, END)

where ARR is an array in which the element FIND is to be searched, BEG refers to the initial index position of the list, END refers to the final position of the list.

Step 1 :   If (BEG > END) Then

        RETURN –1

        [End of If structure]

Step 2 :   Assign MID = (BEG + END) / 2

Step 3 :   If (FIND = ARR[MID]) Then

        RETURN MID

    Else If (FIND < ARR[MID]) Then

        RETURN BINARY_SEARCH (ARR, FIND, BEG, MID–1)

    Else

        RETURN BINARY_SEARCH (ARR, FIND, MID+1, END)

    [End of If structure]

END BINARY_SEARCH()

Program 5.2 :

```c
/* Program to search an element using binary search. */
/* binarys.c */
#include<stdio.h>
int binary_search(int [], int, int, int);
void main()
 {
    int i, n, find, pos = 0, list[100];
    printf("Enter the limit : ");
      scanf("%d", &n);
```

```
        printf("Enter the elements : ");
        for(i = 0 ; i < n ; i++)
           scanf("%d", &list[i]);
        printf("Enter the element to be searched : ");
           scanf("%d", &find);
        pos = binary_search(list, find, 0, n);
        if(pos != -1)
            printf("The element is in the %d position", pos+1);
        else
            printf("The element is not present");
      }
    binary_search(int list[], int find, int beg, int end)
      {
        int mid;
        if(beg > end)
            return(-1);
        mid = (beg + end)/2;
        if(find == list[mid]);
            return(mid);
        else if(find < list[mid])
            return binary_search(list, find, beg, mid-1);
        else
            return binary_search(list, find, mid+1, end);
      }
```

The program displays the following output

RUN 1   Enter the limit : 5

         Enter the elements : 3   12   56   98   987

         Enter the element to be searched : 12
         The element is in the 2 position

RUN 2   Enter the limit : 4

         Enter the elements : 23   34   45   108

         Enter the element to be searched : 78
         The element is not present

The `binary_search()` function used in ***Program 5.2*** takes 4 arguments, an integer array a, the element to be searched, beginning array subscript `beg` and the end array subscript `end`. If the search element is less than the middle element the 4th argument i.e., end subscript is set to `mid-1` and the search continues from the beginning of the array subscript to `mid-1`. If the search element is greater than the middle element, the beginning subscript is set to `mid+1` and the search continues from `mid+1` to the end of the array. This process continues until the search element is equal to the middle element or the list consists of only one element that is not equal to the element to be searched. Since we eliminate elements by half in each pass, the binary search algorithm works faster when compared with the linear search algorithm. Therefore, the running time for average case as well as worst case are almost the same.

5.4 Bubble Sort

The easiest and the most widely used sorting technique among students and engineers is the bubble sort. This sort is also referred as *sinking sort*. *The idea of bubble sort is to repeatedly move the smallest element to the lowest index position in the list.* To find the smallest element, the bubble sort algorithm begins by comparing the first element of the list with its next element and upto the end of the list and interchanges the two elements if they are not in proper order. In either case, after such a pass, the smaller element will be in the lowest index position of the list. The focus then moves to the next smaller element and the process is repeated. Swapping occurs only among successive elements in the list and hence only one element will be placed in its sorted order after each pass. Also, note that once the elements are placed in its sorted order they are not considered for comparison in successive passes. When the focus reaches the end of the list, the smallest element will be `bubbled` from whatever has been its original position to the first index position in the list and the larger values `sink` to the last index position in the list. The sort terminates after a pass in which no elements are interchanged.

Algorithm for Bubble Sort

   BUBBLE (ARR, N)

   where ARR is an array of N elements.

   Step 1 :   Repeat For I = 0, 1, 2, . . . , N–1

   Step 2 :   Repeat For J = I+1 to N–1

   Step 3 :   If (ARR[I] > ARR[J]) Then

                     Interchange ARR[I] and ARR[J]

                     [End of If structure]

   Step 4 :   Increment J by 1

   Step 5 :   [End of **Step 2** For loop]

   Step 6 :   [End of **Step 1** For loop]

   Step 7 :   Print the sorted array ARR.

   END BUBBLE()

Program 5.3 :

```c
/* Program to sort the numbers using bubble sort. */
/* bubble.c */
#include<stdio.h>
void bubble(int a[], int n);
int i, j, n, temp, a[25];
void main()
 {
    printf("Enter the limit : ");
      scanf("%d", &n);
    printf("Enter the elements : ");
    for(i = 0 ; i < n ; i++)
      scanf("%d", &a[i]);
    bubble(a, n);                    /* Calling the function bubble() */
    printf("The sorted list is : ");
    for(i = 0 ; i < n ; i++)
      printf("%d ", a[i]);
 }
void bubble(int a[], int n)
 {
    for(i = 0 ; i <= n-1 ; i++)
      for(j = i+1 ; j <= n-1 ; j++)
        if(a[i] > a[j])
          {
              temp = a[i];
              a[i] = a[j];
              a[j] = temp;
          }
 }
```

The program displays the following output

```
Enter the limit : 6
Enter the elements : 56  91  35  72  48  68
The sorted list is : 35  48  56  68  72  91
```

Unsorted List	56	91	35	72	48	68
	56	91	35	72	48	68
	35	56	91	72	48	68
	35	48	91	72	56	68
	35	48	56	91	72	68
	35	48	56	68	91	72
Sorted List	35	48	56	68	72	91

Fig. 5.1 : Trace of a bubble sort.

The bubble sort algorithm makes (n-1) comparisons, in the first pass to place the smallest element in the first position of the list. The second pass makes (n-2) comparisons, to place the next smallest element in the second position of the list and so on. Thus, the number of comparisons required by the bubble sort is given by

$$[n-1]+[n-2]+\ldots\ldots+2+1 = \frac{n[n-1]}{2} = \frac{[n^2-n]}{2} = o[n^2]$$

Therefore, the time required to execute the binary sort algorithm is proportional to n^2 where n is the number of elements in the list. This sorting method works well only for simple lists. The best case occurs when the input data elements are almost in sorted order. The worst case occurs when the input data elements are in descending order.

5.5 Insertion Sort

The main idea of insertion sort is to consider each element at a time, into the appropriate position relative to the sequence of previously ordered elements, such that the resulting sequence is also ordered. The insertion sort can be easily understood if you know to play cards. Imagine that you are arranging cards after it has been distributed before you in front of the table. As each new card is taken, it is compared with the cards in hand. The card is inserted in proper place within the cards in hand, by pushing one position to the left or right. This procedure proceeds until all the cards are placed in the hand are in order.

Algorithm for Insertion Sort
INSERT (ARR, N)
where ARR is an array of N elements.
Step 1 :   Repeat For I = 1, 2, 3, . . . , N–1
Step 2 :   Assign TEMP = ARR[I]
Step 3 :   Repeat For J = I to 1
Step 4 :   If (TEMP < ARR[J–1]) Then

```
            ARR[J] = ARR[J–1]
       Else
           Goto Step 7
       [End of If structure]
```

Step 5 : Decrement J by 1

Step 6 : [End of **Step 3** For loop]

Step 7 : Assign ARR[J] = TEMP

Step 8 : [End of **Step 1** For loop]

Step 9 : Print the sorted array ARR.

END INSERT()

Program 5.4 :

```c
/* Program to sort the numbers using insertion sort. */
/* insert.c */
#include<stdio.h>
void insert(int a[], int n);
int i, j, n, temp, a[25];
void main()
 {
   printf("Enter the limit : ");
     scanf("%d", &n);
   printf("Enter the elements : ");
     for(i = 0 ; i < n ; i++)
       scanf("%d", &a[i]);
   insert(a, n);                    /* Calling the function insert() */
   printf("The sorted list is : ");
   for(i = 0 ; i < n ; i++)
     printf("%d  ", a[i]);
 }
void insert(int a[], int n)
 {
   for(i = 1 ; i <= n-1 ; i++)
    {
      temp = a[i];
```

```
          for(j = i ; j >= 1 ; j--)
          {
              if(temp < a[j-1])
                a[j] = a[j-1];
              else
                break;
          }
          a[j] = temp;
      }
  }
```

The program displays the following output

```
Enter the limit : 6
Enter the elements : 56   91   35   72   48   68
The sorted list is : 35   48   56   68   72   91
```

Unsorted List	56	91	35	72	48	68
	56	91	35	72	48	68
	35	56	91	72	48	68
	35	48	56	91	72	68
	35	48	56	91	72	68
	35	48	56	68	91	72
	35	48	56	68	72	91
Sorted List	35	48	56	68	72	91

Fig. 5.2 : Trace of an insertion sort.

The number of comparisons, when the list is initially sorted is $O(n^2)$, since only one comparison is made in each pass. In the worst case, (i.e., if the list is arranged in descending order) the number of comparisons required by the insertion sort is given by

$$1 + 2 + 3 + \ldots + [n-2] + [n-1] = \frac{n[n-1]}{2} = \frac{[n^2-n]}{2}$$

Therefore, the number of comparisons is $O(n^2)$. On an average case the number of comparisons is given by

$$\frac{1}{2} + \frac{2}{2} + \frac{3}{2} + \ldots + \frac{[n-2]}{2} + \frac{[n-1]}{2} = \frac{n[n-1]}{4} = \frac{[n^2-n]}{4} = O[n^2]$$

Even though the insertion sort is better than the bubble sort, the number of comparisons are almost the same. Insertion sort is more efficient and takes less time if the list is almost in sorted order. The timing of the insertion sort can be improved by using binary search, rather than a linear search to find the proper position to insert the element in the list. This reduces the total number of comparisons to $O(\log n)$ from $O(n^2)$. Even, if the correct position is found, the sorting time of $O(\log n)$ steps it has to move (n-1)/2 elements forward. Thus, the use of binary search in insertion does not significantly improve time requirements in the sort. Since insertion sort is efficient when 'n' is small, linear search is almost efficient as binary search when n is small.

5.6 Merge Sort

The merge sort algorithm also uses divide and conquer rule for its operation. The first step of the merge sort is to chop the list into two. If the list has even length, split the list into two equal sub lists. If the list has odd length, divide the list into two by making the first sub list one entry greater than the second sub list. Then split both the sub lists into two and go on until each of the sub lists are of size one. Finally, start merging the individual sub lists to obtain a sorted list.

Algorithm for Merge Sort

MERGE_SPLIT (ARR, FIRST, LAST)

where ARR is an array of N elements, FIRST is the initial index of the array

and LAST is the final index of the array.

Step 1 :   If (FIRST < LAST) Then

Step 2 :   Assign MIDDLE = (FIRST + LAST) / 2

Step 3 :   MERGE_SPLIT(ARR, FIRST, MIDDLE)

Step 4 :   MERGE_SPLIT(ARR, MIDDLE+1, LAST)

Step 5 :   MERGE(ARR, FIRST, MIDDLE, MIDDLE+1, LAST)

Step 6 :   [End of If structure]

END MERGE_SPLIT()

MERGE (ARR, F1, L1, F2, L2)

where ARR is an array of N elements, F1 is the initial index of the first sub list,

L1 is the final index of the first sub list, F2 is the initial index of the second sub list,

L2 is the final index of the second sub list.

Step 1 :   Initialize I = F1, J = F2, K = 0

Step 2 :   Repeat While (I <= L1 AND J <= L2)

Step 3 :   If (ARR[I] < ARR[J]) Then

                TEMP[K] = ARR[I]

                  Increment I by 1

            Else

                  TEMP[K] = ARR[J]

                  Increment J by 1

                [End of If structure]

Step 4 :	Increment K by 1
Step 5 :	[End of **Step 2** While loop]
Step 6 :	Repeat While (I <= L1)
Step 7 :	Assign TEMP[K] = ARR[I]
Step 8 :	Increment K by 1
Step 9 :	Increment I by 1
Step 10 :	[End of **Step 6** While loop]
Step 11 :	Repeat While (J <= L2)
Step 12 :	Assign TEMP[K] = ARR[J]
Step 13 :	Increment K by 1
Step 14 :	Increment J by 1
Step 15 :	[End of **Step 11** While loop]
Step 16 :	Assign I = F1, J = 0
Step 17 :	Repeat While (I <= L2 AND J < K)
Step 18 :	Assign ARR[I] = TEMP[J]
Step 19 :	Increment J by 1
Step 20 :	Increment I by 1
Step 21 :	[End of **Step 17** While loop]
Step 22 :	Print the sorted array ARR
END MERGE()	

Program 5.5 :

```c
/* Program to sort the numbers using merge sort. */
/* merge.c */
#include<stdio.h>
void merge_split(int a[], int first, int last);
void merge(int a[], int f1, int l1, int f2, int l2);
int a[25], b[25];
void main()
 {
    int i, n;
    printf("Enter the limit : ");
      scanf("%d", &n);
    printf("Enter the elements : ");
    for(i = 0 ; i < n ; i++)
      scanf("%d", &a[i]);
```

```
      merge_split(a, 0, n-1);          /* Calling the function merge_split() */
      printf("The sorted list is : ");
      for(i = 0 ; i < n ; i++)
        printf("%d ", a[i]);
  }
  void merge_split(int a[], int first, int last)
  {
      int mid;
      if(first < last)
        {
          mid = (first + last)/2;
          merge_split(a, first, mid);
          merge_split(a, mid+1, last);
          merge(a, first, mid, mid+1, last);
        }
  }
  void merge(int a[], int f1, int l1, int f2, int l2)
  {
      int i, j, k = 0;
      i = f1;
      j = f2;
      while(i <= l1 && j <= l2)
        {
          if(a[i] < a[j])
            b[k] = a[i++];
          else
            b[k] = a[j++];
          k++;
        }
      while(i <= l1)
        b[k++] = a[i++];
      while(j <= l2)
        b[k++] = a[j++];
      i = f1;
      j = 0;
      while(i <= l2 && j < k)
        a[i++] = b[j++];
  }
```

The program displays the following output
```
Enter the limit : 6
Enter the elements : 56   91   35   72   48   68
The sorted list is : 35   48   56   68   72   91
```

Unsorted List	56	91	35	72	48	68
	⌊56	91⌋	⌊35⌋	⌊72	48⌋	⌊68⌋
	⌊56⌋	⌊91⌋	⌊35⌋	⌊72⌋	⌊48⌋	⌊68⌋
	⌊56	91⌋	⌊35⌋	⌊48	72⌋	⌊68⌋
	⌊35	56	91⌋	⌊48	68	72⌋
	⌊35	48	56	68	72	91⌋
Sorted List	35	48	56	68	72	91

Fig. 5.3 : Trace of a merge sort.

Unlike quick sort the split stage implemented in the merge sort is singular and the merge stage implementation is complex. The merge sort algorithm passes over the entire list and requires atmost `log n` passes and merges n elements in each pass. Therefore, the total number of comparisons required by the merge sort is given by `O(n log n)`. The main drawback of merge sort is that it requires an additional array (b as in *Program 5.5*) with n elements.

5.7 Quick Sort

Quick sort also referred as *partition exchange sort* was developed by *C.A.R. Hoare.* It is a sorting algorithm, which performs very well on larger lists than any other sorting methods. It employs divide and conquer rule for its operation. *The main idea of the quick sort is to divide the initial unsorted list into two parts, such that every element in the first list is less than all the elements present in the second list.* The procedure is then repeated recursively for both the parts, up to relatively short sequences, which can be sorted until the sequences reduce to length one i.e., we divide the problem into two smaller ones and conquer by solving the smaller ones. The first step of the algorithm requires choosing a pivot value that will be used to divide big and small numbers. Usually the first element of the list is chosen as the pivot value. Once the pivot value has been selected, all elements smaller than the pivot are placed towards the beginning of the set and all the elements larger than the pivot are placed to the right. This process essentially sets the pivot value in the correct place each time. Each side of the pivot is then quick sorted.

Algorithm for Quick Sort

QUICK (ARR, FIRST, LAST)

where ARR is an array of N elements, FIRST refers to the first index of the array and LAST refers to the last index of the array.

Step 1 :   If (FIRST < LAST) Then

Step 2 :   Assign PIVOT = ARR[FIRST], I = FIRST, J = LAST

Step 3 :   Repeat While (I < J)

Step 4 :   Repeat While (ARR[I] <= PIVOT AND I < LAST)

Step 5 :   Increment I by 1

Step 6 :   [End of **Step 4** While loop]

Step 7 :   Repeat While (ARR[J] <= PIVOT AND J > FIRST)

Step 8 :   Decrement J by 1

Step 9 :   [End of **Step 7** While loop]

Step 10 :   If (I < J) Then

                Interchange ARR[I] and ARR[J]

                [End of **Step 10** If structure]

Step 11 :   [End of **Step 3** While loop]

Step 12 :   Interchange ARR[FIRST] and ARR[J]

Step 13 :   QUICK(ARR, FIRST, J–1)

Step 14 :   QUICK(ARR, J+1, LAST)

Step 15 :   [End of **Step 1** If structure]

Step 16 :   Print the sorted array ARR.

END QUICK()

Program 5.6 :

```
/* Program to sort the numbers using quick sort. */
/* quick.c */
#include<stdio.h>
int i, j, n, pivot, a[20];
void quick(int a[], int left, int right);
void swap(int a[], int i, int j);
void main()
 {
    int i, n, a[20];
    printf("Enter the limit : ");
      scanf("%d", &n);
```

```c
      printf("Enter the elements : ");
      for(i = 0 ; i < n ; i++)
        scanf("%d", &a[i]);

      quick(a, 0, n-1);              /* calling the function quick() */
      printf("The sorted list is : ");
      for(i = 0 ; i < n ; i++)
        printf("%d  ", a[i]);
  }
void quick(int a[], int first, int last)
  {
     if(first < last)
       {
         pivot = a[first];
         i = first;
         j = last;
         while(i < j)
           {
             while(a[i] <= pivot && i < last)
               i++;
             while(a[j] >= pivot && j > first)
               j--;
             if(i < j)
               swap(a, i, j);          /* calling the function swap() */
           }
         swap(a, first, j);            /* calling the function swap() */
         quick(a, first, j-1);         /* calling the function quick() */
         quick(a, j+1, last);          /* calling the function quick() */
       }
  }
void swap(int a[], int i, int j)
  {
     int temp;
     temp = a[i];
     a[i] = a[j];
     a[j] = temp;
  }
```

The program displays the following output

```
Enter the limit : 6
Enter the elements : 56   91   35   72   48   68
The sorted list is : 35   48   56   68   72   91
```

Unsorted List	56	91	35	72	48	68
	56	91	35	72	48	68
	35	91	56	72	48	68
	35	72	56	91	48	68
	35	72	56	48	91	68
	35	72	56	48	68	91
	35	68	56	48	72	91
	35	56	68	48	72	91
	35	56	48	68	72	91
Sorted List	35	48	56	68	72	91

Fig. 5.4 : Trace of a quick sort.

The pivot selected should be such that it is smaller than about half the elements and larger than about half the elements. Consider the extreme case where either the smallest or the largest value is chosen as the pivot. When quick sort is called recursively on the values on either side of it, one set of data will be empty while the other would be almost as large as the original data set. If a bad pivot element is chosen, the array will not be evenly, divided and quick sort may be called on an array that is just as big as the original array.

In the best case, quick sort is said to be $O(n \log n)$ and this case occurs when the input array is evenly divided. In the worst case, the timing of the quick sort is $O(n^2)$. This case occurs when the pivot element fails to divide the input array at all (i.e., all elements are greater than or smaller than the pivot element). To improve the efficiency of the sort, there are clever ways to choose the pivot value such that it is extremely unlikely to end up with an extreme value. One such method is called the *median-of-three partitioning*. In this method, three elements are randomly chosen and the median of these three values is chosen as the pivot element.

5.8 Hashing

Hashing is a data structure that allows insertion, deletion and search operations to be carried out in $O(1)$ time. However, hashing is not efficient in operation that requires any ordering information among the elements, such as sorting, finding minimum and maximum element.

Let us assume we need to keep information about 500 students. Hence, we can assign an identification number to each student between 0 and 499 and then use a multidimensional array of 500 elements. Accessing elements in an array is extremely efficient. Array elements are usually accessed by index (also referred as its subscript). To search for a particular record, we have to search the whole array sequentially till we find the required record. Hence, the search time to find a particular data item in the array is proportional to the number of data items in the list, since the item to be searched may be at the first position, any intermediate position or at the last position. Similarly, if we want to store Reg_No (which is supposed to be a 11-digit number) of each student in an array, if we want to directly use the Reg_No as an index, the table should have much more elements than the number of the students, which is a great waste of space. The modern approach is that to make the entries in the table in sorted order of the primary key. Then we apply any of the sorting and searching techniques to get the desired record, which reduces the search time considerably. But the movement of existing records of the table in sorted order is unnecessary and it is a difficult task. To overcome this problem, we generate a table for the records in the file. Then we take the primary key of each record, and apply a function to manipulate it and generate an address value and store the record, at the generated address. Hence, if we map the 500 Reg_Nos of the students and the numbers between 0-499, then we can use the Reg_Nos as search keys and still use an array of 500 elements to store the records. If we can find a mapping between the search keys and indices, we can store each record in the element with the corresponding index. Thus each element would be found with one operation only, such a technique is referred to as *hash addressing* or *hashing*.

This hashing technique is used in searching for a particular record in a file. So we will use the concept of files to analyze this technique. Let us assume a file F having N records. This file has a set of fields SF that forms the primary key PK of the file (i.e., Roll_No field in the students record is unique for every student in the college). This primary key PK is used to uniquely identify a particular record in a file. We assume that this file is maintained by a table kept in memory, which stores the primary key PK and the address part of the record with the primary key PK., which directly refers records in a table by doing arithmetic operations on keys to map them onto the respective table addresses. Such a table which maintains the key (of the item that is mapped to an index value) is referred to as the *hash table*. The hash table is sometimes referred to as *scatter table*, because we are trying to scatter the data throughout the table. Each slot in the array (which is considered as an hash table) is sometimes referred to as a *bucket*.

A hash table is a data structure that works just like an array, except instead of forcing you to use integers as your index you can use any arbitrary data type as your index. Basically a function is needed to map a key K to an integer index i of the hash table (which has indices 0 to n), such a function is referred as the *hashing function.* Ideally, the hash function is used to determine the location (table index) of any record, given its key value.

Hash functions transform the keys into numbers within a predetermined interval (0 to N). These numbers are then used as indices in an array in the hash table to store the records. If the indices are numbers and if N is the size of the array, then h(key) = key % N. This will map all the keys into numbers within the interval (0 to N-1). If the indices are strings of characters, treat the binary representation of a key as a number and then apply them as before. If each character is represented with m bits, then the string can be treated as base-m number.

We should choose a hash function such that it gives us distinct values for different values of primary key. Unfortunately, sometimes, two different values may hash to the same address. This situation, where two different key values hash to the same hash address, is referred to as **collision**.

In this collision condition for inserting the new pair, another possible location has to be found. And if all the buckets are filled then this condition is called **overflow**. so, we should have some resolution methods to avoid such overflow.

The methods for handling overflow are

- Open addressing
- Chaining.

The integer, generated by a hash function between 0 and N-1 is used as an index in a hash table of N elements. Initially all slots in the hash table are blank. To insert values into the hash table, use the hash function to generate an address for each value to be inserted. To search for a key in the table the same hash function is reused. An important consideration in hashing performance is value of the **load factor**, which is denoted by (L) and is defined as the ratio between the number of records to be stored and the size of the table. As the load factor increases, the hashing performance will degrade because of the time required to search collision chains.

The value that the hash function returns for a given key is its hash address. A perfect hash function maps every key into a different table location. It has been proved that if P is a prime number, we obtain better (more even) distribution of the keys over the table. The key obtained by transforming the string into a number using the modulus operation utilizes most of the information in the string. We can use different methods to generate an address from the key value in hash function. Some methods of importance are,

- Truncation method
- Folding method
- Mid square method
- Division method.

5.8.1 Truncation Method

Truncation method is the simplest and easiest method for computing keys using the hash function. Let us consider that you want to generate hash table address for the following keys,

    987456   125978   963294   852137

The truncation method truncates a part of the given keys, depending upon the size of the hash table. If we assume that the size of the hash table is 1000, then the right most (or left most) 3 digits are truncated and used as hash table addresses. The hash table addresses for the given keys are,

456   978   294   137

Since, we are using only the last three digits for computing the hash table address, chances of collisions are more in this method.

5.8.2 Folding Method

In this method, the given keys are broken into groups of digits, and these groups are added to get the hash table address. For example, if the key to be mapped is 143765980, break it into 3 groups of 3 digit numbers (143), (765), (980) and add them up resulting in 1888. You can use the result as it is, as the hash table address. But if the size of the hash table is small, you can truncate the result to last 2 or 3 digits, depending upon the size of the hash table. ***Note that it is not necessary that you have to add the digits in this method, you can also multiply the digits, or multiply each group by a base number, to get the hash address to be stored in the hash table.***

5.8.3 Mid-Square Method

In mid-square method of generating the hash table addresses, after reading the keys, square them and choose the middle digits which are the random address to be stored in the hash table. Let us consider, that you want to generate hash table address for the following keys,

456   978   294   137

Now squaring these keys, will result in the following values,

207936   956484   86436   18769

Now take the two middle digits from each squared value and use that as hash addresses for these keys. The hash table addresses for the given numbers will be,

79   64   43   76

5.8.4 Division Method

This is the one of the best method to get the address, for the key to be mapped. *Take the key, depending upon the size of the hash table choose a prime member and do the modulus operation and store the result of modulus operation as the address of the hash table.* Let us consider, that you want to generate hash table address for the key 987456782. Use any prime number for the modulus operation. Let us consider, the table size is 61 and hence the hash address will be 46 (i.e., 987456883 % 61).

Note that you can also combine all or any combinations of the above methods, to generate the hash address for the corresponding key.

Properties of a good hash function

- The function should compute quickly.
- The function should be easy to compute (with few simple instructions) and understand.
- The function should rely on all or most bits of key.

- The function should distribute the key apparently and randomly.
- It should distribute values uniformly over the address range and avoid collisions as far as possible. To uniformly distribute the values over the address range we have to know the address bound before and define the hash function.

5.9 Overflow Resolution Technique --- Open Addressing

When a data item cannot be placed at the index calculated by the hash function (because of collision), we look for availability of another empty location in the hash table. This process is referred to as *open addressing*. Open addressing is generally used where storage space is large and hence it uses only the space in the hash table and does not use any space outside the hash table for storing the key values. In open addressing, arrays are used as storage for hash tables. Methods of open addressing includes,

- Linear probing (Open adressing).
- Quadratic probing.

5.9.1 Linear Probing

Linear probing method arises when there is no space for new value insertion. In order to solve this, the size of the table is increased and this change in the table size leads to change in the hash function.

For example, Let us use division function method, to find the hash address for the keys specified. Let the keys to be mapped in the hash table be 18, 72, 65, 34 and 13. Using the prime number 7, for the modulus operation, the hash address mapped for the keys are as follows,

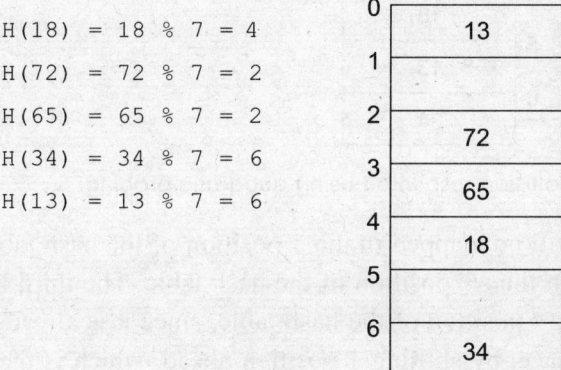

```
H(18) = 18 % 7 = 4
H(72) = 72 % 7 = 2
H(65) = 65 % 7 = 2
H(34) = 34 % 7 = 6
H(13) = 13 % 7 = 6
```

Fig. 5.5 : Collision resolution using linear probing.

The first key (i.e., 18) will be mapped in the 4 position of the hash table. The next key (i.e., 72) will be mapped in the 2nd position of the hash table. The third key (i.e., 65) will try to map the key in the 2nd position of the hash table, since it is already mapped, it will search for the next free place, which is the 3rd position in the hash table. The fourth key (i.e., 34) will be mapped in the 6th position of the hash table. The fifth key (i.e., 13) will try to map the key in the 6th position of the hash table, since it is already mapped, it will search for the next free place, which is the 0th position (since the size of the hash table is 7) in the hash table.

To search an element in the hash table, we check hash address position, corresponding to the key, if the key is not found at that position, then we linearly search the element, following the hash address position. *The main disadvantage* of this resolution technique is the primary clustering problem. That is, if half of hash table is filled or an overflow occurs, it is difficult to find an empty location in the hash table and hence the insertion process takes a longer time.

5.9.2 Quadratic Probing

Quadratic probing is similar to linear probing, except that, instead of looking just one more index ahead each time until we find an empty index, we do the following. On the method we look ahead 1 position, and place the key in the hash table. On the second collision we look 4 (2^2) positions ahead, and on the third we look 9 (3^3) positions ahead and so on. If i is the position in the array in quadratic probing, the step sizes are i+1, i+4, i+9, i+16, and so on. Let us assume, that we want to map the keys considered in the previous example,

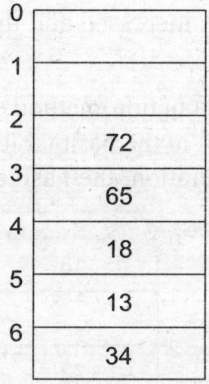

Fig. 5.6 : Collision resolution using quadratic probing.

The first key (i.e., 18) will be mapped in the 4 position of the hash table. The next key (i.e., 72) will be mapped in the 2nd position of the hash table. The third key (i.e., 65) will try to map the key in the 2nd position of the hash table, since it is already mapped, it will search for the next free place, by shifting 1 position ahead, which is the 3rd position in the hash table, and since it is empty, maps the key in that location. The fourth key (i.e., 34) will be mapped in the 6th position of the hash table. The fifth key (i.e., 13) will try to map the key in the 6th position of the hash table, since it is already mapped, it searches for the next free place, by shifting 4 positions ahead, which is the 3rd position in the hash table (since the size of the hash table is 7), since it is already occupied, it searches for the next free place, by shifting 9 positions ahead, which is the 5th position in the hash table and since it is empty, it maps the key in that location.

Quadratic probing is just as easy to implement as linear probing, and has less of a clustering effect than linear probing. The problem with quadratic probing is that it gives rise to secondary clustering. This method will not search all locations in the hash table to find an empty slot. Due to this insertion takes a longer time when compared to linear probing.

5.10 Overflow Resolution Technique --- Separate Chaining

This method uses linked lists for storage as hash tables. It maintains separate chains of elements, which maps to the same hash address. When a data item cannot be placed at the hash address calculated by the hash function, a chain or link is allocated and stores the key element in that chain. This allows an unlimited number of elements to the same hash address and does not require a prior knowledge on number of elements stored in the hash table. This method is referred to as *separate chaining*, because you have a bunch of separate chains in your hash table. In this method, there is no problem of limited storage hence insertions and searching elements are carried out in no time. When we want to map the keys (18, 72, 65, 34 and 13) in the previous example, the hash address mapped for the keys using division function method is 4, 2, 2, 6 and 6. The keys are mapped as shown in *Fig. 5.7*.

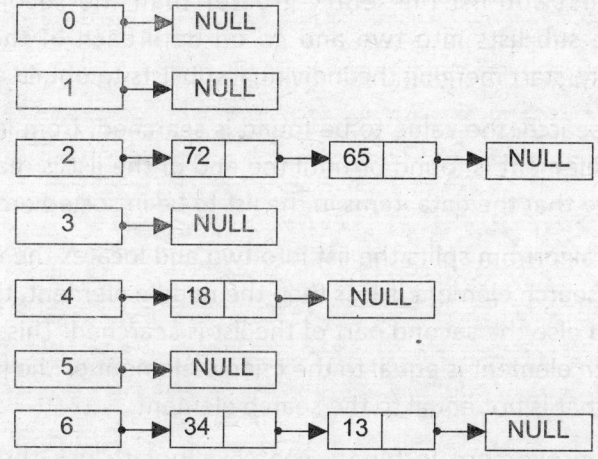

Fig. 5.7 : Collision resolution using separate chaining.

The main advantage of hashing using separate chaining is to get uniform and perfect collision resolution hashing. Since separate chaining is implemented using linked lists, there is no waste of memory (since memory for the hash table is allocated to insert a record when needed). The elements, which have the same memory address will be in the same chain. Hence search operation is fast compared to linear and quadratic probing.

5.11 Summary

- Sorting a group or sequence of elements or data items means rearranging them in either ascending or descending order depending upon the relationship among the data items present in the group.

- Searching is a programming technique that determines whether an element or a data item is present in the given list or not.

- Bubble sort derives its name from the fact that while sorting the data, the smallest data item bubbles to the initial position of the partially sorted list.

- Insertion sort derives its name from the fact that while sorting the data, each data item is inserted in its correct position among the initially placed data items.

- Quick sort divides the initial unsorted list into two parts, such that every element in the first list is less than all the elements present in the second list. The procedure is then repeated recursively for both the parts, up to relatively short sequences, which can be sorted until the sequences, reduces to length one.

- Merge sort algorithm chops the list into two. If the list has even length, split the list into two equal sub lists. If the list has odd length, divide the list into by making the first sub list one entry greater than the second sub list. Then split both the sub lists into two and go on until each of the sub lists are of size one. Finally, start merging the individual sub lists to obtain a sorted list.

- In sequential search, the value to be found is searched, from left to right one by one until the element is found or until the end of the list is reached. This search doesn't require that the data items in the list to be in sorted order.

- Binary search algorithm splits the list into two and locates the middle element in the list. If the search element is less than the middle element, the first part of the list is searched else the second part of the list is searched. This process continues until the search element is equal to the middle element or the list consists of only one element that is not equal to the search element.

- A possible improvement in binary search is not to use the middle element at each step, but to guess more precisely where the key being sought falls within the current interval of interest. This improved version is called fibonacci search.

- A table which maintains key (of the item that is mapped to an index value) is referred to as the hash table.

- Hash table is also referred as scatter table, since we are trying to scatter the data throughout the table.

✍   We basically need a function to map a key K to an integer index i of the hash table (which has indices 0 to n), such a function is referred as the hashing function.

✍   An important consideration in hashing performance is value of the load factor, which is denoted by (L), and is defined as the ratio between the number of records to be stored and the size of the table.

✍   When a data item cannot be placed at the index calculated by the hash function (because of collision), we look for availability of another empty location in the hash table. This process is referred to as open addressing.

5.12 Short-answer Questions

1.   Define sorting.

Sorting a group or sequence of elements or data items means rearranging them in either ascending or descending order depending upon the relationship among the data items present in the group.

2.   What are the factors to be considered while choosing a sorting technique ?

The factors to be considered while choosing a sorting technique are

- Programming time
- Running time of the sorting technique
- Number of comparisons required for sorting the list
- Main or auxiliary memory space needed for the sorting technique.

3.   What is the necessity for sorting techniques ?

All data processing requires accessing records efficiently and quickly. Search techniques are most efficient only when the data items are sorted according to some specified keys. If the list/file is not sorted, searching a record takes more time when the list/file is large.

4.   Mention some of the sorting techniques.

Sorting techniques includes

- Bubble sort
- Merge sort
- Insertion sort
- Quick sort
- Selection sort.

5.   When is sorting method is said to be stable ?

A sorting method is said to be stable when it has minimum number of swaps i.e., if the two data items of matching value are guaranteed not to be rearranged with respect to each other when the algorithm progresses.

6.   List out some of the stable and unstable sorting techniques.

Stable sorting techniques includes

- Bubble sort
- Selection sort
- Insertion sort
- Merge sort.

Unstable sorting techniques includes

- Shell sort
- Radix sort
- Quick sort.

7.   Why is the bubble sort called by that name ?

The bubble sort derives its name from the fact that while sorting the data, the smallest data item bubbles to the initial position of the partially sorted list.

8.   Why is the bubble sort is also called as sinking sort ?

Bubble sort also referred as sinking sort derives its name from the fact that while sorting the data, the largest data item sinks to the final position of the list.

9.   Mention the limitation of insertion sort.

Insertion sort has the limitation, that it compares only the consecutive elements and interchanges the elements by only one space. The smaller elements that are far away require many passes through the sort to properly place them in its correct position.

10.   How sorting is performed in merge sort ?

The merge sort algorithm chops the list into two. If the list has even length, split the list into two equal sub lists. If the list has odd length, divide the list into two by making the first sub list one entry greater than the second sub list. Then split both the sub lists into two and go on until each of the sub lists are of size one. Finally, start merging the individual sub lists to obtain a sorted list.

11.   How sorting is performed in quick sort ?

Quick sort divides the initial unsorted list into two parts, such that every element in the first list is less than all the elements present in the second list. The procedure is then repeated recursively for both the parts, up to relatively short sequences, which can be sorted until the sequences, reduces to length one.

12.   What is median-of-three-portioning method ?

The median-of-three-portioning method is employed in quick sort to find the pivot value. This is done by randomly choosing three elements in the list and finding the median of the elements gives the pivot value.

13. **Define searching.**

Searching is a programming technique that determines whether an element or a data item is present in the given list or not.

14. **Mention some of the searching techniques.**

Searching techniques includes

- Linear search
- Indexed sequential search
- Binary search
- Fibonacci search.

15. **What is meant by binary search ?**

The binary search algorithm is one of the most efficient sorting techniques, which requires the list to be sorted in ascending order. To search for an element in the list, the binary search algorithm splits the list into two and locates the middle element in the list. If the search element is less than the middle element, the first part of the list is searched else the second part of the list is searched. This process continues until the search element is equal to the middle element or the list consists of only one element that is not equal to the search element.

16. **Which is the fastest internal sorting technique & why ?**

As the name indicates, the quick sort technique is the fastest sorting technique. The purpose of quick sort is to move a data item in the correct position just enough to reach its final position in the list. This method therefore reduces unnecessary swaps. The quick sort technique has an advantage of moving a data item to a great distance in one move to place it in its exact position.

17. **Sort the following numbers using bubble sort 30, 5, 95, 6, 3, 15.**

The steps for sorting the numbers using bubble sort includes,

```
30,   5,  95,   6,   3,  15
 3,   5,  95,   6,  30,  15
 3,   5,  95,   6,  30,  15
 3,   5,   6,  95,  30,  15
 3,   5,   6,  15,  30,  95
 3,   5,   6,  15,  30,  95
```

18. **Sort the following numbers using insertion sort 30, 5, 95, 6, 3, 15.**

The steps for sorting the numbers using insertion sort includes,

```
30,   5,  95,   6,   3,  15
 5,  30,  95,   6,   3,  15
 5,  30,  95,   6,   3,  15
 5,   6,  30,  95,   3,  15
 3,   5,   6,  30,  95,  15
 3,   5,   6,  15,  30,  95
```

19.   *Sort the following numbers using quick sort 30, 5, 95, 6, 3, 15.*

The steps for sorting the numbers using quick sort includes,

```
30,   5,  95,   6,   3,  15
15,   5,  95,   6,   3,  30
15,   5,  30,   6,   3,  95
15,   5,   3,   6,  30,  95
 6,   5,   3,  15,  30,  95
 3,   5,   6,  15,  30,  95
```

20.   *Sort the following numbers using merge sort 30, 5, 95, 6, 3, 15.*

The steps for sorting the numbers using merge sort includes,

```
30,   5,  95,   6,   3,  15
 5,  30,  95,   6,   3,  15
 5,  30,  95,   3,   6,  15
 5,  30,  95,   3,   6,  15
 5,  30,  95,   3,   6,  15
 3,   5,   6,  15,  30,  95
```

TECHNICAL Q & A IN C

I – CHOOSE THE CORRECT ANSWER(S) FROM THE CHOICES GIVEN

1. **Which rules determine the meaning of instructions?**
   - (a) semantic rules
   - (b) syntax rules
   - (c) programming rules
   - (d) language rules

2. **An expression that evaluates to true or false is called as _____ expression.**
   - (a) logical (boolean)
   - (b) int
   - (c) char
   - (d) data

3. **An expression that has operands of different data types is called a _____ expression.**
   - (a) unary
   - (b) binary
   - (c) dual
   - (d) mixed

4. **Which of the following is not a legal identifier?**
   - (a) first
   - (b) conversion
   - (c) payRate
   - (d) 'primary

5. **Every character in a string has a relative _____ in the string.**
   - (a) position
   - (b) hierarchy
   - (c) importance
   - (d) area

6. **The maximum number of significant digits is called the _____.**
   - (a) significance
   - (b) position
   - (c) location
   - (d) precision

7. **Preprocessor directives are processed by a program called a(n) _____.**
   - (a) compiler
   - (b) linker
   - (c) preprocessor
   - (d) postprocessor

8. **A set of rules, symbols, and special words that enable you to write code is known as _____.**
   - (a) syntax rules
   - (b) semantic rules
   - (c) programming language
   - (d) syntax

9. **When a value of one data type is automatically changed to another data type a(n) _____ type coercion is said to have occurred.**
   - (a) intrinsic
   - (b) explicit
   - (c) implicit
   - (d) singular

10. **Every C program has a function called _____.**
   - (a) master
   - (b) index
   - (c) main
   - (d) default

11. **Which is a data type that deals with integers, or numbers without a decimal part?**
   - (a) integral
   - (b) enumeration type
   - (c) floating-point
   - (d) numeric

12. **Which of the following is an arithmetic operator?**
   - (a) * multiplication
   - (b) / division
   - (c) % remainder
   - (d) all of the above

13. **The letters that make up a reserved word are always _____.**
   - (a) lowercase
   - (b) uppercase
   - (c) camel case
   - (d) capcase

14. **A sequence of zero or more characters is known as a(n) _____.**
   - (a) character
   - (b) string
   - (c) enumeration
   - (d) set

15. **The EBCDIC character set has _____ values.**
   - (a) 64
   - (b) 128
   - (c) 256
   - (d) 512

16. **A sequence of characters from the computer to an output device is known as a(n) _____.**
   - (a) output stream
   - (b) input stream
   - (c) common output
   - (d) common input

17. **Characters that consist of blanks and certain nonprintable characters, such as tabs, newline characters, are known as _____.**
   - (a) openspace
   - (b) emptyspace
   - (c) blankspace
   - (d) whitespace

18. **One printable character except the blank is valid input for which simple data type?**
   - (a) char
   - (b) int
   - (c) double
   - (d) both b and c

19. **Which logical operator represents not?**
   - (a) !
   - (b) &
   - (c) ||
   - (d) *

20. **The syntax of one-way selection is _____.**
   - (a) if expression statement
   - (b) if(expression) statement
   - (c) expression(if) statement
   - (d) if(statement) expression

21. **Which of the following is a binary operator?**
   - (a) !
   - (b) &&
   - (c) ||
   - (d) both b and c

22. **A two-way selection in C is the _____.**
   - (a) if..else
   - (b) if..then
   - (c) if..only
   - (d) if

23. Since relational and logical operators are evaluated from left to right, the _____ of these operators is said to be from left to right.

    (a) associativity   (b) relativity
    (c) evaluation   (d) calculation

24. An informal mixture of C and ordinary language is known as _____.

    (a) coder   (b) pseudocode
    (c) langcode   (d) program code

25. Which operator represents equal to?

    (a) =   (b) ==
    (c) EQUAL   (d) EQUAL TO

26. The expression used in a switch statement to determine which case to process is called the _____.

    (a) action   (b) selector
    (c) purpose   (d) declaration

27. Which logical operator represents OR operation?

    (a) OR   (b) or
    (c) | (d) ||

28. The statement following the expression in an if statement is sometimes called the _____ statement.

    (a) decision   (b) declarative
    (c) purpose   (d) action

29. Control structures provide alternatives to _____ program execution.

    (a) controlled   (b) sequential
    (c) repetitive   (d) iterative

30. Which of the following operators has the highest order of precedence?

    (a) increment (++)   (b) Unary minus (-)
    (c) decrement (--)   (d) Unary Plus (+)

31. A compound statement consists of a sequence of statements enclosed in _____.

    (a) square braces   (b) parentheses
    (c) curly braces   (d) angle braces

32. Which operator allows you to make comparisons in a program?

    (a) comparison   (b) conditional
    (c) relational   (d) logical

33. A process in which the computer evaluates a logical expression from left to right and stops as soon as the value of the expression is known is called _____.

    (a) circuit evaluation   (b) short-circuit evaluation
    (c) short evaluation   (d) logical evaluation

34. Which operators enable you to combine logical expressions?

    (a) formatted   (b) conditional
    (c) valued   (d) logical (boolean)

35. The primary purpose of a(n) _____ is to simplify the writing of count-controlled loops.

    (a) for loop   (b) if loop
    (c) while loop   (d) if..else loop

36. The do..while loop is useful when it does not make sense to check a condition until after the _____ occurs.

    (a) trigger   (b) action
    (c) condition   (d) opening

37. A flag-controlled while loop uses a _____ variable to control the loop.

    (a) control   (b) loop-control
    (c) boolean   (d) flag

38. In C, do is a(n) _____ word.

    (a) unique   (b) reserved
    (c) action   (d) special

39. What kind of while loop is used when you know how many items of data there are to be read?

    (a) iterative while loop
    (b) counter-controlled while loop
    (c) limited while loop
    (d) specific while loop

40. A(n) _____ statement, when executed in a switch structure, provides an immediate exit from the switch structure.

    (a) continue   (b) break
    (c) halt   (d) end

41. In a while loop, the expression acts as a(n) _____ maker.

    (a) primary key   (b) condition
    (c) expression   (d) decision

42. A break and continue statement alters the flow of _____.

    (a) information   (b) location
    (c) action   (d) control

43. What kind of while loop is used when the sentinel value is not known?

    (a) Counter-controlled while loop
    (b) EOF-(End Of File) controlled while loop
    (c) Flag-controlled while loop
    (d) Unknown sentinel-controlled while loop

44. If the expression in a do..while loop evaluates to _____, the statement executes again.

    (a) true   (b) false
    (c) open   (d) closed

45. In a counter-controlled while loop, the counter is sometimes _____ 0.

   (a) defined as   (b) decremented as
   (c) initialized as   (d) incremented as

46. In a for statement, if the loop condition is omitted, it is assumed to be _____.

   (a) true   (b) false
   (c) open   (d) closed

47. A loop that continues to execute endlessly is called a(n) _____.

   (a) unending loop   (b) endless loop
   (c) infinite loop   (d) elongated loop

48. In a for loop, if the loop condition is initially _____, the loop body does not execute.

   (a) true   (b) false
   (c) open   (d) closed

49. Which of the following is not a looping structure in C?

   (a) if   (b) do..while
   (c) while   (d) for

50. A while loop that uses a sentinel value to end the loop is known as a _____ controlled while loop.

   (a) numerically   (b) counter
   (c) variable   (d) sentinel

51. Using one control structure statement inside another is referred as _____.

   (a) organizing   (b) grouping
   (c) stacking   (d) nesting

52. To call a function, use its name together with the actual _____ list.

   (a) function   (b) module
   (c) parameter   (d) group

53. The header file _____ contains mathematical functions.

   (a) math   (b) cmath
   (c) fmath   (d) stdio

54. Which of the following is not a predefined function?

   (a) pow   (b) islower
   (c) abs   (d) sum

55. The syntax of the actual parameter list is _____.

   (a) expression, variable, ..
   (b) expression variable, variable, ...
   (c) expression, expression or variable, ...
   (d) expression or variable, expression or variable

56. Functions enable you to divide a program into manageable _____.

   (a) tasks   (b) groups
   (c) modules   (d) nodes

57. In C, predefined functions are organized into separate _____.

   (a) divisions   (b) areas
   (c) groups   (d) libraries

58. Functions that have a data type are known as _____ functions.

   (a) user-defined   (b) predefined
   (c) void   (d) value-returning

59. The syntax of the formal parameter list is _____.

   (a) identifier, dataType identifier, dataType, ...
   (b) dataType identifier, identifier, ...
   (c) dataType identifier, dataType identifier, ...
   (d) identifier, dataType identifier, ...

60. An integer that reads forward and backward in the same way is known as a _____.

   (a) unique integer
   (b) bidirectional number
   (c) forward/backward number
   (d) palindrome

61. What lets you divide complicated programs into manageable pieces?

   (a) nodes   (b) subnodes
   (c) functions   (d) packets

62. Which function returns the lowercase value of x if x is uppercase; otherwise, it returns x?

   (a) toupper(x)   (b) tolower(x)
   (c) upper(x)   (d) lower(x)

63. The variable declared in the heading of the function is known as the _____.

   (a) formal parameter   (b) heading parameter
   (c) actual parameter   (d) primary parameter

64. The function heading without the body of the function is known as the _____.

   (a) function header   (b) function prototype
   (c) header prototype   (d) function head prototype

65. The function heading and the body of the function are called the _____ of the function.

   (a) main module   (b) definition
   (c) main group   (d) description

66. The function type is the data type of the _____ value returned by the function.

   (a) final   (b) first
   (c) unique   (d) only

67. When a function exits, the control goes back to the _____.

(a) controller   (b) caller
(c) master   (d) organizer

68. If the function's formal parameter list is empty, it takes the form of _____.

(a) functionType functionName()
(b) functionType Name()
(c) functionType functionParam()
(d) functionType functionList()

69. In a function call, the number of actual parameters and their types must match with the _____ parameters in the order given.

(a) unique   (b) function
(c) specific   (d) formal

70. Functions that do not have a data type are known as _____ functions.

(a) user-defined   (b) predefined
(c) void   (d) value-returning

71. What is the syntax for the formal parameter list of void functions with parameters that are passed by reference?

(a) dataType variable, dataType variable, ...
(b) dataType variable, dataType variable, ...
(c) dataType& variable, dataType& variable, ...
(d) dataType& variable, dataType& variable, ...

72. During data manipulation, the content of the formal parameter directs the computer to manipulate the data of the memory cell indicated by its _____.

(a) location   (b) variable
(c) arguments   (d) content

73. To declare an external variable inside a function use the _____ reserved word.

(a) external   (b) extern
(c) outside   (d) global

74. What is the syntax for the function call of void functions with parameters?

(a) functionName(formal parameter);
(b) functionName(actual parameter list);
(c) functionName(global variables);
(d) functionName(void);

75. A value-returning function returns _____.

(a) a single value   (b) multiple values
(c) void values   (d) control variables

76. Any function that uses global variables is not _____ and typically cannot be used in more than one program.

(a) independent   (b) stand-alone
(c) unique   (d) globally accessible

77. As with value-returning functions, in a function call the number of actual parameters together with their data types must _____ the formal parameters.

(a) match   (b) duplicate
(c) replace   (d) be different from

78. A variable for which memory is allocated at block entry and deallocated at block exit is called a(n) _____ variable.

(a) static   (b) short term
(c) permanent   (d) automatic

79. A formal parameter that receives a copy of the content of the corresponding actual parameter is known as a(n) _____.

(a) reference parameter
(b) open parameter
(c) value parameter
(d) specified parameter

80. Which exactly refer to the accessibility(visibility) of an identifier in a program?

(a) the location of an identifier
(b) the scope of an identifier
(c) the memory address of an identifier
(d) the accessibility of an identifier

81. A variable for which memory remains allocated as long as the program executes is called a(n) _____ variable.

(a) static   (b) short term
(c) permanent   (d) automatic

82. A formal parameter that receives the location (memory address) of the corresponding actual parameter is known as a(n) _____.

(a) reference parameter
(b) open parameter
(c) value parameter
(d) specified parameter

83. Identifiers declared within a function (or block) are known as _____ identifiers.

(a) local   (b) global
(c) functional   (d) unique

84. You can declare a reference (formal) parameter as a constant by using the keyword _____.

(a) con   (b) const
(c) constant   (d) none of the above

85. Identifiers declared outside every function definition are known as _____ identifiers.

(a) local   (b) global
(c) functional   (d) unique

86. You cannot assign a constant value as a default value to a(n) _____ parameter.

(a) reference   (b) automatic
(c) default   (d) formal

87. Which identifiers are not accessible outside the function (block)?
- (a) local
- (b) global
- (c) functional
- (d) unique

88. Because an enumeration is an ordered set of values, the _____ operators can be used with the enumeration type.
- (a) relational
- (b) arithmetic
- (c) numeric
- (d) order

89. The general syntax of the typedef statement is _____.
- (a) typedef existing Name newTypeName;
- (b) typedef existingTypeName newTypeName;
- (c) typedef newTypeName existingTypeName;
- (d) typedef existingTypeName newTypeName

90. In C, a sequence of zero or more characters enclosed in double quotation marks is known as a(n) _____.
- (a) variable
- (b) char
- (c) string
- (d) word

91. A user-defined simple data type is known as a(n) _____ type.
- (a) user-defined
- (b) enumeration
- (c) unique
- (d) specified

92. The enumeration type is a(n) _____ type.
- (a) relational
- (b) integral
- (c) arithmetic
- (d) unique

93. Which of the following statements creates an alias, integer, for the data type int?
- (a) type def integer;
- (b) typedef int integer;
- (c) typedef int real;
- (d) typedef int integer;

94. The values that you specify for the data type must be _____.
- (a) numbers
- (b) variables
- (c) characters
- (d) identifiers

95. Which function returns the number of characters currently in the string?
- (a) strlen()
- (b) strlength()
- (c) stringlen()
- (d) stringlength()

96. What is the syntax for an enumeration type?
- (a) enum typeName{value1, value2, ...};
- (b) enum typeValue{value1, value2, ...};
- (c) enum typeValue{name1, name2, ...};
- (d) enum typeName{object1, object2, ...};

97. You can pass the enumeration type as a(n) _____ to functions just like any other simple data type.
- (a) identifier
- (b) parameter
- (c) enumerated value
- (d) character value

98. If a global identifier in a program has the same name as one of the global identifiers in the header file, the compiler generates a(n) _____ error.
- (a) undefined
- (b) identifier
- (c) syntax
- (d) duplex

99. A collection of a fixed number of components wherein all of the components are of the same data type is known as a(n) _____.
- (a) database
- (b) array
- (c) organization
- (d) area

100. Which statement declares an array list of 10 components?
- (a) int list[9];
- (b) int list[10];
- (c) int list[11];
- (d) int list[12];

101. If either (index < 0) or (index > arraySize - 1), we say that the array is _____ bounds.
- (a) between
- (b) maximized
- (c) in
- (d) out of

102. To copy one array into another array, you must copy it _____.
- (a) all at once
- (b) manually
- (c) in sub groups
- (d) component-wise

103. The address of the first array component is known as the _____ address of an array.
- (a) primary
- (b) memory
- (c) first
- (d) base

104. A sequence of zero or more characters is known as a(n) _____.
- (a) string
- (b) char unit
- (c) alphanumeric
- (d) data group

105. An array in which the components are arranged in a list form is known as a(n) _____ array.
- (a) one-dimensional
- (b) two-dimensional
- (c) multi-dimensional
- (d) list

106. Which of the following is not a basic operation performed on an array?
- (a) initialize
- (b) input data
- (c) output data
- (d) change array bounds

107. C does not check whether the index value is within _____.
- (a) scope
- (b) range
- (c) indices
- (d) formula

108. Other than integers, C allows any _____ type to be used as an array index.

(a) float   (b) char
(c) string   (d) value

109. Which function copies string s2 into string variable s1?

(a) strcopy(s1, s2)   (b) strcopy(s1, s2)
(c) strcpy(s1, s2)   (d) strcpy(s2, s1)

110. The general form of declaring a one-dimensional array is _____.

(a) Type arrayName[intExp];
(b) dataType Name[intExp];
(c) dataType array[intExp];
(d) dataType arrayName[intExp];

111. If the data in an array is numeric, you can find the sum and _____ of the elements of the array.

(a) average   (b) product
(c) quantity   (d) value

112. Like any other simple variable, an array can also be _____ while it is being declared.

(a) initialized   (b) manipulated
(c) resized   (d) deleted

113. Because arrays are passed by reference only, you do not use the symbol _____ when declaring an array as a formal parameter.

(a) *   (b) %
(c) $ (d) &

114. An array whose components are of the type char is known as a(n) _____ array.

(a) char   (b) character
(c) alpha   (d) alphanumeric

115. Which data type has variables of that type which can store only one value at a time?

(a) structured   (b) single
(c) simple   (d) unary

116. To compare struct variables, you compare them _____.

(a) in sorted order   (b) aggregately
(c) group-wise   (d) member-wise

117. The members of a struct are enclosed in _____.

(a) square brackets   (b) angle brackets
(c) curly braces   (d) parentheses

118. The syntax for accessing a struct member is _____.

(a) structMemberName.VariableName;
(b) structVariableName.memberName;
(c) structVariableName.[memberName];
(d) struct[VariableName].memberName;

119. A struct variable can be passed as a(n) _____ either by value or by reference.

(a) variable   (b) control
(c) parameter   (d) object

120. The component employees[50] is the _____ component of the array employees.

(a) 49th   (b) 50th
(c) 51st   (d) 52nd

121. A collection of a fixed number of components in which the components are accessed by name is known as a(n) _____.

(a) object   (b) group
(c) array   (d) struct

122. In arrays, you access a component by using the _____ together with the relative position (index) of the component.

(a) array index   (b) component name
(c) array name   (d) function name

123. As with an array, no aggregate _____ operations are performed on a struct.

(a) binary   (b) relational
(c) Boolean   (d) arithmetic

124. The struct statement must end with a(n) _____.

(a) ampersand   (b) period
(c) colon   (d) semicolon

125. In C, the dot, (.), is an operator, called the _____.

(a) access operator
(b) member access operator
(c) function access operator
(d) member operator

126. An array is a(n) _____ data structure.

(a) open   (b) closed
(c) heterogeneous   (d) homogeneous

127. To convert a number from base 2 to base 10, we first find the _____ of each bit in the binary number.

(a) weight   (b) position
(c) location   (d) amount

128. The body of the recursive function contains a statement that causes the same function to execute before completing the _____ call.

(a) current   (b) last
(c) final   (d) single

129. We can also write an algorithm to find the factorial of a non-negative integer by using a(n) _____ control structure.

(a) iterative   (b) direct
(c) integer   (d) factorial

130. Every recursive call requires the system to allocate _____ space for its formal parameters and (automatic) local variables.

(a) variable   (b) memory
(c) storage   (d) result

131. Every call to a recursive function has its own copy of the _____ of the function.

(a) header   (b) variables
(c) body   (d) footer

132. Every recursive definition must have one or more _____ cases.

(a) open   (b) base
(c) general   (d) specific

133. A very powerful way to solve certain problems for which the solution would otherwise be very complicated is known as _____.

(a) iteration   (b) lineation
(c) circular programming   (d) recursion

134. Solutions that use a looping structure, such as while, for, or do...while, to repeat a set of statements are known as having _____ control structures.

(a) iterative
(b) duplicating
(c) repeating
(d) circular

135. What are the two ways usually used to solve a particular problem?

(a) logically and recursively
(b) logically and iteratively
(c) iteratively and recursively
(d) linearly and recursively

136. When a function terminates, the memory space that is allocated for its formal parameters is then _____.

(a) cleared   (b) deallocated
(c) reallocated   (d) deleted

137. The process of solving a problem by reducing it to smaller versions of itself is called _____.

(a) circular programming   (b) recursion
(c) rounding   (d) fractalization

138. The base case _____ the recursion.

(a) stops   (b) starts
(c) closes   (d) opens

139. When we are inserting an item in a doubly linked list, the insertion of a node in the list requires the adjustment of _____ pointers in certain nodes.

(a) one   (b) two
(c) three   (d) no

140. A list of items, called nodes, in which the order of the nodes is determined by the address stored in each node is known as a(n) _____ list.

(a) linked   (b) object
(c) node   (d) item

141. Building a list in the backward manner, a new node is always inserted at the _____ of the list.

(a) beginning   (b) middle
(c) end   (d) lowest position

142. When building a list backward, the list is empty so the pointer first must be initialized to _____.

(a) open   (b) NULL
(c) closed   (d) the first data record

143. A linked list is a collection of components, called _____.

(a) objects   (b) items
(c) classes   (d) nodes

144. When building a list backward, the new node becomes the _____ node in the list.

(a) end   (b) middle
(c) first   (d) last

145. The function _____ deallocates the memory occupied by each node.

(a) dealloc()   (b) free()
(c) realloc()   (d) freealloc()

146. The function retrieveFirst returns the _____ contained in the first node.

(a) address   (b) link
(c) info   (d) pointer

147. To delete a given item from an ordered linked list, _____.

(a) first we search the list to see whether the item to be deleted is in the list
(b) delete the info component of the node
(c) first sort the nodes
(d) first organize the info sections of the nodes

148. Suppose that current is a pointer of the same type as the pointer head; then the statement _____ copies the value of head into current.

(a) head = current   (b) head = current;
(c) current = head;   (d) current = head

149. We need _____ pointers to build the list forward.

(a) two   (b) three
(c) four   (d) no

150. If a list is _____, then we can insert a new item at either the end or the beginning.

(a) flexible   (b) closed
(c) open   (d) arbitrary

151. **To print the data contained in each node, we must _____ the list starting at the first node.**
    (a) initialize   (b) organize
    (c) preprocess   (d) traverse

152. **When searching an ordered list, we compare the search item with the _____ node in the list.**
    (a) first   (b) middle
    (c) last   (d) current

153. **The length of a list is the number of _____ in the list.**
    (a) objects   (b) nodes
    (c) items   (d) functions

154. **Every node (except the last node) contains the address of the _____ node.**
    (a) first   (b) next
    (c) last   (d) end

155. **When implementing a stack as an array, to keep track of the top position of the array, we can simply declare another _____, called top.**
    (a) class   (b) array
    (c) stack   (d) variable

156. **The queue is empty if front is _____.**
    (a) equal to rear   (b) NULL
    (c) destroyed   (d) set to 9

157. **Because the value of top indicates whether the stack is empty, we can simply set top to _____ to initialize the stack.**
    (a) 0   (b) 1
    (c) 2   (d) 3

158. **The addQueue operation adds a new element at the _____ of the queue.**
    (a) front   (b) middle
    (c) rear   (d) head

159. **Many compilers now first translate arithmetic expressions into some form of _____ notation.**
    (a) postfix   (b) infix
    (c) prefix   (d) refix

160. **If we set the value of top to 0, even though there are elements are in the stack, they are treated as _____.**
    (a) updated information   (b) valuable data
    (c) unlinked data   (d) garbage

161. **A list of homogenous elements, wherein the addition and deletion of elements occurs only at one end, is known as a _____.**
    (a) list   (b) stack
    (c) queue   (d) top

162. **A queue is also called a _____ data structure.**
    (a) Last In Last Out (LILO)
    (b) First In Last Out (FILO)
    (c) Last In First Out (LIFO)
    (d) First In First Out (FIFO)

163. **The end of the stack where additions and deletions occur is known as the _____ of the stack.**
    (a) top   (b) middle
    (c) bottom   (d) initial element

164. **A data structure in which the elements are added at one end, called the rear, and deleted from the other end, called the front, is known as a(n) _____.**
    (a) stack   (b) queue
    (c) tree   (d) list

165. **A stack is also called a _____ data structure.**
    (a) Last In Last Out (LILO)
    (b) First In Last Out (FILO)
    (c) Last In First Out (LIFO)
    (d) First In First Out (FIFO)

166. **A queue is a set of elements of the same type in which the elements are added at one end, called the _____.**
    (a) rear   (b) top
    (c) back   (d) front

167. **Adding, or pushing, an element onto the stack is a _____ step process.**
    (a) one   (b) two
    (c) three   (d) four

168. **Since new items can be added to the stack, we can perform the add operation, called _____.**
    (a) push   (b) pop
    (c) add   (d) pull

169. **We call the add operation on a queue _____.**
    (a) incQueue   (b) enQueue
    (c) deQueue   (d) addQueue

170. **If we try to add a new item to a full stack, the resulting condition is called _____.**
    (a) overflow   (b) underflow
    (c) flow error   (d) flow control

171. **The remove operation for a stack is called _____.**
    (a) push   (b) pop
    (c) add   (d) pull

172. **A sequential search is also called a(n) _____ search.**
    (a) linear   (b) directed
    (c) arrow   (d) binary

173. The algorithms for sequential and binary searches search the list by comparing the target with the list elements and these are called _____ search algorithms.

(a) direction-based   (b) target-based
(c) comparison-based   (d) object-based

174. A sequence of branches from a node x to another node y is called a _____ from x to y.

(a) leaf   (b) branch
(c) path   (d) root

175. If the search item is found in a sequential search, its _____ is returned.

(a) data content   (b) search variable
(c) function   (d) index

176. The most important operation on a list is the _____.

(a) initialization   (b) search
(c) insert   (d) delete

177. The selection sort algorithm sorts a list by selecting the smallest element in the unsorted portion of the list, and then moving this element to the _____ of the list.

(a) bottom   (b) top
(c) middle   (d) next open position

178. Like the quick sort, the merge sort uses the _____ technique to sort a list.

(a) divide-and-conquer
(b) float-to-top
(c) sink-to-bottom
(d) bubble

179. The unique member of an item that uniquely identifies the item in the data set is called the _____ of the item.

(a) index   (b) identifier
(c) object   (d) key

180. The _____ sort algorithm sorts the list by moving each element to its proper place.

(a) insertion   (b) selection
(c) location   (d) place

181. The merge sort and the quick sort differ in how they _____ the list.

(a) link   (b) partition
(c) categorize   (d) update

182. The sequential search always starts at the _____ element in the list.

(a) first   (b) middle
(c) last   (d) index

183. If the list is stored in an array, we can traverse the list in either direction using a(n) _____ variable.

(a) linking   (b) list
(c) traversal   (d) index

184. Once the sublists are sorted, the next step in the merge sort is to _____ the sorted sublists.

(a) further divide   (b) contrast
(c) compare   (d) merge

185. The sequential search continues until either the item is found in the list or _____.

(a) a sentinel value is reached
(b) the search restarts
(c) the entire list is searched
(d) the search list becomes too large

186. If the list is stored in a linked list, we can traverse the list in only one direction starting at the _____ node.

(a) key   (b) controlling
(c) first   (d) last

187. The sequential search is good for _____ lists.

(a) medium sized   (b) very short
(c) single element   (d) very long

188. If the search item is the first element in the list, _____ is/are required.

(a) one comparison
(b) two comparisons
(c) no comparison
(d) multiple comparisons

189. A binary search requires the list _____ to be in order.

(a) elements   (b) variables
(c) functions   (d) operations

190. A list is ordered if its _____ are ordered according to some criteria.

(a) end points   (b) starting point
(c) elements   (d) variables

191. If the binary tree is empty, then the height is _____.

(a) 0   (b) 1
(c) undetermined   (d) pruned

192. The performance of the search algorithm depends on the _____ the binary search tree.

(a) data contained in
(b) shape of
(c) type of variables in
(d) type of search used with

193. A node in the binary tree is called a(n) _____ if it has no left and right children.
   (a) parent   (b) only child
   (c) root   (d) leaf

194. If the item is found in the binary search tree, it returns _____.
   (a) found   (b) identified
   (c) true   (d) false

195. An arrow is usually called a directed edge or a(n) _____.
   (a) directed branch   (b) specific branch
   (c) important branch (d) straight branch

196. In a(n) _____ traversal, the binary tree is traversed as follows: 1. Visit the node. 2. Traverse the left subtree. 3. Traverse the right subtree.
   (a) inorder   (b) preorder
   (c) postorder   (d) levelorder

197. In the diagram of a binary tree, each node of the binary tree is represented as a _____.
   (a) rectangle   (b) triangle
   (c) circle   (d) square

198. A binary tree is a _____ data structure.
   (a) large   (b) variable
   (c) static   (d) dynamic

199. The number of comparisons required to determine whether x is in the binary search tree T is _____ the number of comparisons required to insert x in T.
   (a) one more than   (b) one less than
   (c) the same as   (d) two more than

200. If U and V are two nodes in a binary tree, U is called a(n) _____ of V if there is a branch from U to V.
   (a) child   (b) parent
   (c) sister   (d) brother

201. If the binary search tree is _____, we first compare the search item with the info in the root node.
   (a) empty   (b) nonempty
   (c) open   (d) closed

202. Three lines at the end of an arrow indicate that the subtree is _____.
   (a) open   (b) closed
   (c) empty   (d) full

203. The root node of the binary tree is drawn at the _____ of the drawing.
   (a) top   (b) bottom
   (c) middle   (d) side

204. To make an identical copy of a binary tree, we need to create as many _____ as there are in the binary tree to be copied.
   (a) decision points   (b) nodes
   (c) functions   (d) variables

205. In the inorder traversal of a binary tree, for each node, the _____ subtree is visited first, then the node, and then the _____ subtree.
   (a) primary, secondary (b) secondary, primary
   (c) right, left   (d) left, right

206. After inserting an item in a binary search tree, the resulting binary tree must be _____.
   (a) binary search tree (b) resorted
   (c) updated   (d) relinked

207. Every node in a binary tree has at most _____ children.
   (a) one
   (b) two
   (c) three
   (d) four

208. The path with the smallest weight is known as the _____ path.
   (a) largest   (b) smallest
   (c) longest   (d) shortest

209. Let G be a graph. A graph H is called a _____ of G if V(H) is a subset of V(G) and E(H) is a subset of E(G).
   (a) directed graph   (b) undirected graph
   (c) subgraph   (d) pair graph

210. A graph is empty if the number of _____ is zero.
   (a) parameters   (b) functions
   (c) vertices   (d) variables

211. The set of all elements that are in A or in B are known as the _____.
   (a) vertices   (b) graph
   (c) union   (d) intersection

212. G is called _____ connected if any two vertices in G are connected.
   (a) directly
   (b) indirectly
   (c) strongly
   (d) weakly

213. The weight of the path P is the _____ of the weights of all edges on the path P, which is also called the weight of v from u via P.
   (a) product   (b) sum
   (c) inverse   (d) smallest

214. Vertices u and v are called _____ if there is a path from u to v.

(a) directed   (b) bonded

(c) connected   (d) disconnected

215. A tree in which a particular vertex is designated as a root is called a _____ tree.

(a) minimal   (b) rooted

(c) weighted   (d) spanning

216. When a graph is shown pictorially, the vertices are drawn as _____.

(a) arrows   (b) lines

(c) squares   (d) circles

217. The depth first traversal is similar to the _____ traversal of a binary tree.

(a) preorder   (b) postorder

(c) inorder   (d) levelorder

218. A graph G is a pair where G=(V,E) where V is a finite nonempty set also called a set of _____.

(a) vertices   (b) graph

(c) union   (d) intersection

219. To write programs that process and manipulate graphs, the graphs must be stored or represented in _____.

(a) a variable   (b) computer memory

(c) a function   (d) an array

220. A tree T is a simple graph such that if u and v are _____ vertices in T, there is a uniqpath from u to v.

(a) no   (b) two

(c) three   (d) four

221. A path in which all vertices, except possibly the first and last vertices, are distinct is known as a(n) _____ path.

(a) component   (b) cycle

(c) complex   (d) simple

222. If a weight is assigned to the edges in T, thenT is called a _____ tree.

(a) minimal   (b) rooted

(c) weighted   (d) spanning

223. When a graph is shown pictorially, the label inside the symbol for the vertices represents the _____.

(a) vertex   (b) tangents

(c) union   (d) intersection

224. The breadth first traversal of a graph is similar to traversing a _____ level by level.

(a) function   (b) binary tree

(c) variable   (d) class

BASICS OF C PROGRAM

1.
```c
void main()
{
    int const * ptr = 5;
    printf("%d",++(*ptr));
}
```

2.
```c
void main()
{
    extern int i;
    i = 20;
    printf("%d",i);
}
```

3.
```c
void main()
{
    int i= -1, j= -1, k=0, l=2, m;
    m = i++ && j++ && k++ || l++;
    printf("%d %d %d %d %d",
                    i,j,k,l,m);
}
```

4.
```c
void main()
{
    printf("%x", -1 << 4);
}
```

5.
```c
void main()
{
    int a =- -2;
    printf("c = %d", a);
}
```

6.
```c
#define int char
void main()
{
    int i = 65;
    printf("sizeof(i) = %d",
                    sizeof(i));
}
```

7.
```c
void main()
{
    int i = 10;
    i = !i > 5;
    printf("i = %d",i);
}
```

8.
```c
void main()
{
    printf("\nab");
    printf("\bsi");
    printf("\rha");
}
```

9.
```c
void main()
{
    int i = 5;
    printf("%d %d %d %d %d %d",
            i++,--i,i--,++i,i);
}
```

10.
```c
#define square(x) x*x
void main()
{
    int i = 64 / square(4);
    printf("%d",i);
}
```

11.
```c
#define a 10
void main()
{
    #define a 50
    printf("%d", a);
}
```

12.
```c
#define clrscr() 100
void main()
{
    clrscr();
    printf("%d",clrscr());
}
```

13.
```c
void main()
{
    printf("%p",main);
}
```

14.
```c
void main()
{
    clrscr();
}
clrscr();
```

15.
```c
enum colors{BLACK,BLUE,GREEN};
void main()
{
    printf("%d - %d - %d",
            BLACK,BLUE,GREEN);
    return(1);
}
```

16.
```c
void main()
{
    char far *farther,*farthest,farless;
    printf("%d-%d-%d",sizeof(farther),
    sizeof(farthest),sizeof(farless));
}
```

17.
```c
void main()
{
    char far *a, near *b, huge *c;
    printf("%d - %d - %d",sizeof(a),
            sizeof(b),sizeof(c));
}
```

18.
```c
void main()
{
    int a = 100, b = 200;
    printf("%d - %d");
}
```

```
19. void main()
    {
        char *ptr;
        ptr = "John";
        printf("%c",*&*ptr);
    }

20. void main()
    {
        int i = 5;
        printf("%d",i++ + ++i);
    }

21. void main()
    {
        int i = 5;
        printf("%d",i+++++i);
    }

22. void main()
    {
        int i;
        /* Input value is 234 */
        printf("%d",scanf("%d",&i));
    }

23. #define f(a,b) a##b
    void main()
    {
        int cd = 100;
        printf("%d",f(c,d));
    }

24. void main()
    {
        extern int i;
        i = 20;
        printf("%d",sizeof(i));
    }

25. void main()
    {
        printf("%d",out);
    }
    int out = 100;

26. void main()
    {
        extern out;
        printf("%d",out);
    }
    int out = 100;

27. void main()
    {
        int i = -(-1);
        printf("i = %d, i = %d\n",i,+i);
    }

28. void main()
    {
        char name[10],str[12];
        scanf(" \"%[^\"]\"",str);
    }
    How will the scanf function execute?
```

```
29. void main()
    {
        main();
    }

30. void main()
    {
        char not;
        not = !2;
        printf("%d",not);
    }

31. void main()
    {
        int i = 1;
        printf("%d==1 is ""%s",
               i,i==1?"TRUE":"FALSE");
    }

32. #define MAX 3
    void main()
    {
        typedef char data[MAX];
        data list = {0, 1, 2};
        data name = "John";
        printf("%d %s",list[0],name);
    }

33. #define MAX 3
    #define int data[MAX]
    void main()
    {
        data list = {0, 1, 2};
        printf("%d",list[0]);
    }

34. int i = 10;
    void main()
    {
        extern int i;
        {
            int i = 20;
            {
                const volatile unsigned i=30;
                printf(" %d",i);
            }
            printf(" %d",i);
        }
        printf(" %d",i);
    }

35. void main()
    {
        int i = -1;
        +i;
        printf("i = %d, +i = %d",i,+i);
    }

36. void main()
    {
        int i = -1;
        -i;
        printf("i = %d, -i = %d",i,-i);
    }
```

```
37.#include<stdio.h>
   void main()
   {
      const int i = 4;
      float j;
      j = ++i;
      printf("%d   %f",i,++j);
   }

38.#include<stdio.h>
   void main()
   {
      register i = 4;
      char j[] = "John";
      printf("%s   %d",j,i);
   }

39.void main()
   {
      int i = 5, j = 6, z;
      printf("%d",i+++j);
   }

40.void main()
   {
      char a[] = "12345\0";
      int i = strlen(a);
      printf("The length of the
             variable a is %d\n",++i);
   }

41.void main()
   {
      int a;
      unsigned b=-1;
      printf("%u ",++b);
      printf("%u ",a=--b);
   }

42.void main()
   {
      int i=i++, j=j++, k=k++;
      printf("%d %d %d",i,j,k);
   }

43.void main()
   {
      static int i=i++, j=j++, k=k++;
      printf("i = %d j = %d k = %d",i,j,k);
   }

44.#define prod(a,b) a*b
   void main()
   {
      int x = 3, y = 4;
      printf("%d",prod(x+2,y-1));
   }

45.void main()
   {
      float a = 2;
      printf("%f\n", a >> 2);
   }
```

```
46.void main()
   {
      enum { i = 10, j = 20, k = 50 };
      printf("%d\n", ++k);
   }

47.void main()
   {
      float a = 2;
      printf("%f\n", a << 2);
   }

48.void main()
   {
      float a = 2, b = 4;
      printf("%lf\n", a % b);
   }

49.void main()
   {
      float a = 5, b = 2;
      printf("%lf\n",fmod(a,b));
   }

50.typedef enum errorType
   {
      warning, error, exception
   }error;
   void main()
   {
      error e;
      e = 1;
      printf("%d",e);
   }

51.typedef struct error
   {
      int warning, error, exception;
   }error;
   void main()
   {
      error e;
      e.error=1;
      printf("%d",e.error);
   }

52.#define something 100
   #ifdef something
   int some = 0;
   #endif
   void main()
   {
      int thing = 0;
      printf("%d %d",some,thing);
   }

53.#ifdef something
   int some = 0;
   #endif
   void main()
   {
      int thing = 0;
      printf("%d %d",some,thing);
   }
```

```
54. #ifdef something
    #if something == 0
    int some = 0;
    #endif
    void main()
      {
        int thing = 0;
        printf("%d %d",some,thing);
      }

55. void main()
      {
        int arr[2][2];
        printf("%d\n",((arr==*arr)
            &&(*arr==arr[0]))));
      }

56. void main()
      {
        int i = 5;
        printf("%d",++i++);
      }

57. void main()
      {
        int i = 5;
        printf("%d",i=++i==6);
      }

58. void main()
      {
        char text[] = "%d\n";
        text[1] = 'c';
        printf(text, 65);
      }

59. void main()
      {
        int i=5, j=10;
        i = i &= j && 10;
        printf("%d %d",i,j);
      }

60. void main()
      {
        int i=5, j=10;
        i = i |= j || 10;
        printf("%d %d",i,j);
      }

61. void main()
      {
        int i=5, j=10;
        j = i = j + 10;
        printf("%d %d",i,j);
      }

62. void main()
      {
        int i=5, j=10;
        j = i = j > 1;
        printf("%d %d",i,j);
      }
```

```
63. void main()
      {
        int i = 4, j = 7;
        j=j||i++&&printf("YOU CAN WIN");
        printf("%d %d",i,j);
      }

64. void main()
      {
        register int a = 2;
        printf("Address of a = %d",&a);
        printf("Value of a = %d",a);
      }

65. void main()
      {
        extern i;
        printf("%d\n",i);
        {
          int i = 20;
          printf("%d\n",i);
        }
      }

66. #define DIM(array,type)
            sizeof(array)/sizeof(type)
    void main()
      {
        int arr[10];
        printf("The dimension of the
            array is %d",DIM(arr,int));
      }

67. void main()
      {
        printf("sizeof(void *) = %d\n",
            sizeof(void*));
        printf("sizeof(int *) = %d\n",
            sizeof(int*));
        printf("sizeof(double *) = %d\n",
            sizeof(double *));
        printf("sizeof(struct unknown*)
        =%d",sizeof(struct unknown*));
      }

68. void main()
      {
        char a[4] = "HELLO";
        printf("%s",a);
      }

69. void main()
      {
        char a[4] = "HELL";
        printf("%s",a);
      }

70. void main()
      {
        extern int i;
        {
          int i = 20;
          {
            const volatile unsigned i=30;
            printf("\t%d", i);
          }
          printf("\t%d", i);
        }
        printf("\t%d", i);
      }
```

❖ ❖ ❖ ❖ ❖ ❖ ❖

IF STATEMENTS

```
1. void main()
   {
       float a=123.4;
       double b=123.4.;
       if(a == b)
         printf("INDIA will win match");
       else
         printf("INDIA may win match");
   }

2. void main()
   {
       static int a=5;
       printf("%d ",a--);
       if(a)
         main();
   }

3. void main()
   {
       static int i = 5;
       if(--i)
       {
         main();
         printf("%d ",i);
       }
   }

4. void main()
   {
       int i=3;
       switch(i)
       {
         default: printf("Zero");
                 break;
         case 1: printf("One");
                 break;
         case 2: printf("Two");
                 break;
         case 3: printf("Three");
                 break;
       }
   }

5. void main()
   {
       int a=1,b=2;
       switch(b)
       {
         case 1: printf("Good Morning");
                 break;
         case b: printf("Bad Morning");
                 break;
       }
   }

6. #define FALSE -1
   #define TRUE 1
   #define NULL 0
   void main()
   {
       if(NULL)
         puts("NULL");
       else if(FALSE)
         puts("TRUE");
       else
         puts("FALSE");
   }
```

```
7. void main()
   {
       int year;
       /* Input value is 10 */
       scanf("%d",&year);
       if((year%4==0&&year%100!=0)
                   ||year%100==0)
         printf("%d is a leap year");
       else
         printf("%d is not a leap year");
   }

8. void main()
   {
       int i=0, j=0;
       if(i&&j++)
         printf("%d - %d",i++,j);
       printf("%d - %d",i,j);
   }

9. void main()
   {
       int a = 0;
       int b = 20;
       char c = 1;
       char d = 10;
       if(a, b, c, d)
         printf("Programming");
   }

10. void main()
    {
        char str[]="\0";
        if(printf("%s\n",str))
          printf("Good Morning");
        else
          printf("Good Night");
    }

11. void main()
    {
        int a, b = 100,c;
        if(a = b % 2)
        {
            c=100;
        }
        printf("%d - %d - %d",a,b,c);
    }

12. void main()
    {
        float a=1.5;
        switch(a)
        {
          case 1: printf("1");
                  break;
          case 2: printf("2");
                  break;
          default: printf("0");
                   break;
        }
    }
```

❖ ❖ ❖ ❖ ❖ ❖ ❖

LOOPING STATEMENTS

1.
```c
void main()
{
    char str[] = "man";
    int i;
    for(i = 0; str[i]; i++)
    printf("\n%c%c%c%c",
    str[i],i[str],*(str+i),*(i+str));
}
```

2.
```c
void main()
{
    int arr[] = {2.2,3,4.4,5,6.6};
    int i, *ptr1=arr, *ptr2=arr;
    puts("");
    for(i = 0; i < 5; i++)
    {
        printf("%d ",*arr);
        ++ptr2;
    }
    for(i = 0; i < 5; i++)
    {
        printf("%d ",*ptr1);
        ++ptr1;
    }
}
```

3.
```c
void main()
{
    char *str="I Love You",*ptr;
    ptr = str;
    while(*str != '\0')
    ++*str++;
    printf("%s    %s",str,ptr);
}
```

4.
```c
void main()
{
    int a = 1;
    while(a <= 5)
    {
        printf("%d",a);
        if(a > 3)
        goto INDIA;
        a++;
    }
}
int func()
{
    here:
    printf("I LOVE INDIA");
}
```

5.
```c
void main()
{
    int i;
    char *ptr;
    static char Lang[4][20]
    ={"Pascal","C","C++","Java"};
    ptr = Lang[2];
    Lang[2] = Lang[3];
    Lang[3] = ptr;
    for(i = 0; i <= 3; i++)
    puts(Lang[i]);
}
```

6.
```c
void main()
{
    int a = 0;
    for(; a++; printf("%d",a));
    printf("%d",a);
}
```

7.
```c
int i;
void main()
{
    unsigned int i;
    for(i = 1; i > -3; i--)
    printf("c aptitude");
}
```

8.
```c
void main()
{
    while(1)
    {
        if(printf("%d",printf("%d")))
        break;
        else
        continue;
    }
}
```

9.
```c
void main()
{
    unsigned int i = 5;
    while(i-- >= 0)
    printf("%u",i);
}
```

10.
```c
void main()
{
    unsigned int i = 65000;
    while(i++ != 0);
    printf("%d",i);
}
```

11.
```c
void main()
{
    int i = 0;
    while(+(+i--) != 0)
    i-=i++;
    printf("%d",i);
}
```

12.
```c
void main()
{
    signed char i = 0;
    for(; i >= 0; i++);
    printf("%d",i);
}
```

13.
```c
void main()
{
    unsigned char i = 0;
    for(; i >= 0 ; i++);
    printf("%d",i);
}
```

14.
```c
void main()
{
    char i = 0;
    for(; i >= 0; i++);
    printf("%d",i);
}
```

```
15.void main()
   {
      while(strcmp("John", "John\0"))
       printf("Strings Are Not Equal");
   }
16.void main()
   {
      char str1[]={'J','o','h','n'};
      char str2[]={'J','o','h','n','\0'};
      while(strcmp(str1,str2))
       printf("Strings Are Not Equal");
   }
17.void main()
   {
      int i = 3;
      for(; i+=0 ;)
         printf("%d",i);
   }
18.void main()
   {
      static int i;
      while(i <= 10)
         (i > 2) ? i++ : i--;
      printf("%d",i);
   }
19.void main()
   {
      char ch;
      for(ch=0;ch<=127;ch++);
         printf("%c  %d",ch,ch);
   }
20.void main()
   {
      int l1 = 1;
      l1:
      printf("Testing in C");
      while(l1)
         break;
      goto l1;
   }
```

❖ ❖ ❖ ❖ ❖ ❖ ❖

FUNCTIONS

```
1. void main()
   {
      char str[] = "Hello World";
      display(str);
   }
   void display(char *str)
   {
      printf("%s",str);
   }
2. void main()
   {
      show();
   }
   void show()
   {
      printf("Fashion Show");
   }
3. void main()
   {
      int i = _func(10);
      printf("%d",--i);
   }
   int _func(int i)
   {
      return(i++);
   }
4. void main()
   {
      char c = ' ', x, Convert(z);
      x = Convert(c);
      printf("%c",x);
   }
   Convert(z)
   {
      return z;
   }
5. int sum(int m, int n);
   void main(int argc, char **argv)
   {
      printf("Enter a Character : ");
      getchar();
      sum(argv[1], argv[2]);
   }
   int sum(int m, int n)
   {
      return(m+n);
   }
6. a()
   {
      printf("Hai");
   }
   b()
   {
      printf("Hello");
   }
   c()
   {
      printf("Bye");
   }
   void main()
   {
      int (*ptr[3])();
      ptr[0] = a;
      ptr[1] = b;
      ptr[2] = c;
      ptr[2]();
   }
```

```
7. void main()
   {
     int i;
     i = abc();
     printf("%d",i);
   }
   abc()
   {
     _AX = 1000;
   }
```

8. The following notations of defining functions are known as?

```
   i. int abc(int a,float b)
      {
         // some code
      }
   ii. int abc(a,b)
      int a; float b;
      {
         // some code
      }
```

```
9. void main()
   {
     char a[100];
     a[0] = 'a';
     a[1] = 'b';
     a[2] = 'c';
     a[3] = 'd';
     func(a);
   }
   func(char a[])
   {
     a++;
     printf("%c",*a);
     a++;
     printf("%c",*a);
   }
```

```
10. func(int a, int b)
    {
      return(a =(a == b));
    }
    void main()
    {
      int process(), func();
      printf("The value of process
      is %d",process(func,3,6));
    }
    process(int(*pf)(),int val1,int val2)
    int(*pf)();
    {
      return((*pf) (val1,val2));
    }
```

```
11. int func(int);
    void main()
    {
      int a = func(sizeof(float));
      printf("Value is %d",++a);
    }
    int func(int func)
    {
      func += 2.5;
      return(func);
    }
```

```
12. void pascal func(int, int, int);
    void main()
    {
      int i = 10;
      func(i++, i++, i++);
      printf(" %d",i);
    }
    void pascal func(integer:i,
                integer:j, integer:k)
    {
      write(i, j, k);
    }
```

```
13. void pascal fun1(int i,int j,int k)
    {
      printf("%d %d %d",i,j,k);
    }
    void cdecl fun2(int i,int j,int k)
    {
      printf("%d %d %d",i,j,k);
    }
    void main()
    {
      int i = 5;
      fun1(i++, i++, i++);
      printf(" %d\n",i);
      i = 5;
      fun2(i++, i++, i++);
      printf(" %d",i);
    }
```

14. Describe the C statement.
```
    void(* abc(int,void(*def)()))();
```

```
15. int DIM(int array[])
    {
      return sizeof(array)/sizeof(int);
    }
    void main()
    {
      int arr[10];
      printf("The array dimension is
                 %d",DIM(arr));
    }
```

❖ ❖ ❖ ❖ ❖ ❖ ❖

POINTERS

```
1. void main()
   {
       char *ptr;
       printf("%d %d",sizeof(*ptr),
                    sizeof(ptr));
   }

2. void main()
   {
       char s[] = {'a','b','c','\n',
                        'c','\0'};
       char *ptr,*str;
       ptr = &s[3];
       str = s;
       printf("%c",++*ptr + ++*str-9);
   }

3. void main()
   {
       int arr[2][2][2] = {{10,2,3,4},
                        {5,6,7,8}};
       int *ptr,*qtr;
       ptr = &arr[2][2][2];
       *qtr = ***arr;
       printf("%d--%d",*ptr,*qtr);
   }

4. void main()
   {
       int a[2][3][2] =
                {
                {{1,2},{3,4},{5,6}},
                {{9,8},{7,6},{5,4}}
                };
       printf("%u - %u - %u - %d \n",
                    a,*a,**a,***a);
       printf("%u - %u - %u - %d",
             a+1,*a+1,**a+1,***a+1);
   }

5. void main()
   {
       int a[] = {10,20,30,40,50};
       int i, *ptr;
       for(i = 0; i < 5; i++)
         {
           printf("%d",*a);
           a++;
         }
       ptr = a;
       for(i = 0; i < 5; i++)
         {
           printf("%d",*ptr);
           ptr++;
         }
   }

6. void main()
   {
   static int a[]={0,1,2,3,4};
   int *p[] = {a,a+1,a+2,a+3,a+4};
   int **ptr = p;
   ptr++;
   printf("\n %d %d %d",
                ptr-p,*ptr-a, **ptr);
   *ptr++;
   printf("\n %d %d %d",
                ptr-p,*ptr-a,**ptr);
   *++ptr;
   printf("\n %d %d %d",
                ptr-p,*ptr-a,**ptr);
   }

7. void main()
   {
       int arr[3][3]  =   {
                        {1, 2, 3},
                        {4, 5, 6},
                        {7, 8, 9},
                        };
       printf("%d", *(*(arr)));
   }

8. void main()
   {
       void *vp;
       char ch = 'A', *cp = "Hello";
       int j = 20;
       vp = &ch;
       printf("%c",*(char*)vp);
       vp = &j;
       printf("%d",*(int*)vp);
       vp = cp;
       printf("%s", (char*)vp+3);
   }

9. void main()
   {
       static char *s[]  =
       {"Black","White","Green","Red"};
       char **ptr[]={s+3,s+2,s+1,s};
       char ***p;
       p = ptr;
       **++p;
       printf("%s",*--*++p+3);
   }

10. void main()
    {
        int  i, n;
        char *x = "John";
        n = strlen(x);
        *x = x[n];
        for(i = 0; i < n; ++i)
          {
            printf("%s\n",x);
            x++;
          }}

11. void main()
    {
        char *cptr, c = 10;
        void *vptr, v = 0;
        cptr = &c;
        vptr = &v;
        printf("%c %v", c, v);
    }

12. void main()
    {
        char *str1 = "abcd";
        char str2[] = "abcd";
        printf("%d %d %d",sizeof(str1),
           sizeof(str2),sizeof("abcd"));
    }
```

```
13.void main()
   {
     int *j;
      {
        int i = 10;
        j = &i;
      }
     printf("%d",*j);
   }

14.void main()
   {
     char *p;
     int *q;
     long *r;
     p = q = r = 0;
     p++;
     q++;
     r++;
     printf("%p - %p - %p",p,q,r);
   }

15.int arr[] = {1, 2, 3};
   void main()
   {
     int *ptr;
     ptr = arr;
     ptr += 3;
     printf("%d", *ptr);
   }

16.void main()
   {
     char *ptr;
     ptr = "%d\n";
     ptr++;
     ptr++;
     printf(ptr-2,300);
   }

17.void main()
   {
     void *v;
     int integer = 2;
     int *i = &integer;
     v = i;
     printf("%d",(int*)*v);
   }

18.void main()
   {
     int a[10];
     printf("%d",*a+1-*a+3);
   }

19.void main()
   {
     char *ptr = "abcde";
     printf("%c",++*(ptr++));
   }

20.void main()
   {
     char *p = "abcde";
     printf("%c", ++*p++);
   }
```

```
21.void main()
   {
     int *mptr =(int*)malloc(sizeof(int));
     int *cptr=(int*)
     calloc(sizeof(int),1);
     printf("%d\n",*mptr);
     printf("%d",*cptr);
   }

22.void main()
   {
     int a = 2, *ptr1, *ptr2;
     ptr1 = ptr2 = &a;
     *ptr2 += *ptr2 += a += 2.5;
     printf("%d %d %d",a,*ptr1,*ptr2);
   }

23.void main()
   {
     char *ptr = "GOOD";
     char arr[ ] = "GOOD";
     printf("%d %d %d %d %d",
            sizeof(ptr),sizeof(*ptr),
            strlen(ptr),sizeof(arr),
            strlen(arr));
   }

24.int swap(int *a,int *b)
   {
     *a = *a + *b; *b = *a - *b;
     *a = *a - *b;
   }
   void main()
   {
     int x = 10, y = 20;
     swap(&x, &y);
     printf("x= %d y = %d\n", x, y);
   }

25.void main()
   {
     void swap();
     int x = 10, y = 20;
     swap(&x, &y);
     printf("x = %d y = %d",x,y);
   }
   void swap(int *a, int *b)
   {
     *a ^= *b; *b ^= *a;
     *a ^= *b;
   }

26.void main()
   {
     int i = 257; int *ptr = &i;
     printf("%d %d",*((char*)ptr),
                    *((char*)ptr+1));
   }

27.void main()
   {
     int i = 258; int *ptr = &i;
     printf("%d %d",*((char*)ptr),
                    *((char*)ptr+1));
   }

28.void main()
   {
     int i = 300;
     char *ptr = &i;
     *++ptr = 2;
     printf("%d",i);
   }
```

```
29.void main()
   {
       char *str = "hello";
       char *ptr = str, ch = 127;
       while(*ptr++)
          ch = (*ptr < ch) ? *ptr:ch;
       printf("%d", ch);
   }
```

30.What is the subtle error in the following code segment?

```
   void fun(int n, int arr[])
   {
       int *ptr = 0, i = 0;
       while(i++ < n)
         ptr = &arr[i];
       *ptr = 0;
   }
```

```
31.void main()
   {
       int *i = 0x400;
       /* i points to the address 400 */
       *i = 0;
       /*sets value of memory location pointed by i*/
   }
```

```
32.void main()
   {
       int i = 10, j = 2;
       int *ip = &i, *jp = &j;
       int k = *ip / *jp;
       printf("\n\n\n%d",k);
   }
```

33.Is this code legal?
```
   int *ptr;
   ptr = (int *) 0x400;
```

```
34.char *someFun()
   {
       char *temp = "String constant";
       return temp;
   }
   int main()
   {
       puts(someFun());
   }
```

```
35.char *someFun1()
   {
       char temp[] = "string";
       return temp;
   }
   char *someFun2()
   {
       char temp[]={'s','t','r','i','n','g'};
       return temp;
   }
   int main()
   {
       puts(someFun1());
       puts(someFun2());
   }
```

❖ ❖ ❖ ❖ ❖ ❖ ❖

STRUCTURES & UNIONS

```
1. void main()
   {
       struct X
       {
           int x = 2;
           char name[] = "John";
       };
       struct X *s;
       printf("%d",s->x);
       printf("%s",s->name);
   }
```

```
2. void main()
   {
       struct X
       {
           int x;
           struct Y
           {
               char s;
               struct X *p;
           };
           struct Y *q;
       };
   }
```

```
3. void main()
   {
       struct X
       {
           int x;
           char name[5];
       };
       struct X *s=malloc(sizeof(struct X));
       printf("%d",s->x);
       printf("%s",s->name);
   }
```

```
4. void main()
   {
       struct X
       {
           int x;
           struct Y
           {
               char s;
               struct X *p;
           };
           struct Y *q;
       };
   }
```

```
5. struct A_A
   {
       struct A_A *prev;
       int i;
       struct A_A *next;
   };
   void main()
   {
       struct A_A abc,def,ghi,jkl;
       int x = 100;
       abc.i = 0;
       abc.prev = &jkl;
       abc.next = &def;
       def.i = 1;
       def.prev = &abc;
       def.next = &ghi;
       ghi.i = 2;
```

```
        ghi.prev = &def;
        ghi.next = &jkl;
        jkl.i = 3;
        jkl.prev = &ghi;
        jkl.next = &abc;
        x=abc.next->next->prev->next->i;
        printf("%d",x);
    }
```

6.
```
struct A_A
    {
        int x, y;
    };
    struct A_A origin, *ptr;
    void main()
    {
        ptr = &origin;
        printf("Value of ptr is %d - %d
            \n",(*ptr).x,(*ptr).y);
        printf("Value of ptr is %d - %d
                \n",ptr->x,ptr->y);
    }
```

7.
```
void main()
    {
        struct student
            {
                char name[30];
                struct date dob;
            }stud;
        struct date
            {
                int dd, mm, yy;
            };
        scanf("%s %d %d %d", stud.name,
        &student.dob.dd,&student.dob.mm,
                    &student.dob.yy);
    }
```

8.
```
void main()
    {
        struct date;
        struct student
            {
                int rollno;
                char name[30];
                struct date dob;
            }stud;
        struct date
            {
                int dd, mm, yy;
            };
        scanf("%s %d %d %d", stud.name,
        &student.dob.dd,&student.dob.mm,
                    &student.dob.yy);
    }
```

9. There were 10 records stored in "somefile.dat" but the following program printed 11 names. What went wrong?
```
void main()
    {
        struct student
            {
                char name[30];
                int rollno;
            }stud;
        FILE *fp =
            fopen("C;\somefile.dat","r");
        while(!feof(fp))
            {
```
```
                fread(&stud,sizeof(stud),1,fp);
                puts(stud.name);
                printf("%s -- %d",
                        stud.name,stud.rollno);
            }
    }
```

10. Is the following code legal?
```
struct a
    {
        int x;
        struct a b;
    };
```

11. Is the following code legal?
```
struct a
    {
        int x;
        struct a *b;
    };
```

12. Is the following code legal?
```
typedef struct a
    {
        int x;
        Type *b;
    }Type;
```

13. Is the following code legal?
```
typedef struct a Type;
struct a
    {
        int x;
        Type *b;
    };
```

14. Is the following code legal?
```
void main()
    {
        typedef struct a Type;
        Type someVariable;
        struct a
            {
                int x;
                Type *b;
            };
    }
```

15.
```
void main()
    {
        struct X
            {
                int x;
                float y;
            };
        struct X v1, v2;
        v1. x = 10;
        v1.y = 20.5;
        v2 = v1;
        printf("\t%d\t%f\t%d\t%f",
                v1.x,v1.y,v2.x,v2.y);
    }
```

❖ ❖ ❖ ❖ ❖ ❖ ❖

III – PREDICT THE OUTPUT FOR THE FOLLOWING QUESTIONS

1. Which of the following are valid identifiers?

(a) circle area   (b) char

(c) INT   (d) book's

(e) no_of_side   (f) no-constraint

2. Which of the following are valid expression?

(a) b*b - 4.0*a*c = d;   (b) 5.54 = e;

(c) f = +g--h;   (d) f = +g- -h;

(e) i = 'j' + 2;   (f) k += 8;

3. What is the type of each of the following integer constants?

(a) 1234   (b) 1234u

(c) 1234s   (d) 1234L

(e) 1234us   (f) 1234uL

4. What is the type of each of the following constants?

(a) 12   (b) -34.56

(c) "34.56"   (d) '3'

(e) '\n'   (f) "Son\'s"

5. Which of the following are valid character constants?

(a) 'a'   (b) 'abc'

(c) "Z_1234"   (d) '\t'

6. Which of the following are valid string constants?

(a) 3 + "Program"   (b) 'book'

(c) "1234.56789"   (d) "\n\t"

7. What are the results of the following statements if, a = 2, b = 3 and c = 4?

(a) ++a + a++;   (b) c *= c + 1;

(c) a + b * c / a;   (d) c -= a += 5;

(e) 'a' + a;   (f) sizeof(a) + a;

8. What is the equivalent C statement for the expression?

(a) b*b - 4*a*c / 2*a

(b) b*b - 4*a*c / (2*a)

(c) ((b*b) - (4*a*c)) / 2*a

(d) ((b*b) - (4*a*c)) / (2*a)

9. What are the results of the following expressions if, a = 2, b = 3 and c = 4?

(a) a & b | c   (b) ~a ^ b & c

(c) a | b & c   (d) ~~b

(e) a << b >> c   (f) a ^ b ^ c

10. What are the results of following conditional expressions if, a = 2, b = 0 and c = 4?

(a) a && b || c   (b) !!b

(c) a || b && c   (d) a && b || !c

11. Simplify the expressions by removing the parenthesis and the logical not operator.

(a) !(a)   (b) !(a > b)

(c) !(!a >= b)   (d) !(a != !b)

12. What are the results of following conditional expressions if a = 1, b = 0, c = 3 and d = 4?

(a) !(a)

(b) (a || b && c || d)

(c) (!a && b || c && !d)

(d) !(a && (!b || !c))

13. What is the result of the following C statements?

(a) 5 / 2   (b) 5.0 / 2

(c) 5 % 2   (d) 5 % 2.0

14. What is the value of the following C statements.

(a) 38 * 1000   (b) 38.0 * 1000

(c) 'A' * 1000   (d) 'A' * 1000.0

15. What is the value of a in the following C statements?

(a) char a = '\123';   (b) char a = 0x53;

(c) char a = 0123;   (d) char a = 83;

16. Which of the following is not a data type in C?

(a) integer   (b) character

(c) void   (d) boolean

17. Specify the precedence order of the operators.

(a) arithmetic, relational, logical, assignment

(b) arithmetic, logical, relational, assignment

(c) relational, arithmetic, logical, assignment

(d) logical, arithmetic, relational, assignment

18. What are the operators which have left to right associativity?

(a) unary

(b) binary

(c) ternary

(d) comma

19. Specify the type of operators used.

(a) sizeof(double)

(b) abc + x_y_z

(c) return((a>b) ? a:b)

(d) a += b;

20. Which of the following is a valid binary expression?

(a) -3

(b) a + b

(c) c += 25

(d) 4 - e

21. Which of the following is a valid assignment statement?

(a) i = (j++, k++);

(b) i = j = k;

(c) i += k;

(d) 5000 = k;

22. What is the value of x after the execution of the statements, when x = 5 initially?

(a) x %= 10;

(b) x = x+++3;

(c) x >>= 1;

(d) x <<= 1;

(e) x >>= -1;

(f) x <<= -1;

23. What is the value of z after the execution of the statements, when x = 5 and y = 7 initially?

(a) z = (x++, y++);

(b) z = x += ++y;

(c) z = x++ - --y;

(d) z = x + y-- -x + x++;

24. Which of the following statements are true about mixed mode operation?

(a) Implicit cast generated by the compiler

(b) Compiler automatically carries constant casting

(c) Explicit cast used to change expression type

(d) Explicit cast changes variable type in memory

25. What is the purpose of the header files in C?

(a) Function decalration

(b) Function definition

(c) Both (a) and (b)

(d) None of the above

26. What does the following scanf() format specifiers indicate?

(a) %d

(b) %x

(c) %u

(d) %o

(e) %lu

(f) %ld

27. What does the following scanf() format specifier indicate?

(a) %e

(b) %f

(c) %g

(d) %lf

28. What does the following printf() format specifiers indicat?

(a) %n

(b) %X

(c) %p

(d) %o

(e) %i

(f) %hx

29. What is the justification, decimal fraction and width of the format specifier %-9.2f?

(a) Right justify, 2, 9

(b) Left justify, 2, 9

(c) Right justify, 9, 2

(d) Left justify, 9, 2

30. What is the output of the C statement?

printf("%d", printf("ProgramminG"));

(a) 11ProgramminG

(b) ProgramminG11

(c) Compile time error

(d) Warning

31. Which of the following file accessing modes are used in C?

(a) Text

(b) Binary

(c) Text & Binary

(d) None of the above.

32. Which of the following specifies an infinite loop?

(a) for(;;);

(b) while(1);

(c) for(;i=4;);

(d) while(true);

(e) for(;!0;);(f)

while(a = 0);

33. How many times the following loop executes, when the value of i = 0 initially?

(a) for(;i < 5; i--);

(b) while(i);

(c) for(;i=4;i++);

(d) while(! i);

(e) do{ }while(i = 0);

(f) do{ }while(!i);

34. What is the following C statement used to read the one line input from the keyboard?

(a) scanf("%s",str)

(b) scanf("%[^\n]",str)

(c) gets(str)

(d) getchar()

35. Which is the simplest file structure?

(a) Sequential

(b) Indexed

(c) Random

(d) All of the above

36. What are the sizes of character, integer and float pointers.

(a) 1, 1, 1   (b) 2, 2, 2

(c) 4, 4, 4   (d) 1, 2, 4

37. Which one is not a preprocessor directive?

(a) #define   (b) #zigma

(c) #line   (d) #pragma

38. Which operator is used to concatenate (combine) two separate strings into one single string in a macro definition?

(a) #   (b) ##

(c) +   (d) strcat

39. Argument list in the scanf() is the example of

(a) Call by value   (b) Call by reference

(c) Call by pointer (d) Both (b) and (c)

40. Argument list in the printf() is the example of

(a) Call by value   (b) Call by reference

(c) Call by address (d) Both (b) and (c)

41. How many argument(s) are available in the main() function?

(a) 0   (b) 1

(c) 2   (d) 3

42. Which of the following is a valid main() function?

(a) main(int)   (b) main(int,char*)

(c) main(int,char*,char*) (d) main()

43. In switch-case statement, which accept the following as its input.

(a) Integer variable   (b) Float variable

(c) Character variable   (d) String variable

(e) Character constant   (f) Integer constant

44. What is the use of break statement in a switch-case control statement?

(a) Control pass to after switch block

(b) Control pass to next switch case statement

(c) Control pass to first switch case statement

(d) Control pass to default switch case statement

45. Which of the following character is not a terminating character in the string?

(a) NULL   (b) '\0'

(c) 0   (d) ';'

46. Evaluate the value of the following expressions.

(a) fabs(5)   (b) fabs(-5)

(c) fabs(5.5)   (d) fabs(-5.44)

(e) fabsl(5.5)   (f) fabsl(-5.44)

47. Evaluate the value of the following expressions.

(a) floor(5.4)   (b) ceil(5.4)

(c) floor(5.5)   (d) ceil(5.5)

(e) floor(5.6)   (f) ceil(5.6)

48. Evaluate the value of the following expressions, when x = 1.234, y = 5.678

(a) floor(x*10 + y*5)

(b) ceil(x*10 + y*5)

(c) floor(x+0.05) / 10

(d) ceil(x+0.05) / 10

(e) floor(x+y*0.5)*50

(f) ceil(x + y * 0.5) * 50

49. In the following tree structures, which is efficient interms of space and time complexities?

(a) Full Binary Tree

(b) Incomplete binary tree

(c) Complete binary tree

(d) Right skewed binary tree

50. Which of the following statements about switch-case statements is true?

(a) Two case labels can have the same value

(b) Condition in switch control expression must evaluate to an integer type

(c) It can have two default statements

(d) Break statement after every case is compulsary

51. Which of the following situations creates dangling problem in if-else statements?

(a) Any nested if statement

(b) Nested if statement without a true statement

(c) Nested if statement without a false statement

(d) Nested if statement without true or false statement

52. What is the value of the following expressions?

(a) tolower('A')   (b) toupper('A')

(c) tolower('g')   (d) toupper('g')

(e) tolower('5')   (f) toupper('5')

53. Which of the following statements about for loop is true?

(a) Initialization part is executed first

(b) Condition part is executed first

(c) If the condition is false, the loop is exited in the first check itself

(d) Multiple statements are allowed in a for loop

(e) Initialization part and increment / decrement part are compulsary in a for loop

54. Which of the following statements about while loop is true?

(a) Multiple statements are allowed in the loop

(b) It is a pre test loop

(c) If the condition is false, the loop is not exited in the first check itself

(d) While statement is terminated with a semicolon

(e) Initialization is not required in the loop

55. Which of the following is not a jump statement?

(a) break   (b) continue

(c) goto   (d) move

(e) jump   (f) return

56. Which of the following header file contains the function prototypes for functions used in file operations?

(a) stdio.h   (b) stdlib.h

(c) stdarg.h   (d) stddef.h

(e) file.h   (f) stdfile.h

57. Which of the following statement is correct, when a file is opened in r mode?

(a) Opens the file for reading, and sets the file pointer at the beginning of the file

(b) Opens the file for reading, and sets the file pointer at the end of the file

(c) Opens the file for reading, and clears all the text from the file

(d) Returns an error if the file doesn't exist

58. Which of the following statement is correct, when a file is opened in w mode?

(a) Opens the file for writing, and sets the file pointer at the beginning of the file

(b) Opens the file for writing, and sets the file pointer at the end of the file

(c) Opens the file for writing, and clears all the text from the file

(d) Returns an error if the file doesn't exist

59. If the file is not opened sucessfully, what happens in C?

(a) Returns an error if file doesn't exist

(b) Terminates the program

(c) Continues execution of next statement in the program

(d) Displays an error message

60. Find the odd man out in the following options.

(a) malloc()   (b) calloc()

(c) realloc()   (d) free()

61. Find the odd man out in the following options.

(a) getch()   (b) getche()

(c) getchar()   (d) kbhit()

62. Which function is used to convert string to double?

(a) atoi()   (b) atof()

(c) atol()   (d) strtod()

(e) fcvt()   (b) gcvt()

63. What is the purpose of functions exec..(), spawn..() and system() in C?

(a) Enable the program into run the other file

(b) Enable program into run the other child process

(c) Enable the program into include other files

(d) All of the above

64. When an array is passed as an argument to a function, C uses _____.

(a) Call by value   (b) Call by reference

(c) Call by pointer   (d) Call by address

65. What is the equivalent of the statement a[r]?

(a) r[a]   (b) *(a+r)

(c) *(&a[r])   (d) a+r

66. Which of the following statements are true regarding pointers?

(a) String constants can be assigned to character pointer

(b) String constatnts can be assigned to character array

(c) Possible to cast a pointer to int as a pointer to float

(d) Implicit cast is done, when pointers of different types are assigned

67. What is the equivalent of the statement a[r][c]?

(a) *(*(a+r)+c)   (b) (*(a+r)+c)

(c) *(&a[r][0]+c)   (d) *(a[r]+c)

68. Which of the following statements are true regarding array?

(a) Assign one array variable to another

(b) Array contains different type of elements

(c) Array index starts from -1

(d) Array index may be a character constant

69. Which of the following statement defines a pointer variable to a float or one dimensional float array?

(a) float *ptr   (b) float &ptr

(c) float **ptr   (d) float &&ptr

70. What is the incremented value for the following options when char *p, int *q, float *r, double *s?

(a) p++   (b) q++

(c) r++   (d) s++

71. What is the size of the following pointer variables?

(a) char *ptr   (b) char near *ptr

(c) char far *ptr   (d) char huge *ptr

72. The declaration float **p defines p is,

(a) pointer of pointer to float variable

(b) pointer of float variable

(c) pointer to address of float variable

(d) two dimesional pointer

73. Sorting is not possible by using which of the following methods?

(a) insertion   (b) selection

(c) exchange   (d) deletion

74. How many different trees are possible with 10 nodes ?

(a) 1004   (b) 1014

(c) 1024   (d) 1034

75. In construction of a tree which is the suitable and efficient data structure?

(a) array   (b) linked list

(c) stack   (d) queue

ANSWERS --- I

(1) (c)	(2) (a)	(3) (d)	(4) (d)	(5) (a)	(6) (d)	(7) (c)
(8) (c)	(9) (c)	(10) (c)	(11) (b)	(12) (d)	(13) (a)	(14) (b)
(15) (c)	(16) (a)	(17) (d)	(18) (d)	(19) (a)	(20) (b)	(21) (d)
(22) (a)	(23) (a)	(24) (b)	(25) (b)	(26) (b)	(27) (d)	(28) (a)
(29) (a)	(30) (b)	(31) (c)	(32) (c)	(33) (b)	(34) (d)	(35) (a)
(36) (c)	(37) (c)	(38) (b)	(39) (b)	(40) (b)	(41) (d)	(42) (d)
(43) (a)	(44) (a)	(45) (c)	(46) (a)	(47) (c)	(48) (b)	(49) (a)
(50) (d)	(51) (d)	(52) (c)	(53) (a)	(54) (d)	(55) (d)	(56) (a)
(57) (d)	(58) (d)	(59) (c)	(60) (d)	(61) (c)	(62) (b)	(63) (a)
(64) (b)	(65) (b)	(66) (c)	(67) (b)	(68) (a)	(69) (c)	(70) (c)
(71) (c)	(72) (b)	(73) (b)	(74) (b)	(75) (a)	(76) (b)	(77) (a)
(78) (c)	(79) (c)	(80) (b)	(81) (a)	(82) (a)	(83) (a)	(84) (b)
(85) (b)	(86) (d)	(87) (a)	(88) (c)	(89) (b)	(90) (c)	(91) (b)
(92) (b)	(93) (b)	(94) (d)	(95) (a)	(96) (a)	(97) (b)	(98) (c)
(99) (b)	(100) (b)	(101) (d)	(102) (b)	(103) (d)	(104) (a)	(105) (d)
(106) (d)	(107) (b)	(108) (b)	(109) (c)	(110) (d)	(111) (a)	(112) (a)
(113) (a)	(114) (a)	(115) (c)	(116) (d)	(117) (c)	(118) (b)	(119) (c)
(120) (c)	(121) (c)	(122) (a)	(123) (d)	(124) (d)	(125) (b)	(126) (c)
(127) (a)	(128) (a)	(129) (a)	(130) (b)	(131) (b)	(132) (b)	(133) (d)
(134) (a)	(135) (b)	(136) (b)	(137) (b)	(138) (a)	(139) (b)	(140) (a)
(141) (a)	(142) (d)	(143) (d)	(144) (c)	(145) (b)	(146) (a)	(147) (a)
(148) (c)	(149) (a)	(150) (a)	(151) (d)	(152) (a)	(153) (b)	(154) (b)
(155) (d)	(156) (b)	(157) (a)	(158) (c)	(159) (a)	(160) (d)	(161) (b)
(162) (d)	(163) (a)	(164) (b)	(165) (c)	(166) (a)	(167) (b)	(168) (a)
(169) (b)	(170) (a)	(171) (b)	(172) (a)	(173) (b)	(174) (c)	(175) (d)
(176) (b)	(177) (a)	(178) (a)	(179) (d)	(180) (a)	(181) (b)	(182) (a)
(183) (d)	(184) (d)	(185) (a)	(186) (c)	(187) (b)	(188) (a)	(189) (a)
(190) (c)	(191) (c)	(192) (a)	(193) (d)	(194) (d)	(195) (c)	(196) (a)
(197) (b)	(198) (c)	(199) (d)	(200) (c)	(201) (b)	(202) (b)	(203) (c)
(204) (a)	(205) (b)	(206) (d)	(207) (a)	(208) (b)	(209) (d)	(210) (c)
(211) (c)	(212) (c)	(213) (c)	(214) (b)	(215) (c)	(216) (d)	(217) (d)
(218) (c)	(219) (a)	(220) (d)	(221) (b)	(222) (b)	(223) (c)	(224) (a)
(225) (b)						

ANSWERS --- I I

BASICS OF C PROGRAM

1. **ANSWER :**

   Compiler error : Cannot modify a constant value.

   EXPLANATION :

   The variable ptr is a pointer to a "constant integer". But we tried to change the value of the "constant integer", which is impossible, hence the error.

2. **ANSWER :**

   Linker Error : Undefined symbol i

   EXPLANATION :

   The extern storage class declaration (extern int i) specifies to the compiler that the memory for i is allocated in some other program and that address will be used in the current function at the time of linking. But since linker doesn't finds anyother variable of name i in the program with memory space allocated for it. Hence a linker error has occurred .

3. **ANSWER :**

   0 0 1 3 1

   EXPLANATION :

   Always keep in mind, that logical operations always gives a result of 1 or 0. Logical AND (&&) operator has higher priority over the logical OR (||) operator. So the expression ("i++ && j++ && k++") is executed first. The result of this expression is 0 (since, -1 && -1 && 0). Now the next expression (i.e., 0 || 2) is evaluated. whose result evaluates to 1 (because OR operator always gives 1 except for '0 || 0' combination). So the value of m is 1. The values of other variables are also incremented by 1, hence you get the above output.

4. **ANSWER :**

   fff0

   EXPLANATION :

   The constant -1 is internally represented as all 1's. When left shifted four times (since << 4) the least significant 4 bits are filled with 0's.The %x format specifier specifies that the integer value be printed as a hexadecimal value.

5. **ANSWER :**

   a = 2;

   EXPLANATION :

   During initialization of a, unary minus (or negation) operator is used twice. As per maths, minus * minus= plus, same rule applies in programming also hence you get the above output.. Also note that you cannot give --2, since -- operator can only be applied to variables for decrement operation as --a).

6. **ANSWER :**

   sizeof(i) = 1

   EXPLANATION :

   Since #define replaces the string int by the macro definition char.

7. **ANSWER :**

   i = 0

   EXPLANATION :

   In the expression !i>5 , NOT (!) operator has more precedence than ' >' symbol. ! is a unary logical operator.

   !i (!10) is 0 (not of true is false). 0 > 5 is false (zero).

8. **ANSWER :**

   hai

   EXPLANATION :

   \n - newline character

   \b - backspace character

   \r - linefeed character.

   The first printf() function prints the string, "ab" in a new line. The second printf() function initially moves backspace by one character (i.e., \b) and removes the character b, and then prints the string, "si" in the same line. After the execution of the second line, the output is "asi". The third printf() function initially moves to the first position of the same line (i.e., \r),and then prints the string, "ha" in the same line. After the execution of the third line, the output is "hai".

9. **ANSWER :**

   4 4 6 6 5

   EXPLANATION :

   The arguments in a printf() function call are pushed into the stack from left to right. The evaluation is by popping out from the stack and the evaluation takes place from right to left, hence you get the above output..

10. **ANSWER :**

    64

    EXPLANATION :

    The macro definition called square(4) will substituted by 4 * 4 so the expression becomes i = 64 / 4 * 4. Since / and * has equal priority the expression will be evaluated as (64/4)*4 (i.e., 16 * 4 = 64).

11. **ANSWER :**

    50

    EXPLANATION :

    The preprocessor directives can be redefined anywhere in the program and hence the most recently assigned value will be taken.

12. **ANSWER :**

    100

    EXPLANATION :

    Preprocessor executes as a seperate pass before the execution of the compiler. So textual replacement of clrscr() to 100 occurs.The input program to compiler looks like this,

    ```
    void main()
      {
        100;
        printf("%d\n", 100);
      }
    ```

    Note that 100; is a perfectly executable statement but with no action, hence it doesn't rise any error.

13. **ANSWER :**

    Some address will be printed.

    EXPLANATION :

    Function names are just addresses (similar as array names are addresses). We know that main() is also a function. So the address of main() function will be printed. The format specifier %p in printf() function specifies that the argument is an address, which is printed as hexadecimal numbers.

14. **ANSWER :**

No output / error.

EXPLANATION :

The first declaration of clrscr() occurs inside the main() function, so it becomes a function call. In the second usage, clrscr(); is a function declaration (because it is not inside any function).

15. **ANSWER :**

0 - 1 - 2

EXPLANATION :

The enum data type assigns numbers to its enumeration list starting from 0, if not explicitly defined, hence you get the above output.

16. **ANSWER :**

4 - 2 - 1

EXPLANATION :

The first output (i.e., 4) represents the size of the character far pointer. The second output (i.e., 2) represents the size of the character pointer. The third output (i.e., 1) represents the size of the character.

17. **ANSWER :**

4 - 2 - 4

EXPLANATION :

The first output (i.e., 4) represents the size of the character far pointer. The second output (i.e., 2) represents the size of the character near pointer. The third output (i.e., 4) represents the size of the character huge pointer.

18. **ANSWER :**

200 - 100

EXPLANATION :

The printf() function takes the values of the first two assignments of the program in reverse order. Any number of printf's may be given. All of them uses only the first two assignment values. If more number of assignments are given in the program, then the printf() function will take garbage values.

19. **ANSWER :**

J

EXPLANATION :

* is a dereferencing operator and & is a reference operator. They can be used any number of times with a variable provided it is meaningful. Here ptr points to the first character in the string "John". *ptr dereferences it and so its value is 'J'. Again & references it to an address and * dereferences it to the value 'J'.

20. **ANSWER :**

12

EXPLANATION :

The input statement to the compiler looks like this (5 + 7), and hence you get the above output.

21. **ANSWER :**

Compiler Error.

EXPLANATION :

The expression i+++++i is parsed as i ++ ++ + i, which is an illegal combination of operators.

22. **ANSWER :**

1

EXPLANATION :

The scanf() function returns number of input items successfully read. Here 10 is given as input which should have been scanned successfully. So number of items read is 1, hence you get the above output.

23. **ANSWER :**

100

EXPLANATION :

The ## used in the macro definition, is a concatenation operator, used to concatenate two seperate strings into one string. In the program, we concatenate the variable name 'c' and 'd, as 'cd'. The value for the variable cd is 100, hence you get the above output.

24. **ANSWER :**

Linker error: undefined symbol 'i'.

EXPLANATION :

The variable i declared using the keyword extern specifies global declaration that the variable i is defined somewhere else. The compiler passes the external variable to be resolved by the linker, so compiler doesn't find an error. During linking, the linker searches for the definition of i. Since it is not found the linker raises an error.

25. **ANSWER :**

Compiler error: undefined symbol out in function main.

EXPLANATION :

The rule is that a variable is available for use, only from the point of declaration, hence you get an error. Even when out is a global variable, it is not available in the main() function. Hence an error.

26. **ANSWER :**

100

EXPLANATION :

This is the correct way of writing the previous program.

27. **ANSWER :**

i = 1, i = -1

EXPLANATION :

The expression -(-1) evaluates to 1, and hence you get the above output.

28. **ANSWER :**

First it checks for the leading white space and discards it.Then it matches with a quotation mark and then it reads all character upto another quotation mark.

29. **ANSWER :**

Runtime error : Stack overflow.

EXPLANATION :

The main() function calls itself again and again recursively. Each time the function is called its return address is stored in the call stack. Since there is no condition to terminate the function call, the call stack overflows at runtime and it terminates the program resulting in an run time error.

30. **ANSWER :**

    0

    EXPLANATION :

    The character ! is used to denote a logical NOT operator. In C, the value 0 is considered to be the boolean value for FALSE, and any non-zero value is considered to be the boolean value for TRUE. Here 2 is a non-zero value so TRUE. !TRUE is FALSE so it prints 0.

31. **ANSWER :**

    1==1 is TRUE

    EXPLANATION :

    When two strings are placed together or separated by whitespace they are concatenated (this is called as "stringization" operation). So the string is as if it is given as "%d==1 is %s". Thus conditional operator(?:) evaluates to "TRUE".

32. **ANSWER :**

    0 John

    EXPLANATION :

    The typedef is used for declaring new types, hence typedef assigns 'data' as character array of size [MAX], and hence you get the above output.

33. **ANSWER :**

    Compiler Error.

    EXPLANATION :

    The #define preprocessor is used for textual replacement, hence the variable data is declared in the preprocessor directive and not in the function, hence you get errors.

34. **ANSWER :**

    30 20 10

    EXPLANATION :

    Each open and closed flower braces in the program introduces new block and hence a new scope. In the innermost block i is declared as, const volatile unsigned which is a valid declaration. i is assumed to be of type int and hence the printf() function prints the value of the variable 30. In the next block, i has value 20 and hence the printf() function prints the value of the variable 20. In the outermost block, i is declared as extern, so no storage space is allocated for it. After compilation is over the linker resolves it as global variable i (since it is the only variable visible there)and hence the printf() function prints the value of the variable 10.

35. **ANSWER :**

    i = -1, +i = -1

    EXPLANATION :

    Unary + is the only dummy operator in C, whenever it comes you can just ignore it just because it has no effect in the unary expressions.

36. **ANSWER :**

    i = -1, -i = 1

    EXPLANATION :

    -i is executed and this execution doesn't affect the value of i. In printf first you just print the value of i. After that the value of the expression -i = -(-1) is printed.

37. **ANSWER :**

    Compiler Error.

    EXPLANATION :

    The variable i is declared as a constant ane hence you cannot change the value of constant variable i.

38. ANSWER :

John 4

EXPLANATION :

If you declare i as a register variable, the compiler will treat it as ordinary integer and it will take as integer value. The variable i may be stored either in register or in memory.

39. ANSWER :

11

EXPLANATION :

The expression i+++j is evaluated as (i++ + j), and hence you get the above output.

40. ANSWER :

The length of the variable a is 6

EXPLANATION :

The char array 'a' will hold the initialized string, whose length will be counted from 0 till the null character. Hence the variable 'i' will hold the value equal to 5, after the pre-increment operation, the value of i is incremented to 6, which is printed as output.

41. ANSWER :

0 65535

EXPLANATION :

The value -1 initialized to the unsigned variable 'b' is represented in 2's complement as all 1's in 16 bit format, which evaluates to 65535. The variable 'b' when incremented by 1, results in all 0's in 16 bit format, which evaluates to 0. The variable 'b' when decremented by 1, results in all 1's again in 16 bit format, which evaluates to 65535, hence you get the above output.

42. ANSWER :

Garbage values.

EXPLANATION :

An identifier is available to use in program code from the point of its declaration,,hence expressions such as i = i++ are valid statements. The variables i, j and k are automatic or local variables and hence they contain some garbage value, which will be displayed as output.

43. ANSWER :

i = 1 j = 1 k = 1

EXPLANATION :

Since static variables are initialized to zero by default, you get the above output.

44. ANSWER :

10

EXPLANATION :

The macro expands and evaluates as (x + 2 * y - 1) = (x + (2 * y) - 1) = 10

45. ANSWER :

Compiler Error : Cannot apply right shift to float.

EXPLANATION :

Bitwise operators cannot be applied on float values, hence you get the above error.

46. **ANSWER :**

Compiler Error: Lvalue required.

EXPLANATION :

Enumeration constants cannot be modified, hence you cannot apply ++ operation with it.

47. **ANSWER :**

Compiler Error : Cannot apply left shift to float.

EXPLANATION :

Bitwise operators cannot be applied on float values, hence you get the above error.

48. **ANSWER :**

Compiler Error : Cannot apply mod to float.

EXPLANATION :

Modulus operators cannot be applied on float values, hence you get the above error.

49. **ANSWER :**

1.000000

EXPLANATION :

The fmod() function is to find the modulus values for floats similar to modulus (%) operator used for integers.

50. **ANSWER :**

Compiler error: Multiple declaration for error.

EXPLANATION :

The name error is used in the two meanings. One means that it is a enumerator constant with value 1. The another use is that it is a type name (due to typedef) for enum errorType. Given a situation the compiler cannot distinguish the meaning of error to know in what sense the error is used, hence you get the above error.

51. **ANSWER :**

1

EXPLANATION :

The three usages of the identifier error can be distinguishable by the compiler at any instance, hence it is perfectly valid since they are in different namespaces. Note that this code is given here to just explain the concept behind. In real programming don't use such overloading of names. It reduces the readability of the code.

52. **ANSWER :**

0 0

EXPLANATION :

This is a simple example for conditional compilation. The macro named 'something' is defined, hence the conditonal compilation of ifdef evaluates to true and declares the global variable 'some' and initialize the variable with a value of zero, which gets printed in the output.

53. **ANSWER :**

Compiler error : undefined symbol some

EXPLANATION :

The macro named 'something' is not defined, hence the conditonal compilation of ifdef evaluates to false, hence the declaration, int some = 0; is effectively removed from the source code and hence you get the following output.

54. ANSWER :

0 0

EXPLANATION :

This code is to show that preprocessor expressions are not the same as the ordinary expressions. If a name is not known the preprocessor treats it to be equal to zero, hence you get the above output.

55. ANSWER :

1

EXPLANATION :

The variable arr is made up of a 2 single arrays that contains 2 integers each. The variable 'arr' refers to the beginning of all the 2 arrays. *arr refers to the start of the first 1D array (of 2 integers) that is the same address as 'arr'. So the expression (arr == *arr) evaluates to true. Similarly, *arr is nothing but *(arr + 0), adding a zero doesn't change the value/meaning and again arr[0] is the another way of telling *(arr + 0). So the expression (*(arr + 0) == arr[0]) is true (1) evaluates to true. Since both parts of the expression evaluates to true the result is true and you get the above output.

56. ANSWER :

Compiler Error: Lvalue required in function main

EXPLANATION :

++i yields an rvalue. For evaluating the postfix expression ++, an lvalue is required.

57. ANSWER :

1

EXPLANATION :

The expression can be treated as i = (++i==6), because == is of higher precedence than = operator. In the inner expression, ++i is equal to 6 yielding true(1), hence you get the above output.

58. ANSWER :

A

EXPLANATION :

Initially the string text is initialized to "%d\n". Due to the assignment text[1] = 'c' the string becomes, "%c\n". Since this string becomes the format string for the printf() function and ASCII value of 65 is 'A' you get the above output.

59. ANSWER :

1 10

EXPLANATION :

The expression is evaluated as i = (i &= (j && 10)); The inner expression (j && 10) evaluates to 1 because j==10. i is 5. i = 5 & 1 is 1, hence you get the above output.

60. ANSWER :

5 10

EXPLANATION :

The expression is evaluated as i = (i |= (j || 10)); The inner expression (j || 10) evaluates to 1 because j==10. i is 5. i = 5 | 1 is 5, hence you get the above output.

61. ANSWER :

20 20

EXPLANATION :

The expression (j = i = j + 10;) is an example of a simple assignement statement, which evaluates as j = i = 20, hence you get the following output.

62. ANSWER :

1 1

EXPLANATION :

The expression (j = i = j > 10;) evaluates as j > i is 1, hence you get the following output.

63. **ANSWER :**

4 1

EXPLANATION :

The boolean expression needs to be evaluated only till the truth value of the expression is not known. j is not equal to zero, means that the expression's truth value is 1. Because it is followed by || and true || (anything) evaluates to true, hence the remaining expression is not evaluated and so the value of i remains the same. Similarly when && operator is involved in an expression, when any of the operands become false, the whole expression's truth value becomes false and hence the remaining expression will not be evaluated, since false && (anything) evaluates to false.

64. **ANSWER :**

Compiler Error: '&' on register variable

EXPLANATION :

The address of (&) operator cannot be applied on register variables.

65. **ANSWER :**

Linker Error : Unresolved external symbol i

EXPLANATION :

The identifier i is available in the inner block and so using extern has no use in resolving it.

66. **ANSWER :**

10

EXPLANATION :

The size of integer array of 10 elements is 10 * sizeof(int). The macro expands to sizeof(arr)/sizeof(int) = 10 * sizeof(int) / sizeof(int) = 10.

67. **ANSWER :**

sizeof (void *) = 2
sizeof (int *) = 2
sizeof (double *) = 2
sizeof(struct unknown *) = 2

EXPLANATION :

The pointer to any data type are of same size.

68. **ANSWER :**

Compiler Error : Too many initializers.

EXPLANATION :

The array a is of size 4 but the string constant requires 6 bytes (including the terminating NULL character) to get stored.

69. **ANSWER :**

HELL%@!~@!@???@~~!

EXPLANATION :

The character array has the memory just enough to hold the string "HELL" and doesnt have enough space to store the terminating null character. So it prints the HELL correctly and continues to print garbage values till it accidentally comes across a NULL character. Note that you may get any combination of characters, after the string "HELL" each time you evaluate.

70. **ANSWER :**

30   20   0

EXPLANATION :

The first printf() function represents the value of const volatile unsigned variable 'i'. Second printf() function represents the value of local variable 'i'. Third printf() function represents the value of extern variable 'i' which is declared outside the main function.

IF STATEMENTS

1. **ANSWER :**

   INDIA may win match

   EXPLANATION :

   Floating point variables declared using float, double and long double values cannot be predicted exactly. Depending on the number of bytes, the precision with of the value represented varies from compiler to complier. Float takes 4 bytes and long double takes 10 bytes. So float stores the value 123.4 with less precision than the long double value. ***Note that never compare using floating point numbers with relational operators.***

2. **ANSWER :**

   5 4 3 2 1

   EXPLANATION :

   The static variable 'a' is initialized only once, and any change in the value of the static variable 'a' is retained between function calls. The main() function is treated as an ordinary function, which can be called recursively, hence you get the above output.

3. **ANSWER :**

   0 0 0 0

   EXPLANATION :

   The static variable 'i' is initialized only once, and any change in the value of the static variable 'a' is retained between function calls. The main() function is treated as an ordinary function, which is called recursively, unless I becomes equal to 0 and hence you get the above output.

4. **ANSWER :**

   Three

   EXPLANATION :

   A simple C program using switch case statement. The default case can be placed anywhere inside the loop. It is executed only when all other cases doesn't match, hence you get the above output.

5. **ANSWER :**

   Compiler Error: Constant expression required.

   EXPLANATION :

   The case statement can have only constants or constant expressions (this implies that we cannot use variable names directly hence the above error).

6. **ANSWER :**

   TRUE

   EXPLANATION :

   The input program to the compiler after processing by the preprocessor is,

   ```
   void main()
   {
      if(0)
        puts("NULL");
      else if(-1)
        puts("TRUE");
      else
        puts("FALSE");
   }
   ```

   The preprocessor replaces the values given only in the conditional statements and doesn't replace the values given inside the double quotes in the puts() function. The if condition (NULL) evaluates to a boolean value zero, which is indicated as false, hence the condition will never become true, so it goes to else if condition. In else if condition evaluates to a boolean value one, hence the condition will always be true, and exits without executing the else conditon, hence you get the above output.

7. **ANSWER :**

1000 is a leap year

EXPLANATION :

A simple C program to check the given year is leap year or not.

8. **ANSWER :**

0 - 0

EXPLANATION :

Logical operators employ a technique referred as lazy evaluation, for their operators. These operators evaluate their operand first, and then evaluate the right hand operator only if it is required. Clearly false && any operand is always false, and true || is always true. In such cases, the second and subsequent operand are not evaluated Hence in the above case, the value of i is 0, and hence the printf statement following the if statement is not executed. The values of i and j remain unchanged and gets printed.

9. **ANSWER :**

Programming

EXPLANATION :

The comma operator has the associativity from left to right. Only the rightmost value is returned and the other values are ignored. Thus the value of last variable y is returned to check in the if statement. Since it is a non zero value, the if condition becomes true and hence the printf statement following the if statement is executed and, hence you get the above output.

10. **ANSWER :**

Good Morning

EXPLANATION :

The printf() function returns the number of characters it prints. Hence printing a null character returns 1 (note that null character is considered as a single character) which makes the if statement evaluated to true, hence you get the above output.

11. **ANSWER :**

0 - 100 - 9874

EXPLANATION :

The value of (b % 2) is 0, is assigned to a. The if condition reduces to if(a) or in other words if(0) and the statement(s) inside the flower braces will not be executed, hence the value of c will not be initialized. The value of c which is printed, is a garbage value. You will get different values for the same program, if you execute it in different systems and different times.

12. **ANSWER :**

Compiler Error: switch expression not integral

EXPLANATION :

Switch statements can be applied only to integral types.

LOOPING STATEMENTS

1. ANSWER :

mmmm
aaaa
nnnn

EXPLANATION :

str[i], i[str], *(i+str), *(str+i) are all different ways of expressing the same location in an array. Generally array name is the base address for that array. Here str is the base address for the array and i is the index number from thebase address. So, indirecting it with * is same as str[i]. i[str] may be surprising. But in the case of C it is same as str[i].

2. ANSWER :

2 2 2 2 2 2 3 4 5 6

EXPLANATION :

Initially pointer arr is assigned to both ptr1 and ptr2. In the first loop, since only ptr2 is incremented and not arr, the value 2 will be printed 5 times. In second loop ptr itself is incremented, hence you get the above output.

3. ANSWER :

J!Mpwf!Zpv

EXPLANATION :

++*str++ will be parse in the following order.

- *str is the value at each location currently pointed by str will be considered.
- ++*str the retrieved value will be incremented.
- When ; is encountered the location will be incremented that is str++ will be executed.

Hence you get the above output.

4. ANSWER :

Compiler error: Undefined label 'INDIA'

EXPLANATION :

The scope of the labels is limited to functions. The label 'INDIA' is available only in function func() Hence it is not visible in main() function.

5. ANSWER :

Compiler error: Lvalue required

EXPLANATION :

Array names are pointer constants. So it cannot be modified.

6. ANSWER :

1

EXPLANATION :

Before entering into the for loop the condition is evaluated, which evaluates to 0 (i.e., false) and exits from the loop (Note the semicolon at the end of for loop), and a is incremented by 1, hence you get the above output.

7. ANSWER :

No Output.

EXPLANATION :

Inside the main() function i is an unsigned integer. It is compared with a signed value (i.e., -3). Since both data types doesn't match, the condition becomes false and control exits from loop, hence the printf() function will not be executed and you will not get any output from the program.

8. **ANSWER :**

01

EXPLANATION :

The inner printf executes first to print some garbage value. The printf returns no of characters printed and this value also cannot be predicted. Still the outer printf prints something and so returns a non-zero value. So it encounters the break statement and comes out of the while statement.

9. **ANSWER :**

5 4 3 2 1 0 65535 65534 65533 65534 . . .

EXPLANATION :

Since i is an unsigned integer it can never become negative. So the expression (i-- >= 0) will always be true, hence an infinite loop is executed and hence you get the above output.

10. **ANSWER :**

1

EXPLANATION :

Note the semicolon after the while statement. When the value of i becomes 0 it exits from the while loop. The value 1 is printed due to the post-increment operation on i, hence the value of i, increments by 1, and hence you get the above output.

11. **ANSWER :**

-1

EXPLANATION :

Unary + is the only a dummy operator in C. So it has no effect on the expression and now the while loop is, while(i-- != 0) which is false and so exits from the while loop. The value −1 is printed due to the post-decrement operation on i, hence the value of i, decrements by 1, and hence you get the above output.

12. **ANSWER :**

-128

EXPLANATION :

Note the semicolon after the for statement. The initial value of i is set to 0. The inner loop executes to increment the value from 0 to 127 (the positive maximum positive range of char) and then it rotates to the negative minimum value of -128. The condition in the for loop fails and so exits from the for loop. It prints the current value of i that is -128.

13. **ANSWER :**

Infinite loop.

EXPLANATION :

The difference between the previous question and this one is that the char is declared to be unsigned. So the i++ can never yield negative value and hence the condition (i >= 0) never becomes false so it forms an infinite for loop.

14. **ANSWER :**

Behavior is implementation dependent.

EXPLANATION :

The char is signed/unsigned by default is implementation dependent. If the implementation treats the char to be signed by default the program will print −128 and terminate. On the other hand if it considers char to be unsigned by default, it goes to infinite loop.

15. **ANSWER :**

No output.

EXPLANATION :

Ending the string constant with \0 explicitly makes no difference. So "John" and "John\0" are equivalent. So, the strcmp() function returns 0 (false) and hence the printf() function will not be executed and you will not get any output from the program.

16. **ANSWER :**

No output.

EXPLANATION :

Initializing the string constant with \0 explicitly makes no difference. So "John" and "John\0" are equivalent. So, the strcmp() function returns 0 (false) and hence the printf() function will not be executed and you will not get any output from the program.

17. **ANSWER :**

Compiler Error: Lvalue required.

EXPLANATION :

As we know that increment operators return rvalues and hence it cannot appear on the left hand side of an assignment operation.

18. **ANSWER :**

32767

EXPLANATION :

Since i is a static variable it is initialized to 0 by default, hence the condition in the while loop evaluates to true. Inside the while loop, the conditional operator evaluates to false, executing i--. This continues till the integer value rotates to positive value (32767). The while condition becomes false and hence, comes out of the while loop, printing the current value of i.

19. **ANSWER :**

Implementaion dependent

EXPLANATION :

The char data type may be signed or unsigned by default. If it is signed then ch++ is executed. When ch reaches 127 it rotates back to -128. Thus char is always smaller than 127.

20. **ANSWER :**

Infintely prints the string "Testing in C".

EXPLANATION :

Here the declaration int l1 = 1 represents l1 is a integer variable. But l1: represents label value which is used by the goto statement later. Both the things are different in C. The statement goto l1 tranfer the control flow again into label statement without any condition. So the program is infinitely prints the string "testing in C".

FUNCTIONS

1. **ANSWER :**

   Compiler Error : Type mismatch in redeclaration of function display.

   EXPLANATION :

   When the function call statement (i.e., display(str)) is encountered, the compiler doesn't know anything about the display() function. It assumes the arguments and return type of the display() function to be of type integers, (which is the default type). When it sees the actual function display(), the arguments and type contradicts with what it has assumed previously, hence you get the above error.

   The solutions to avoid such errors are as follows,

   - Use function declaration statements for all functions used in the program, above the main() function.
   - Use function declaration statement atleast before its use in the function.

2. **ANSWER :**

   Compiler Error: Type mismatch in redeclaration of function show.

   EXPLANATION :

   When the function call statement (i.e., show(str)) is encountered, the compiler doesn't know anything about the show() function. It assumes the arguments and return type of the show() function to be of type integers, (which is the default type). When it sees the actual function show(), the arguments and type contradicts with what it has assumed previously, hence you get the above error.

   The solutions to avoid such errors are as follows,

   - Declare void show() in main() function.
   - Define void show() before main() function.

3. **ANSWER :**

   9

   EXPLANATION :

   The return statement (i++) will first return i and then increments it. The returned value (i.e., 10) is first decremented and then printed in the printf() function.

4. **ANSWER :**

   Compiler Error.

   EXPLANATION :

   Declaration of the function convert() in the variable declaration statement is wrong.

5. **ANSWER :**

   Compiler Error. Type Mismatch in call to fucntion sum().

   EXPLANATION :

   The arguments argv[1] & argv[2] are strings. They are passed to the function sum without converting it to integer values.

6. **ANSWER :**

   Bye

   EXPLANATION :

   The varaible ptr is an array of pointers to functions of return type int. ptr[0] is assigned to address of the function a(). Similarly ptr[1] and ptr[2] for the functions b() and c() respectively. ptr[2]() is in effect of writing c(), since ptr[2] points to the function c().

7. **ANSWER :**

1000

EXPLANATION :

Normally the return value from the function is through the information from the accumulator. Here _AX is the pseudo global variable denoting the accumulator. Hence, the value of the accumulator is set 1000 so the unction returns value 1000.

8. **ANSWER :**

i. ANSI C notation

ii. Kernighan & Ritche notation

9. **ANSWER :**

bc

EXPLANATION :

The base address is modified only in function and as a result the variable 'a' points to 'b' then after incrementing points to 'c', hence you get the above output.

10. **ANSWER :**

The value if process is 0 !

EXPLANATION :

The function 'process' has 3 parameters - 1, a pointer to another function 2 and 3, integers. When this function is invoked from main(), the following substitutions for formal parameters take place: func for pf, 3 for val1 and 6 for val2. This function returns the result of the operation performed by the function 'func'. The function func has two integer parameters. The formal parameters are substituted as 3 for a and 6 for b. since 3 is not equal to 6, a==b returns 0. therefore the function returns 0 which in turn is returned by the function 'process'.

11. **ANSWER :**

Value is 7

EXPLANATION :

In the function definition int func(int func), the function name and the argument name can be the same. The function func() is called in which the sizeof(float) (i.e., 4 is passed), after the first expression the value in func will be 6, as 'func' is an integer variable hence the value stored in 'func' will have implicit type conversion from float to int. The value of the variable 'func' is returned to main() is printed after preincrement operation, hence you get the above output.

12. **ANSWER :**

Compiler Error : Unknown type integer or integer cannot start a parameter declaration.

Compiler Error : Undeclared function write().

EXPLANATION :

The use of pascal keyword doesn't mean that pascal code can be used. It means that the function follows pascal argument passing mechanism in calling functions.

13. **ANSWER :**

5 6 7 8

5 6 7 8

EXPLANATION :

Pascal argument passing mechanism forces the arguments to be called from left to right, where as cdecl is the normal C argument passing mechanism where the arguments are passed from right to left.

14. **ANSWER :**

abc is a ptr to a function which takes 2 argument, an integer variable and a pointer to a funtion which returns void. The return type of the function abc() is of type void.

EXPLANATION :

Use clock-wise rule to find the result.

15. **ANSWER :**

The array dimension is 1

EXPLANATION :

Arrays cannot be passed to functions as arguments and only pointers can be passed, hence the argument which is equivalent to int *array (this is one of the very few places where [] and * usage are equivalent). The return statement evluates as, sizeof(int *)/ sizeof(int), which happens to be equal in this case.

POINTERS

1. **ANSWER :**

1 2

EXPLANATION :

The sizeof() operator gives the number of bytes of its operand. The ptr variable is a character pointer, which needs one byte for storing its value (a character), hence sizeof(*ptr) returns a value of 1. Since it needs two bytes to store the address of the character pointer sizeof(ptr) returns 2.

2. **ANSWER :**

d

EXPLANATION :

The statement (ptr = &s[3]), indicates that the variable ptr is pointing to character '\n'. The statement (str=s), indicates that the variable str is pointing to character 'a'. The expression ++*ptr, indicates that, ptr (pointing to '\n') is incremented by one. The ASCII value of '\n' is 10, which is then incremented to 11. The value of ++*ptr is 11. The expression ++*str, indicates that, str (pointing to 'a') is incremented by one. The ASCII value of '\a' is 97, which is then incremented to 98. The value of ++*str is 98. Now performing (11 + 98 - 9), we get 100, which is the ascii value of "d", hence we get the above output.

3. **ANSWER :**

SomeGarbageValue--1

EXPLANATION :

The variable declaration arr[2][2][2], in the first line declares only 2 two-dimensional arrays. But in the statement ptr = &arr[2][2][2], you are trying to access the 3rd two-dimensional array (which you are not declared) so it will take some garbage values. If you print *ptr, it will print some garbage value. The statement *qtr = ***a, assigns the starting address of the array 'arr' to an integer pointer 'qtr'. Now the variable 'qtr' is pointing to starting address of the array 'arr'. If you print *qtr, it will print first element of array 'arr'.

4. **ANSWER :**

100 - 100 - 100 - 1
114 - 106 - 104 - 2

EXPLANATION :

The given array is a 3-dimesional array. Let us assume the initial address of the array is 100, which means the subsequent array values are stored in contiguous memory locations. Thus, the first printf() function a, *a, **a prints the address of first element of the array, and the indirection pointer ***a gives the value of the first location of the array. The second printf() function a+1 increases in the third dimension and thus points to value at the address 114, *a+1 increments in second dimension and thus points to 106, **a+1 increments the first dimension and thus points to 104 and ***a+1 gets the value at the second location (since first location is incremented by 1) and points to the value in the second location (i.e., 2), hence you get the above output.

5. **ANSWER :**

   Compiler Error: lvalue required.

   EXPLANATION :

   Error is in statement a++. The operand must be an lvalue and may be of any of scalar type for any operator, but array name only when subscripted becomes an lvalue. Simply using an array name is unalterable lvalue.

6. **ANSWER :**

   1 1 1
   2 2 2
   3 3 3

   EXPLANATION :

   The given array is a 3-dimesional array. Let us assume the initial address of the array is 1000, which means the subsequent array values are stored in contiguous memory locations. The statement **ptr = p assigns the starting address of the array 'p' to the pointer variable 'ptr'. After execution of the instruction ptr++, the value in ptr becomes 1002, if scaling factor for integer is 2 bytes. The expression (ptr-p) indicates (value in ptr - the value in the starting location of array 'p'), (i.e., (1002 - 1000) / (scaling factor)) = 1. The expression (*ptr-a) indicates (value at address pointed by ptr - starting value of array 'a'), (i.e., (1002 - 1000) / (scaling factor)) = 1. The expression **ptr indicates the value stored in the location pointed by the pointer of ptr, (i.e., the value pointed by 1002) = 1. Hence the output of the first printf() function is 1 1 1. Similarly for all other printf() functions.

7. **ANSWER :**

   1

   EXPLANATION :

   The given array is a 2-dimesional array, which points to its first element and hence the output.

8. **ANSWER :**

   A20lo

   EXPLANATION :

   Since a void pointer (i.e., *vp) is used it can be type casted to any other type pointer. The statement (vp = &ch) stores the address of char 'ch' and the next statement prints the value stored in the pointer vp after type casting it to the proper data type pointer, whose output is 'A'. The statement (vp = &j) stores the address of integer 'j' and the next statement prints the value stored in the pointer vp after type casting it to the proper data type pointer, whose output is 20. The statement (vp = cp) stores the address of string 'cp' and the next statement prints the value stored in the pointer vp after type casting it to the proper data type pointer, whose output is "lo".

9. **ANSWER :**

   ck

   EXPLANATION :

   In this problem we have an array of character pointers 's' pointing to the start of 4 strings. Then we have 'ptr' which is a pointer to a pointer of type char and a variable 'p' which is a pointer to a pointer to a pointer of type char. The vaiable 'p' holds the initial value of ptr, (i.e., p = s+3). The next statement increments value in p by 1, hence it becomes s+2. In the printf() function the expression *++p is first evaluated, thus incrementing the value of p to s+1, then the predecrement operation is evaluated, and we get s+1 - 1 = s. The indirection operator now gets the value from the array of s and adds 3 to the starting address. Thus the string is printed from the third position, hence we get the output as 'ck'.

10. **ANSWER :**

   (blank line)
   ohn
   hn
   n

   EXPLANATION :

   Here a string (a pointer to char) is initialized with the value "John". The strlen() function returns the length of the string, thus n has a value 4. The next statement assigns value at the nth location (i.e., '\0') to the first location.

Now the string becomes "\0ohn". Now the printf statement prints the string after each iteration it increments it starting position. Loop starts from 0 to 4. The first time x[0] = "\0" hence it prints nothing and pointer value is incremented. The second time it prints from x[1] i.e., "ohn" and the third time it prints "hn" and the last time it prints "n" and the loop terminates.

11. **ANSWER :**

Compiler Error : size of v is Unknown.

EXPLANATION :

You can create a variable of type void * but not of type void, since void is an empty data type. In the second line you are creating variable vptr of type void* and v of type void hence an error.

12. **ANSWER :**

2 5 5

EXPLANATION :

The sizeof(str1) is a character pointer so it gives you the size of the pointer variable. The next sizeof(str2) indicates the name of the array whose size is 5 (which includes the '\0' NULL character). The last sizeof("abcd") is similar to the second one, hence you get the above output.

13. **ANSWER :**

10

EXPLANATION :

The variable i is a block level variable and the visibility is inside that block only. But the lifetime of j is lifetime of the function so it lives upto the exit of the main() function. Since i is has its allocated space, *j prints the value stored in i (since j points i), hence you get the above output.

14. **ANSWER :**

0001 - 0002 - 0004

EXPLANATION :

The ++ operator when applied to pointers increments address according to their corresponding data-types.

15. **ANSWER :**

Garbage value.

EXPLANATION :

The pointer ptr is pointing out of array range of 'arr'.

16. **ANSWER :**

300

EXPLANATION :

The pointer points to % since it is incremented twice and again decremented by 2, it points to "%d\n" and 300 is printed.

17. **ANSWER :**

Compiler Error. We cannot apply indirection on type void*.

EXPLANATION :

Void pointer is a generic pointer type. No pointer arithmetic can be done on it.

18. **ANSWER :**

4

EXPLANATION :

The expressions *a and -*a cancels out. The result is as simple as (1+3) = 4.

19. **ANSWER :**

b

EXPLANATION :

The expression (ptr++) assigns the first value to ++*(ptr++) expression, and then gets incremented, the expression ++* increments the value in the address stored pointed by the pointer.

20. **ANSWER :**

b

EXPLANATION :

There is no difference between the expression ++*(p++) and ++*p++. Note that parenthesis is a visual clue for the reader to see which expression is first evaluated.

21. **ANSWER :**

Some garbage value
0

EXPLANATION :

The memory space allocated by the malloc() function is uninitialized, whereas the calloc() function returns the allocated memory space initialized to zeros.

22. **ANSWER :**

16 16

EXPLANATION :

The pointer variables ptr1 and ptr2 both refer to the same memory location a. So changes through ptr1 and ptr2 ultimately affects only the value of a.

23. **ANSWER :**

2 1 4 5 4

EXPLANATION :

The expression sizeof(ptr) is similar to sizeof(char*), which is 2. The expression sizeof(*ptr) is similar to sizeof(char), which is 1. The expression sizeof(arr) is size of the character array, which is 1. When sizeof() operator is applied to an array it returns the sizeof the array and it is not the same as the sizeof the pointer variable, since the sizeof(arr) where arr is the character array and the size of the array is 5 because the space necessary for the terminating NULL character should also be taken into account.

24. **ANSWER :**

x = 20 y = 10

EXPLANATION :

This is one way of swapping two values, using pointers.

25. **ANSWER :**

x = 20 y = 10

EXPLANATION :

Just another way of swapping two values, using bitwise operators.

26. **ANSWER :**

1 1

EXPLANATION :

The integer value 257 is stored in the memory as, 00000001 00000001 (in 16 bits), so the individual bytes are taken by casting it to char* and gets printed.

27. **ANSWER :**

2 1

EXPLANATION :

The integer value 258 is stored in the memory as, 00000001 00000010 (in 16 bits), so the individual bytes are taken by casting it to char* and gets printed.

28. **ANSWER :**

556

EXPLANATION :

The integer value 300 in binary notation is, 00000001 00101100. It is stored in memory as, 00101100 00000001.

Result of the expression *++ptr = 2 makes the memory representation as, 00101100 00000010. So the integer corresponding to it is 00000010 00101100 is 556.

29. **ANSWER :**

0

EXPLANATION :

After 'ptr' reaches the end of the string the value pointed by 'str' is '\0'. So the value of 'str' is less than that of 'ch'. So the value of 'ch' is 0.

30. **Answer & EXPLANATION :**

If the body of the loop never executes ptr is assigned no address so ptr remains NULL. When *ptr = 0, may result in problem (which gives rise to runtime error or NULL pointer assignment and terminate the program).

31. **ANSWER :**

Undefined behavior.

EXPLANATION :

The second statement results in undefined behavior because it points to some location whose value may not be available for modification. This type of pointer in which the non-availability of the implementation of the referenced location is referenced as 'incomplete types'.

32. **ANSWER :**

5

EXPLANATION :

The output is simple, it divides the two integer pointer values, hence you get the above output.

33. **ANSWER :**

Yes.

EXPLANATION :

The pointer ptr will point at the integer in the memory location 0x400.

34. **ANSWER :**

String constant

EXPLANATION :

The program suffers no problem and gives the output correctly because the character constants are stored in code/data area and not allocated in the stack, so this doesn't lead to dangling pointers.

35. **ANSWER :**

Garbage values.

EXPLANATION :

Both the functions suffer from the problem of dangling pointers. In someFun1() temp is a character array and so the space for it is allocated in heap and is initialized with character string "string". This is created dynamically as the function is called, and is also deleted dynamically on exiting the function so the string data is not availabl in the calling main() function leading to print some garbage values. The function someFun2() also suffers from the same problem but the problem can be easily identified in this case.

STRUCTURES & UNIONS

1. **ANSWER :**

   Compiler Error.

   EXPLANATION :

   You cannot initialize variables in structure declaration.

2. **ANSWER :**

   Compiler Error.

   EXPLANATION :

   The structure Y is nested within structure X. Hence, the elements of structure Y are to be accessed through the instance of structure X, which needs an instance of Y to be known. If the instance is created after defining the structure the compiler will not know about the instance relative to X. Hence for nested structure Y a member have to be declared.

3. **ANSWER :**

   Compiler Error.

   EXPLANATION :

   Initialization should not be done for structure members inside the structure declaration.

4. **ANSWER :**

   Compiler Error.

   EXPLANATION :

   In the end of nested structure Y a member have to be declared.

5. **ANSWER :**

   2

   EXPLANATION :

   All the above statements form a circular doubly linked list. The statement, abc.next -> next -> prev -> next -> i; points to "ghi" node, the value at that particular node is 2.

6. **ANSWER :**

   Value of ptr is 0 - 0
   Value of ptr is 0 - 0

   EXPLANATION :

   The variable ptr is a pointer to structure variable. We can access the elements of the structure either with an arrow mark or with indirection operator. Since structure A_A is globally declared, its members x & y are initialized as zeroes by default.

7. **ANSWER :**

   Compiler Error: Undefined structure date

   EXPLANATION :

   Inside the struct definition of 'student' the member of type struct date is given. The compiler doesn't have the definition of date structure before it is used, hence you get the above error.

8. **ANSWER :**

   Compiler Error: Undefined structure date.

   EXPLANATION :

   Only declaration of struct date is available inside the structure definition of 'student' but to have a variable of type struct date the definition of the structure is required.

9. **EXPLANATION :**

The fread() function reads 10 records and prints the names successfully. It will return EOF only when the fread() function tries to read another record and fails while reading EOF. So it prints the last record again. After this only the condition feof(fp) becomes false, hence exits from the while loop.

10. **ANSWER :**

No.

EXPLANATION :

Is it not legal for a structure to contain a member of the same type as in this case. Because this will cause the structure declaration to be recursive without end.

11. **ANSWER :**

Yes.

EXPLANATION :

*b is a pointer to type struct a and so is legal. The compiler knows, the size of the pointer to a structure even before the size of the structure is determined (as you know the pointer to any type is of same size). This type of structures is known as self-referencing structure.

12. **ANSWER :**

No.

EXPLANATION :

The typename 'Type' is not known at the point of declaring the structure (i.e., forward references are not made for typedefs).

13. **ANSWER :**

Yes.

EXPLANATION :

The typename Type is known at the point of declaring the structure, because it is already typedefined.

14. **ANSWER :**

No.

EXPLANATION :

When the declaration, (typedef struct a Type;) is encountered body of struct a is not known, these are referred as incomplete types.

15. **ANSWER :**

10   20.5   10   20.5

EXPLANATION :

Structure variable assignments are equivalent to the simple data type variable assignments. So the values of v1.x and v1.y into the v2.x and v2.y.

ANSWERS --- III

1. (c) and (e)
2. (d), (e) and (f)
3. (a) Integer
   (b) Unsigned integer
   (c) Signed integer
   (d) Long integer
   (e) Unsigned short integer
   (f) Unsigned long integer
4. (a) Integer
   (b) Double
   (c) String
   (d) Character
   (e) Character
   (f) String
5. (a) and (d)
6. (a), (c) and (d)
7. (a) 6
   (b) 20
   (c) 8
   (d) -3
   (e) 99
   (f) 4
8. (d)
9. (a) -6
   (b) -3
   (c) 2
   (d) 3
   (e) 1
   (f) 5
10.(a) 1
   (b) 0
   (c) 1
   (d) 0
11.(a) a == 0
   (b) a <= b
   (c) (a == 0) < b
   (d) a == b == 0
12.(a) 0
   (b) 1
   (c) 0
   (d) 0
13.(a) 2

       (b) 2.5
       (c) 1
       (d) Compile Error: Illegal use of floating point
14. (a) -27536
     (b) 38000.000
     (c) -536
     (d) 65000.000
15. (a) 'S'
     (b) 'S'
     (c) 'S'
     (d) 'S'
16. (a), (b) and (d)
17. (a)
18. (b) and (d)
19. (a) size of operator
     (b) Arithmetic operator
     (c) Ternary operator
     (d) Assignment operator
20. (b), (c) and (d)
21. (a), (b) and (c)
22. (a) 5
     (b) 9
     (c) 2
     (d) 10
     (e) 0
     (f) 0
23. (a) 7
     (b) 15
     (c) 7
     (d) 24
24. (a), (b) and (c)
25. (a)
26. (a) Signed integer
     (b) Hexadecimal integer
     (c) Unsigned integer
     (d) Octal integer
     (e) Unsigned long integer
     (f) Signed long integer
27. (a) Floating point value in exponent form
     (b) Float type
     (c) Floating point value in exponent type or
         float type depending on value
     (d) Double type

28. (a) Pointer to integer
     (b) Unsigned Hexadecimal integer
         (With A, B, C, D, E, F)
     (c) Pointer
     (d) Unsigned octal integer
     (e) Signed decimal integer
     (f) Unsigned Hexadecimal integer
         (With a, b, c, d, e, f)
29. (b)
30. (b)
31. (c)
32. (a), (b), (c) and (e)
33. (a) 32769
     (b) 0
     (c) Infinite Loop
     (d) Infinite Loop
     (e) 1
     (f) Infinite Loop
34. (b) and (c)
35. (a)
36. (b)
37. (b)
38. (b)
39. (b)
40. (a)
41. (a), (c) and (d)
42. (b), (c) and (d)
43. (a), (c), (e) and (f)
44. (a)
45. (d)
46. (a) 5.0
     (b) 5.0
     (c) 5.5
     (d) 5.44
     (e) 0.0 (ANSI C does not support fabsl())
     (f) -0.0 (ANSI C does not support fabsl())
47. (a) 5
     (b) 6
     (c) 5
     (d) 6
     (e) 5
     (f) 6
48. (a) 40.0
     (b) 41
     (c) 0.1
     (d) 0.2
     (e) 200
     (f) 250.00

49. (a)
50. (b)
51. (a)
52. (a) a
     (b) A
     (c) g
     (d) G
     (e) 5
     (f) 5
53. (a), (c) and (d)
54. (a), (b) and (e)
55. (d) and (e)
56. (a)
57. (a) and (d)
58. (a), (c) and (d)
59. (a), (c) and (d)
60. (d)
61. (d)
62. (b) and (d)
63. (d)
64. (b)
65. (a), (b) and (c)
66. (a), (b) and (c)
67. (a), (c) and (d)
68. (d)
69. (a)
70. (a) 1
     (b) 2
     (c) 4
     (d) 8
71. (a) 2
     (b) 2
     (c) 4
     (d) 4
72. (a) and (d)
73. (d)
74. (c)
75. (b)

❖ ❖ ❖ ❖ ❖ ❖ ❖

I. ASCII TABLE

Table lists the ASCII characters and its ASCII codes in decimal. The first 32 ASCII characters and the last character are control characters. These characters are non-printable characters, which are enclosed in parentheses (*e.g.*, (nul)).

Character	ASCII code	Character	ASCII code	Character	ASCII code	Character	ASCII code
(nul)	0	(sp)	32	@	64	`	96
(soh)	1	!	33	A	65	a	97
(stx)	2	"	34	B	66	b	98
(etx)	3	#	35	C	67	c	99
(eot)	4	$	36	D	68	d	100
(enq)	5	%	37	E	69	e	101
(ack)	6	&	38	F	70	f	102
(bel)	7	'	39	G	71	g	103
(bs)	8	(	40	H	72	h	104
(ht)	9	)	41	I	73	i	105
(nl)	10	*	42	J	74	j	106
(vt)	11	+	43	K	75	k	107
(np)	12	,	44	L	76	l	108
(cr)	13	−	45	M	77	m	109
(so)	14	.	46	N	78	n	110
(si)	15	/	47	O	79	o	111
(dle)	16	0	48	P	80	p	112
(dc1)	17	1	49	Q	81	q	113
(dc2)	18	2	50	R	82	r	114
(dc3)	19	3	51	S	83	s	115
(dc4)	20	4	52	T	84	t	116
(nak)	21	5	53	U	85	u	117
(syn)	22	6	54	V	86	v	118
(etb)	23	7	55	W	87	w	119
(can)	24	8	56	X	88	x	120
(em)	25	9	57	Y	89	y	121
(sub)	26	:	58	Z	90	z	122
(esc)	27	;	59	[	91	{	123
(fs)	28	<	60	\	92			124
(gs)	29	=	61	]	93	}	125
(rs)	30	>	62	^	94	~	126
(us)	31	?	63	_	95	(del)	127

II. PRECEDENCE TABLE

Description	Operator	Associativity
Function expression	()	Left to right
Array expression	[]	Left to right
Structure operator	->	Left to right
Structure operator	,	Left to right
Unary minus	–	Right to left
Increment/Decrement	++ ––	Right to left
One's complement	~	Right to left
Negation	!	Right to left
Address of	&	Right to left
Value at address	*	Right to left
Type cast	(type)	Right to left
Size in bytes	sizeof	Right to left
Multiplication	*	Left to right
Division	/	Left to right
Addition	+	Left to right
Subtraction	–	Left to right
Left shift	<<	Left to right
Right shift	>>	Left to right
Less than	<	Left to right
Less than or equal to	<=	Left to right
Greater than	>	Left to right
Greater than or equal to	>=	Left to right
Equal to	==	Left to right
Not equal to	!=	Left to right
Bitwise AND	&	Left to right
Bitwise exclusive OR	^	Left to right
Bitwise inclusive OR	|	Left to right
Logical AND	&&	Left to right
Logical OR	||	Left to right
Conditional	?:	Right to left
Assignment	=	
	*= /= %=	Right to left
	+= –= &=	Right to left
	^= |=	Right to left
	<<= >>=	Right to left
Comma	,	Right to left

III. HEADER FILES

This appendix is a summary of the header files defined by the ANSI (American National Standards Institutions) standard. The header files contain information about various library functions that are functionally same. Header files may be included in any order in the program. A header file must be included outside any external declaration or definition and before the use of library functions present in it.

The library functions, constants, global variables, and macros of the standard library are declared in the following header files.

alloc.h	assert.h	bios.h	conio.h	ctype.h
dir.h	dos.h	errno.h	fcntl.h	float.h
graphics.h	io.h	limits.h	malloc.h	math.h
mem.h	process.h	setjmp.h	share.h	signal.h
stdarg.h	stddef.h	stdio.h	stdlib.h	string.h
time.h	values.h			

Some header files consist only of constants, global variables, and macros, such header files are of less importance. Hence they are not included in the description.

alloc.h

The header <alloc.h> declares functions for dynamic allocation (also reallocation) of memory, and clearing of used memory before the execution of program is over.

Example : calloc(), malloc(), etc.,

assert.h

The assert macro is used to add diagonstics to programs. If NDEBUG is defined at the time of including <assert.h> the assert macro is ignored.

bios.h

The header <bios.h> declares functions for interrupt enabling / disabling functions, which are used for RS-232 communications and bios interrupts.

Example : bioscom(), biostime(), etc.,

conio.h

The header <conio.h> declares functions which deals with the text mode window.

Example : crscr(), clreol(), etc.,

ctype.h

The header <ctype.h> declares functions for testing characters. For each function the argument is an int, whose value must be EOF or representable as an unsigned char, and the return value is an int. These functions return non-zero value, if the argument C satisfies the condition. zero otherwise.

dir.h

This header file <dir.h> defines library functions that are used to do functions with the directories existing in the drive. The functions present in the header file can be used to rename, delete, copy , move a directory from one place to another and so on.

Example : chdir(), rmdir(), etc.,

dos.h

This header file <dos.h> defines library functions that are used to do functions with the operating system. The functions present in the header file can be used to get, set, data, time, and so on. The functions can also be used to suspend execution of a time for some time also.

Example : delay(), gettime(), etc.,

errno.h

Many of the functions in the library set status indicators when error or end of file occurs. These indicators may be set and tested explicitly. The integer expression errno may contain an error number that gives further information about the most recent error.

fcntl.h

This header file < fcntl.h >defines "open flags" for open and similar library functions.

float.h

The subset of the header file <float.h> are constants related to floating-point arithmetic. When a value is given, it represents the minimum magnitude for the corresponding quantity, Each implementation defines appropriate values.

Example : FLI_RADIX, FLT_DIG, etc.,

graphics.h

The header file <graphics.h> consists of library functions used to do all graphics effects in graphical window. This is one of the header files, which has lot of library functions in it.

Example : circle(), detectgraph(), etc.,

io.h

The header file <io.h> defines library functions used for input and output functions that are not present in the stdio.h header file.

Example : chmod(), eof(), etc.,

limits.h

The header file <limits.h> defines constants for the size upto integral types. The values below are acceptable minimum magnitudes. Larger values may also be used.

Example : CHAR_BIT, CHR_MAX, etc.,

math.h

The header file <math.h> declares mathematical functions and macros. All those functions returns a datatype double. Angles for trignometric functions are expressed in radians.

Example : sin, cos, exp, etc.,

setjmp.h

The declarations in <setjmp.h> provide a way to avoid the normal function call and return sequence, typically to permit an immediate return from a deeply nested function call.

Example : longjmp, setjmp, etc.,

signal.h

The header file <signal.h> provides facilities for handling exceptional conditions that arise during execution, such as an interrupt signal from an external source or an error in the execution. The initial state of signals is implementation defined. The group of signal functions have their names beginning with SIG.

Example : SIGABRT, SIGFPE, etc.,

stdarg.h

The header file <stdarg.h> facilitates for stepping through a list of function arguments of unknown number and type. The macro can be called once, after the arguments have been processed.

Example : va_list, va_start, etc.,

stdlib.h

The header file <stdlib.h> declares function for number conversion, storage allocation and similar tasks.

Example : ralloc, malloc, exit, etc.,

stdio.h

The input and output functions, types and macros defined in <stdio.h> represent nearly one third of the library. It is one of the commonly used header file.

Example : scanf, printf, etc.,

string.h

The header file <string.h> is used to perform functions in strings. There are two group of string functions defined in the header file <string.h>. They have names beginning with str and also have names beginning with mem.

Example : strcpy, strcmp, memcpy, etc.,

time.h

The header file <time.h> declares types and functions for manipulating data and time. Some functions process local time, which may differ from calendar time.

Example : fm_sec, fm_min, fm_hour, etc.,

IV. LIBRARY FUNCTIONS

MEMORY MANAGEMENT FUNCTIONS (include file is alloc.h)

Function	Purpose
calloc(n, size)	Allocates n multiple blocks of (n * size) bytes and initializes the memory to 0.
coreleft(void)	Returns a measure of RAM memory not in use.
farcalloc(n, size)	Allocates n multiple blocks of (n * size) bytes larger than 64K.
farcoreleft(void)	Returns a measure of RAM memory not in use beyond the highest allocated block.
farfree(*ptr)	Releases a block of memory previously allocated by farcalloc(), farmalloc(), & farrealloc() pointed by ptr.
farmalloc(size)	Allocates a block of (size) bytes larger than 64K.
farrealloc(*ptr, size)	Reallocates the size of the allocated block to size pointer by ptr previously alocated by farcalloc() / farmalloc(), copying the contents to a new location, if necessary.
free(*ptr)	Releases a block of memory previously allocated by calloc(), malloc(), & realloc() pointed by ptr.
malloc(size)	Allocates a block of (size) bytes.
realloc(*ptr, size)	Reallocates the size of the allocated block to size pointer by ptr previously allocated by calloc() / malloc(), copying the contents to a new location, if necessary.

INTERRUPT ENABLING / DISABLING FUNCTIONS (include file is bios.h)

bioscom()	Used for RS-232 communications (serial I/O).
biosdisk()	Used for interrupt 0x13 to issue disk operations directly to the BIOS.
biosequip()	Used for BIOS interrupt 0x11 to return an integer describing the equipment connected to the system.
bioskey()	Performs various keyboard operations using BIOS interrupt 0x16.
biosmemory()	Used for BIOS interrupt 0x12 to return the size of RAM.
biosprint()	Used for BIOS interrupt 0x17 to perform various printer functions on the printer identified by port.
biostime()	Used for BIOS interrupt 0x1A to either read or set the BIOS timer.

TEXT MODE WINDOW FUNCTIONS (include file is conio.h)

clrscr()	Clears the text mode window.
clreol()	Clears to end of line in text window.
cgets(char*str)	Reads a string of characters from the console and stores the string (and the string length) in the location *str.

cputs(char*str)	Writes the null-terminated string str to the current text window. It does not append a newline character.
cprintf(..)	Displays formatted output to the text window on the screen.
cscanf(..)	Reads formatted input from the text window.
getch()	Reads a character from console but does not echo to the screen.
getche()	Reads a character from console and echoes to the screen.
gotoxy()	Positions cursor in text window.
movetext()	Copies text on screen from one rectangle to another.
gettext()	Copies text from text-mode screen to memory.
puttext()	Copies text from memory to text-mode screen.
kbhit()	Checks for currently available keystrokes.
putch()	Displays the character to the current text window.
textbackground()	Selects a new text background color.
window()	Defines active text-mode window.

CHARACTER CLASSIFICATION FUNCTIONS (Include file is ctype.h)

isalnum(c)	Determine if argument is alphanumeric. Return non-zero value if true; 0 otherwise.
isalpha(c)	Determine if argument is alphabetic. Return non-zero value if true; 0 otherwise.
isascii(c)	Determine if argument is an ASCII character. Return non-zero value if true; 0 otherwise.
iscntrl(c)	Determine if argument is an ASCII character. Return non-zero value if true; 0 otherwise.
isdigit(c)	Determine if argument is a decimal digit character. Return non-zero value if true; 0 otherwise.
isgraph(c)	Determine if argument is a graphic ASCII character. Return non-zero value if true; 0 otherwise.
islower(c)	Determine if argument is lowercase. Return non-zero value if true; 0 otherwise.
isodigit(c)	Determine if argument is graphic ASCII character. Return non- zero value if true; 0 otherwise.
isprint(c)	Determine if argument is a printing ASCII character. Return non-zero value if true; 0 otherwise.
isspace(c)	Determine if argument is a white space character. Return non -zero value if true; 0 otherwise.
isupper(c)	Determine if argument is uppercase. Return non-zero value if true; 0 otherwise.
isxdigit(c)	Determine if argument is hexadecimal digit. Return non-zero value if true; 0 otherwise.
tolower(c)	Tests the charater and converts to lowercase if uppercase.
toupper(c)	Tests the charater and converts to uppercase if lowercase.

DIRECTORY FUNCTIONS (include file is dir.h)

chdir(char *path)	Causes the directory specified by path to become the current working directory.
findfirst(char *path)	Uses the DOS system call 0x4E to begin a search of a disk directory.
findnext(char *path)	Finds subsequent files that match the pathname argument of findfirst.
getcurdir()	Gets current directory for specified drive.
getcwd()	Gets the current working directory.
mkdir(char *path)	Creates a new directory from the given path.
rmdir(char *path)	Deletes the directory whose path is given by path.
searchpath(char *p)	Searches the DOS path (p) for a file.

DISK OPERATING SYSTEM FUNCTIONS (include file is dos.h)

absread()	Uses DOS interrupt 0x25 to read specific disk sectors.
abswrite()	Uses DOS interrupt 0x26 to write specific disk sectors.
allocmem()	Uses the DOS system call 0x48 to allocate a block of free memory and return the segment address of the allocated block.
delay()	Suspends execution of a program for an interval in milliseconds.
disable()	Used to disable interrupts.
dostounix()	Converts date and time to UNIX time.
enable()	Used to enable interrupts.
freemem()	Frees a previously allocated DOS memory block.
geninterrupt()	Macro that generates a software interrupt.
getdate()	Gets DOS system date.
gettime()	Gets DOS system time.
inport()	Reads a word from a hardware port.
inportb()	Reads a byte from a hardware port.
int86()	Used to execute an 8086 software interrupt.
intdos()	Used to execute DOS interrupt 0x21 to invoke a specified DOS function.
intr()	An alternate interface for executing software interrupts, that generates an 8086 software interrupt.
keep()	Exits the current program and remains resident.
nosound()	Turns the PC speaker off.
outport()	Sends a word to a hardware port.
outportb()	Sends a byte to a hardware port.
setdate()	Sets DOS system date.
settime()	Sets DOS system time.
sleep()	Suspends execution for an interval in milliseconds.
sound()	Turns the PC speaker on at the specified frequency.
unixtodos()	Converts date and time from UNIX to DOS format.

INPUT / OUTPUT FUNCTIONS (include file is io.h)

access()	Determines the accessibility of a file.
chmod()	Sets file access permissions (read, write & execute).
chsize()	Changes the size of the file.
close()	Closes a file previously opened by open().
creat()	Creates a new file or overwrite an existing one.
creatnew()	Creates and opens a new file for reading & writing in binary mode.
eof()	Checks for end-of-file.
filelength()	Gets file size in bytes.
lock()	Sets or resets file-sharing locks.
mktemp()	Makes a unique file name.
open()	Opens a file.
read()	Uses DOS function 0x3F (read system call) to read bytes from a file into a buffer.
remove()	Macro that removes a file.
rename()	Changes the name of a file from oldname to newname.
tell()	Gets current position of file pointer.
write()	Uses DOS function 0x40 to write bytes from a buffer to a file.

ARITHMETIC FUNCTIONS (Include file is math.h)

cos(d)	Returns the cosine value of double d.
cosh(d)	Returns the hyberbolic cosine value of double d.
ceil(d)	Returns a value rounded up to the next higher integer.
acos(d)	Returns the arc cosine of d.
asin(d)	Raises the exponential e to the power d.
exp(d)	Returns the value of power d.
fabs(d)	Returns the absolute value of d.
floor(d)	Returns the value rounded down to the next lower integer.
fmod(d_1,d_2)	Returns the remainder of d_1/d_2.
hypot(d)	Returns the hypotenuse of right triangle.
log(d)	Returns the natural logarithm of d.
log10(d)	Returns the natural logarithm of d.
modf(d)	Returns the argument into integer and fractional parts.
pow(d_1,d_2)	Returns d_1 raised to the d_2 power.
sin(d)	Returns the sine of d.
sinh(d)	Returns the hyperbolic sine of d.
sqrt(d)	Returns the square root of d.
tan(d)	Returns the tangent of d.
tanh(d)	Returns the hyperbolic tangent of d.

MEMORY FUNCTIONS (include file is mem.h)

memccpy()	Copies a block of n bytes from source to destination.
memchr()	Searches for a character in a string of n bytes.
memcmp()	Compares the first n bytes of strings s1 and s2.
memcpy()	Copies a block of n bytes from source to destination.
memicmp()	Compares the first n bytes of strings s1 and s2, ignoring case.
memmove()	Moves a block of n bytes from source to destination.
movmem()	Moves a block of length bytes from source to destination.
setmem()	Sets a block of length bytes in memory.

PROCESS FUNCTIONS (include file is process.h)

abort()	Abnormally terminates a process.
execl()	Enable your program to load and run other files (child processes).
exit()	Terminates the program.
getpid()	Gets the process ID of the program.
spawnl()	Enable your programs to run other files (child processes) and returns control to your program when the child processes finish.
system()	Invokes the DOS command interpreter file from inside an executing C program to execute a DOS command.

JUMP FUCTIONS (include file is setjmp.h)

longjmp()	Performs a nonlocal goto.
setjmp()	Sets up for a nonlocal goto.

SIGNAL RECEIVING / SENDING FUNCTIONS (include file is signal.h)

raise()	Sends a software signal to the executing program.
signal()	Specifies signal-handling actions.

DATA CONVERSION FUNCTIONS (Include file is stdlib.h)

atof(s)	Converts string to float.
atoi(s)	Converts string to int.
atol(s)	Converts string to long.
abs(i)	Returns the absolute value of an integer.
ecvt(d)	Converts double to string.
fcvt(d)	Converts double to string.
gcvt(d)	Converts double to string.
itoa(i)	Converts int to string.
strtod(s)	Converts long to string.
strtal(s)	Converts string to double.
strtoul(s)	Converts string to an unsigned long integer.
ultoa(l)	Converts unsigned long integer to string.

STRING MANIPULATION FUNCTIONS (Include file is string.h)

strcats(s_1, s_2)	Concatenate string s_2 at the end of string s_1 and terminates with '\0' return s_2.
strchr(s_1, c)	Compare characters of s_1 with characters starting from the headofstring s_1. Return pointer to the first occurence of c in s1. If c is not present in s_1 returns NULL.
strcmp(s_1, s_2)	Compares two strings lexicographically with regard to case. Returns a negative value if $s_1 < s_2$, 0 if s_1 and s_2 are identical and a positive value if $s_1 > s_2$.
strcpy(s_1, s_2)	Copies string s_2 to string s_1.
strdup(s)	Duplicates string s.
strcmpi(s_1, s_2)	Same as strcmp.
strlen(s)	Returns the number of characters of strings.
strlwr(s)	Converts a string to lowercase.
strncat(s_1, s_2)	Concatenates string s_2 at the end of string s_1 return s_2 and terminates s_2 with '\0'.
strncmp(s_1, s_2, n)	Same as strcmp but it compares a portion of string s_2 to string s_1 upto position n.
strncpy(s_1, s_2, n)	Same as strcpy but it copies a portion of string s_2 to string s_1 upto position n.
strnicmp(s_1, s_2, n)	Same as strncmp but it compares a portion of string s_1 and string s_2 upto position n.
strrchr(s_1, c)	Compare characters of s_1 with character c starting from the tail of string s_1. Returns pointer to the last occurence of c in s_1. If c is not present in s_1 it returns NULL.
strrev(s)	Reverses the string s.
strset(s_1, s_2)	Set all characters within s_1 to s_2 (excluding the null character).
strstr(s_1, s_2)	Return pointer to first occurence of whole string s_2 in s_1. If string s_2 is not present in s_1 returns NULL.
strupr(s)	Converts a string to uppercase.
strpbrk(s_1, s_2)	Return pointer to first occurence in string s_1 of any character of string s_2 and if not character of s_2 is present return NULL.

STANDARD INPUT/OUTPUT FUNCTIONS (Include file is stdio.h)

fclose(f)	closes a file f, return 0 if file is sucessfully closed.
feof(f)	Determine if and end-of-file condition has been reached. If so, return a non-zero value; 0 otherwise.
fgetc(f)	Reads a single character from file f.
fgetchar(f)	Reads a single character from keyboard.

| fgets(s,n,f) | Reads a string s, containing n characters, from file f. |
| fopen(ft,s) | Opens a file named f of types s. Returns a pointer to the file. |

TIME FUNCTIONS (INCLUDE FILE IS TIME.H)

asctime()	Converts date and time to ASCII.
clock()	Returns number of clock ticks since program start.
ctime()	Converts date and time to a string.
difftime()	Computes difference between two times.
gmtime()	Converts date and time to Greenwich Mean Time (GMT).
localtime()	Converts date and time to a structure.
mktime()	Converts time to calendar format
stime()	Sets system date and time
time()	Get time of the day.

NOTE

*	denotes a pointer.
c	denotes a character-type argument.
d	denotes a double-precision argument.
f	denotes a file argument.
i	denotes an integer argument.
l	denotes a long integer argument.
s	denotes a string argument.

V. COMMON PROGRAMMING ERRORS

This appendix is written for C programmers for finding the bugs (errors) while compiling a program. It is advisable to keep track of such errors and to see that these known errors are not present in the program. I have made a list of more common programming mistakes, which will help you while writing programs. They are not arranged in any particular order.

1. Forgetting to include stdio.h

Forgetting to include the header file stdio.h in the program, which reads input from the user through the keyboard or displays the output in the screen. In such cases, the compiler issues an error message.

2. Forgetting to include header files

Forgetting to include the header file in the program, which uses one or more functions from a header file. In such cases, the compiler issues an error message.

3. Missing semicolons

Every C statement must end with a semicolon. Missing a semicolon may cause confusion to the compiler and results in error messages.

Example : x = a + b + c

The above statement should be

        x = a + b + c;

4. Using a keyword as an identifier

Another common mistake is to use a keyword as an identifier for naming variables, arrays, functions, structures, or unions, etc., In such cases, the compiler issues an error message.

5. Placing semicolons at the end of function definitions

It is common error to place a semicolon after the right paranthesis enclosing the argument list of a function definition.

6. Missing semicolons at the end of function declarations

It is common error to omit a semicolon after the right paranthesis enclosing the argument list of a function declaration (i.e., function prototype).

7. Ending an if statement with a semicolon

Another common mistake is to put a semicolon at the end of the if statement after the paranthesis. The semicolon may cause the if statement to omit its body, and hence it will not perform any action regardless of whether the condition is true or false.

8. Ending a loop with a semicolon

Another common mistake is to put a semicolon in a wrong place i.e., in a loop. A semicolon may cause a loop to be an indefinite loop and the programmer may get a different output.

Example :

```
#include<stdio.h>
main()
```

```
{
    int x = 1;
    while(x<=50);
    {
        printf("ERROR");
        x++;
    }
}
```

In the above program the value of x will not be incremented and always remains less than 50. Therefore the program segment should be

```
while(x <= 50)
{
    printf("ERROR");
    x++;
}
```

9. Improper comment characters

Every comment should start with a /* and end with a */. Anything between them is ignored by the compiler. If we miss out the opening /* or the closing /* then the compiler searches in the program treating all lines as comments. Hence, we will get an error message. One opening /* should be always accompanied with a closing */. Otherwise, an error will be occurred.

10. Undeclared variables

C requires every variable to be declared for its type, before it is used. During the development of a large program, one may use a variable, which has not been declared. Hence, it gives an error.

11. Using spaces in between operators

It is a common error to leave spaces in between operators like ==, !=, >=, <=.

12. Reversing the order of operators

It is a common error to reverse the order of operators like =! , =>, =<.

13. Confusing the precedence of various operators

Expressions are evaluated according to the precedence of operators. It is common among beginners to confuse the precedence of operators.

Example : `while(sum=multiply() <= 50)`.

         `sub = 5 - sum;`

The function call multiply() returns the product of three numbers, which is compared to 50. If the value is equal to or less than 50, the relational test is true, and 1 is assigned to sum, otherwise 0 will be assigned. But this is not the output the programmer wanted.

The expression should be

```
while(sum=multiply()) <= 50
    sub = 5 - sum;
```

14.   Omitting the ampersand (&) before the variables used in scanf()

Example :

```
int x;
scanf("%d",x);
```

The statement should be

```
int x;
scanf("%d",&x);
```

15.   Using the operator = instead of operator ==

It is quite possible to forget the use of operator == and use the operator = when we perform a relational test.

Example :

```
#include<stdio.h>
main()
  {
      int x = 1;
      while(x = 5)
      {
          printf("Hello");
          x++;
      }
  }
```

The statement is syntactically valid and the variable x assigns 5 and then, because x = 5 is true, the message is printed once and the control will come out of the loop since x becomes 6. Hence we have fallen in an indefinite loop. Similar mistakes can occur in other control statements such as for and if. Such a mistake in the loop control statements might cause infinite loops.

The statement should be

```
while(x == 5)
```

16.   Omitting the break statement at the end of a case in a switch statement

Example :

```
#include<stdio.h>
main()
  {
      int x;
      printf("ENTER AN INTEGER : ");
      scanf("%d", &x);
      switch(x)
      {
          case 1 :
                  printf("Hello");
```

```
        case 2 :
                printf ("Welcome");
        case 3 :
                printf ("Hai");
        }
    }
```

Remember that if a break statement is not included at the end of a case then execution will continue to the next case. Since in the above program we have not used break after the printf() in case1, case2 and case3, the control prints all the messages.

The program segment should be

```
switch (x)
{
    case 1 :
        printf ("Hello");
        break;
    case 2 :
        printf ("Welcome");
        break;
    case 3 :
        printf ("Hai");
        break;
}
```

17.   Using a continue statement in a switch statement

It is a common error to use the continue statement instead of break statement in a switch statement. Note that the continue statement works only in loops, and never with a switch statement.

18.   Crossing the bounds of an array
Example :

```
#include<stdio.h>
main ()
    {
    int x[50], i, sum = 0;
    for (i = 1; i < 100; i++)
        sum += x[i];
    }
```

In the array x there is no such element as x[50], since array counting begins with 0 and not 1. Compiler would give warning if the program exceeds the bounds.

Note that all indices starts from 0 and not 1.

19.   Missing braces at the end of a compound statement

It is common error to forget a closing brace when using a loop. It will be usually detected by the compiler because an opening brace should have a closing brace. However, if we put a matching brace in a wrong place, the compiler won't notice the mistake and the program will produce unexpected results.

20.   Inserting a semicolon at the end of macro definition

A macro definition must not end with a semicolon. A semicolon may cause confusion to the compiler and results in error messages.

Example :

```
# define SIZE 40;
```

The above macro definition should be

```
# define SIZE 40
```

21.   Confusing a character constant and a character string

In the statement

```
ch = 'x';
```

a single character is assigned to ch. In the statement

```
str = "x";
```

a pointer to the character string "x" is assigned to str.

Note that in the first case, the declaration of ch would be

```
char ch;
char *str;
```

22.   Forgetting to reserve an extra location in a character array (string) for the null terminator

Remember that each character array ends with a null character ('\0'), therefore its dimension should be declared big enough to hold the characters in the array as well as the null character.

Example : For storing a string "ANITHRA" the dimension of the array should be 8

```
i.e., char str[8] = "ANITHRA";
```

23.   Missing quotes

Every string must be enclosed in double quotes, while a single character constant must be enclosed in single quotes. If we miss the quotes, the string or the character would be referred to as a variable name.

Example :

```
if(output == NO)

mark = B;
```

In the above example NO and B are treated as variables and therefore an error message "undefined variable B", "undefined variable NO" will occur.

24.   Passing the wrong argument type to a function

For example,

```
abc = fleet_sum(5);
```

If the fleet_sum function is expecting a floating-point argument, then the above statement will produce a result with errors, since an integer value is being passed. We can use a typecast operator to explicitly convert a value that is passed to a function.

25. Omitting return type declarations

Omitting the return type of a function is a syntax error, if the function returns any other value other than int.

For example,

```
abc = sum (a);
```

If the return type for a function is not declared explicitly then the compiler will assume that is function returns a value of type int. Hence, it is a good programming practice to use void as a return type.

26. Mismatch of arguments in the function definitions and the function call

It's a syntax error to call a function, which doesn't match with the arguments present in function definition or function declaration.

27. Defining a function inside another function

It's a syntax error to define a function inside another function. However, you can call a function from any other function.

28. Confusing the operator (->) with the operator(.) when referencing structure variables.

Note that, the dot or period operator (.) is used for structure variables, while the arrow operator -> is used for structure pointer variables. So if **'a'** is a structure variable then the notation **a.b** is used to refer the member **b** of **a**, where as if **'a'** is a pointer to a structure, then the notation **a -> b** is used to refer the member **b** of the structure pointed to by **b**.

29. Leaving a blank space between the name of a macro and its argument list in the #define statements.

For example,

```
#define    PAY   (m, n)    (30, 80)
```

The above macro definition is incorrect, as the preprocessor considers the first blank space after the defined name as the start of the definition of that name.

The definition should be

```
# define PAY(M, N)    (30, 80)
```

30. Omitting parenthesis around arguments in macro definition

```
#define SQR(x)   x * x
main()
{
    int y;
    y = 16/SQR(4);
    printf("%d",y);
}
```

In the above example we would expect the value of y to be 1, but the value will be 16. this happens because on preprocessing the arithmetic statements takes the form of

```
y = 16 / 4 * 4 ;
```

the arithmetic statement should be

```
y = 16/(SQR(4));
```

VI. KEYS USED IN C COMPILER

MENU BAR KEYS

Keys used	Menu item	Function
Alt+Spacebar	Menu	Takes you to the (System) menu
Alt+C	Compile menu	Takes you to the Compile menu
Alt+D	Debug menu	Takes you to the Debug menu
Alt+E	Edit menu	Takes you to the Edit menu
Alt+F	File menu	Takes you to the File menu
Alt+H	Help menu	Takes you to the Help menu
Alt+O	Options menu	Takes you to the Options menu
Alt+P	Project menu	Takes you to the Project menu
Alt+R	Run menu	Takes you to the Run menu
Alt+S	Search menu	Takes you to the Search menu
Alt+W	Window menu	Takes you to the Window menu
Alt+F4 (Alt+x)	File quit	Exits C compiler.

GENERAL IDE KEYS

Keys used	Menu item	Function
F1	Help	Displays a help screen.
F2	File-Save	Saves the file that's in the active edit window.
F3	File-Open	Brings up a dialog box so you can open a file.
F4	Run/Go to cursor	Runs your program to the line where the cursor is positioned.
F5	Zoom window	Zooms the active window.
F6 (ctrl+F6)	Next window	Cycles through all open windows.
F7	Run/Trace Into	Runs your program in debug mode, tracing into functions.
F8	Run/Step Over	Runs your program in debug mode, stepping over function calls.
F9	Compile & Make	Invokes the project manager to make an .exe file.
F10	None	Takes you to the menu bar.

EDITING KEYS

Keys used	Menu item	Function
Ctrl+Del	Edit-Clear	Removes selected text from window; doesn't put it in Clipboard
Ctrl+Ins	Edit-Copy	Copies selected text to Clipboard
Shift+Del	Edit-Cut	Places selected text in Clipboard, deletes selection
Shift+Ins	Edit-Paste	Pastes text from Clipboard into the active window
Alt+Bksp	Edit-Undo	Restores text in active window to previous state
Ctrl+L	Search/Search again	Repeats last Find or Replace command.

WINDOW MANAGEMENT KEYS

Keys used	Menu Item	Function
Alt+#		Displays a window, where # is the number of the window you want to view
Alt+0	List All	Displays a list of open windows
Ctrl+F4 (Alt+F3)	Close window	Closes the active window
Shift+F5	Tile window	Tiles all open windows
Alt+F5	Debug/Inspect	Opens an inspector window
Shift+F5	User Screen	Displays user screen
F5	Zoom Window	Zooms/unzooms the active window
Ctrl+F6 (F6)	Window³Next	Switches the active window
Ctrl+F5		Changes size or position of active window.

ONLINE HELP KEYS

Keys used	Menu Item	Function
F1	Help-Contents	Opens a context-sensitive help screen
F1 F1	Help-Using Help	Brings up help on help. (Just press F1 when you're already in the help system.)
Shift+F1	Help-Index	Brings up help index
Alt+F1	Help-Previous Topic	Displays previous help screen
Ctrl+F1	Help-Topic Search	Calls up language-specific help in the active edit window.

DEBUGGING / RUNNING KEYS

Keys used	Menu Item	Function
Alt+F5	Debug/Inspect	Opens an Inspector window
Alt+F7	Search-Previous Error	Takes you to previous error
Alt+F8	Search-Next Error	Takes you to next error
Alt+F9	Compile	Compiles to .obj
Ctrl+F2	Run/Program Reset	Resets running program
Ctrl+F3	Debug/Call Stack	Brings up call stack
Ctrl+F4	Debug/Evaluate/Modify	Evaluates an expression
Ctrl+F5 (ctrl+F7)	Debug/Add Watch	Adds a watch expression
Ctrl+F8	Debug/Toggle Breakpoint	Sets or clears conditional breakpoint
Ctrl+F9	Run	Runs program
F4	Run/Go To Cursor	Runs program to cursor position
F7	Run/Trace Into	Executes tracing into functions
F8	Run/Step Over	Executes skipping function calls
F9	Compile/Make	Makes (compiles/links) program

VII. GLOSSARY

access time	The time taken to retrieve information from memory.
address	It is the value that points to a location in memory.
address bus	Unidirectional line or pathway, which is used to identify the storage location in memory where the next instruction to be executed or the next piece of data will be found.
adjacency matrix	The most frequently used graph representation scheme.
algorithm	It is a step-by-step recipe for solving an instance of a problem.
analog computer	Device that performs operations on data that are represented within the device by continuous variables having a physical resemblance to the dependent quantities being represented by using one kind of physical quantity to represent another.
application software	Software used to accomplish specific tasks other than just running the computer system.
arguments	List of values provided when a function is called. They are specified within paranthesis following the function name.
array	It is a group of elements that share a common name and that are differentiated from one another by their positions within the array.
ASCC	Automatic Sequence Controlled Calculator.
ASCII	American Standard Code for Information Interchange.
Assembler	A translator program that translates the program written in assembly language into machine language.
assignment	It is the process of copying of one variable into another.
associativity	The direction (either left to right or right to left) used to evaluate an expression when a statement contains number of operators.
auto	It is a storage class specifier for local variables.
auto variable	It is visible only in the block or function in which it is declared. All the variables are of type auto by default.
auxiliary memory	See *secondary memory*.
base-10 number system	Also referred as decimal number system since it uses ten separate symbols, 0 to 9 to represent its numbers.
base-16 number system	Also referred as hexadecimal number system since it uses sixteen separate symbols, 0 to 9 and A to F to represent its numbers.
base-2 number system	Also referred as binary number system since it uses only two separate symbols, 0 and 1 to represent its numbers.
base-8 number system	Also referred as octal number system since it uses eight separate symbols, 0 to 7 to represent its numbers.
BCD	Binary-coded decimal.
binary	A numbering system with a base of 2. The binary digits are 0 and 1.

binary-coded decimal	Simplest method used to represent the decimal digits 0 through 9 using binary digits.
binary search	It is a method used to search an ordered list. The search begins in the middle of a table and determines whether the argument is in the upper or lower half of the table, and then continues to divide that portion of the table in which the argument being sought is located until the argument is found.
binary tree	It is a tree that has nodes either empty or not more than two child nodes, each of which may be a leaf node.
bit	Binary digit. It is the basic storage unit of a computer with the capability of storing the values in the form by 0 and 1.
bitwise operator	Operators that operate on individual bits within a word of memory.
Blaise Pascal	A French mathematician who invented the first mechanical calculator.
boolean	It is a variable or an expression that can assume the value of true or false.
branch	A link between a parent node and its child node.
break	It is used to terminate or to exit from a loop or from a switch statement.
bubble sort	Compares the top element of an array with its successor, swapping (interchanging) the two if they are not in proper order.
buffer	A storage location used for temparary storage of data transferred from disk.
bug	It is an error in a program.
bus	Group of lines interconnecting central processing unit and I/O devices.
byte	Eight binary digits (i.e eight bits). A character represented in ASCII requires a byte of storage.
call by reference	It is the process of calling a function using pointers to pass the address of variables.
call by value	It is the process of passing the actual value of variables.
calloc	This function allocates a block of memory (in bytes) dynamically and it returns a pointer variable. The block of memory is initialized with zeroes.
card reader	An input device that uses punched cards to read input information into the computer.
case	It is used to provide labels in a switch statement.
CD-ROM	Compact Disk-Read Only Memory. A common type of portable secondary storage device that can store large volumes of data in the range 700 MB to 1.4 GB.
cell	It is the smallest part of memory that can be accessed by the CPU to store data. Each cell has a capability to store one bit of information. Each cell stores a binary digit referred to as bit.
central processing unit	Heart of the computer system, mounted on the motherboard of every computer, responsible for the interpretation and execution of instructions.
char	It is a primary datatype which is used to declare character variables and arrays.
Charles Babbage	Father of computers, who invented the analytical engine.

Chinese Abacus	First calculating device invented some 2000 years ago in china.
child	*See child node.*
child node	Each link in the root node.
circular queue	It is another form of a linear queue in which the last position is connected to the first position of the list.
comma operator	Operator which connects multiple statements in a single expression.
command	A directive to the operating system to provide a specific function.
command line	The entire command issued to the operating system.
command line arguments	These are arguments that are passed on to main function at the command prompt.
compiler	It is a software program that translates a high-level language program into a machine language program at a stretch.
computing	Study of processes that describe and transform data and instructions. It is a set of techniques to define and solve problems, a way to complete an information-based task adaptable to a specific problem.
computer	A programmable electronic data processing machine, which receives, stores, correlates, performs arithmetic and logical operations & outputs a large volume of data with high speed as per the instruction given.
complete binary tree	It is a binary tree in which every non-leaf node has exactly two children not necessarily to be on the same level.
complete graph	*See strongly connected graph.*
connected graph	It is a graph if there exists a path from any vertex to any other vertex.
constants	It refers to fixed values that does not change during execeution of a program.
contiguous	A storage characteristic that specifies that the values are stored in consecuetive locations either in memory or on disk.
continue	It is used to transfer the control to the begining of the loop, but does not terminate a loop.
control bus	Bi-directional line or pathway, which is used to pass the controlling signals from control unit to other units in the system and vice versa.
cycle	It is a path containing at least three vertices such that the starting and the ending vertices are the same.
data field	It is a field in the node of a linked list that contains the actual data (i.e., value) of the element to be stored in the list.
data bus	Line or pathway, which is used to transfer data from one unit to another in both directions.
data structure	It defines the way of organizing all data items that includes not only the elements stored, but also stores the relationship between the elements.
data type	It is the named set of values and operations defined to manipulate the character set.
debugging	It is the process of removing errors from the program.

default   It is the default label in the switch statement. The control is transferred to the default statement when none of the case labels match with the expressions in the switch.

dereference operator   It is the asterisk (*) symbol to indicate a pointer's contents (an address).

degree   The number of sub trees in a node.

depth   *See height.*

deque   It is another form of a queue in which insertions and deletions are made at both the front and rear ends of the queue.

dequeue   It is an operation in queue which is used to remove an element from queue at the front end.

derived data type   It is a composite data type constructed from other types, such as array pointer, structure, union,etc.,

digital computer   Device that reads input, stores data, processes and outputs discrete signals.

digraph   *See directed graph.*

directory   The index that stores file locations.

directed cycle   It is the cycle in a digraph.

directed graph   It is a graph in which an edge between any two nodes is directionally oriented.

directed path   In a digraph, all edges are in same direction, which follows a path always moving in the same direction indicated by arrows.

directory   The index that stores file locations.

do   It is a control statement that creates a loop of operations. It is used with while keyword.

double   It is a floating-point datatype specifier, which is used to double the number of digits after decimal point of a floating-point value.

double word   It is a collection of bits on 16-bit boundary.

doubly linked list   It is a linked list containing two pointers, one to its predecessor and another to Its successor, thus allowing traversal of the list both backwards and forwards.

dynamic data structure   A data structure formed when the number of data items are not known in advance.

DRAM   Dynamic Random Access Memory. Type of physical read/write memory used in most personal computers and holds data only if it is continuously/ constantly refreshed or reenergized or it will lose its contents.

E²PROM   Electrically Erasable Programmable Read Only Memory.

EAPROM   Electrically Alterable Programmable Read Only Memory.

EDSAC   Electronic Delay Storage Automatic Calculator

EDVAC   Electronic Discrete VAriable Computer

EEPROM   Electrically Erasable Programmable Read Only Memory. Type of ROM that can be erased using electrical pulses and reprogrammed, byte-by-byte, without removing it from the computer system.

edge   It is a connection between two nodes. Also referred as a link.

else	It is used to specify an alternative path in a two-way branch control of execution. It is used with if statement.
ENIAC	Electronic Numeric Integrator And Calculator.
enum	It is used to create a user-defined integer datatype.
enqueue	It is an operation in queue which is used to add a new element in to a queue at the rear end.
EOF	End-Of-File, the condition that occurs when a read operation attempts to read after it has processed the last piece of data.
EPROM	Erasable Programmable Read Only Memory. Type of information in which the information written can be erased by exposing EPROM to high intensity ultraviolet rays for 5 to 10 minutes, and the new information can be written.
expression	It is the sequence of operators and operands that reduces to a single value.
extern	It is a storage class specifier.
extern variable	It is a variable that is visible to all functions and all parts of the program. It is usually declared at the start of the program before main function.
external address	It is the address of the first node in the list, which is stored in the head pointer of the list.
fibonacci numbers	Numbers that are the sum of the preceding two numbers in a sequence. For example : 1 , 1 , 2 , 3 , 5 , 8 and so on.
fibonacci search	A possible improvement in binary search is not to use the middle element at each step, but to guess more precisely a fibonacci number where the key being sought falls within the current interval of interest.
FIFO lists	*see queue.*
file	It is a place in the disk where a group of related data is stored permanently, so that the information is accessed and altered whenever necessary.
file pointer	Pointer to structure that contains information about a file.
float	It is a primary datatype which is used to declare a variable to store a single-point precision value.
for	It is a control statement which is used to create a loop of iterative operations.
free	It deallocates the block of memory already allocated by malloc or calloc.
front end	The end at which deletions are made in a queue.
full binary tree	It is a binary tree in which all the leaves are on the same level and every non-leaf node has exactly two children.
function	It is a self contained program segment (block of statements) that performs some specific well defined task.
function call	It is a statement that invokes another function.
function declaration	It provides the information needed to call a function. This gives information about the name of the function, its return type, and the type of each argument passed in the function.
function prototype	*see function declaration.*
garbage collection	The effective recovery of memory that is no longer in use.

garbage value	It is an unpredictable value which is taken by an auto or register variable when it is not initialized.
getch	It is similar to getchar which reads a single character the instant it is typed without waiting for the ENTER key to be hit. It will not display the character in the screen.
getchar	It is a function which is used for getting a single character from the standard input device. It does not require any arguments.
getche	It is similar to getch which reads a single character the instant it is typed without waiting for the ENTER key to be hit, but it will echo or displays the character that the user have typed, in the screen.
gets	It stands for *"get string"*. This function is used to get a string as input and automatically assigns a null-character to the end of the string.
global variable	*see extern variable*.
goto	It is a transfer statement that enables us to skip a group of statements unconditionally.
graph	It consists of a set of non-empty vertices (referred as nodes in case of trees), together with a set of edges, and each edge joins two different vertices.
graphics tablet	A graphical input device used by graphics designers, which can produce much more accurate drawings on the screen than could do with a mouse or any other pointing device.
gray code	It is a binary code with the property that only one-bit changes between any two consecutive elements.
hard disk	Commonly used secondary storage device made up of a rigid metal disk covered by a magnetic material used to store large volumes of data.
hardware	Various functional units of the computer system.
head pointer	It is a pointer in the linked list, which points to the first node in the list that stores the address of the first node of the list.
header file	It is a file containing the declarations that are to be used in one or more files. This file is normally included in a program with an # include directive.
height	It is the maximum level of any node in the tree.
hybrid computer	Combination of both analog and digital computers, which posses all the good qualities of analog and digital computers.
if	It is a control statement that is used to test an expression and transfer the control to a particular statement or a group of statements depending upon the value of expression.
impact printers	Type of printers that produce their output by making a contact (by using hammers, which strike against a ribbon and paper) to print the text on the paper to be printed.
Input unit	Comprises of various input devices, which helps to read the program, data, information, and various operating commands into the computer.
incidence matrix	*See adjacency matrix*.
indegree	It is the number of edges entering from the vertex in a digraph.

Infix notation	It is the normal way of expressing mathematical expressions. In this notation the operator is present in between the operands.
insertion sort	It is a type of sorting by inserting the i *th* element in the i *th* pass in its correct place.
int	It is a primary data type which is used to declare a variable that stores integer values.
Joseph-Marie Jacquard	Inventor of an automatic loom referred as the **Jacquard loom**, with a punched-card system.
joysticks	Input device used for playing computer games and controlling computer simulations.
keyword	It is a word that has a specific meaning in certain contexts.
Lady Ada Lovelace	First programmer who programmed for computer to run on the Babbage's Analytical Engine.
leaf node	See terminal node.
library	It is a collection of programs or functions that can be utilised by several programmers.
LIFO lists	*see stack.*
linear data structure	A data structure having a linear relation ship between its adjacent elements
linear search	*See sequential search.*
link	*See edge.*
link field	It is a field in the node of a linked list also referred as the next address field that contains the address of the next node in the list.
linked list	It is a linear dynamic data structure that can grow and shrink during its execution time.
left sub trees	All the nodes to the left of a given node in a binary tree.
left-skewed binary tree	It is a binary tree, which has only left child nodes.
long	It is a data type modifier that can be applied to some of the basic data types to increase their size.
LSB	Least Significant Bit. The rightmost bit in a number system with bit position zero.
macro	A statement or series of statements inserted in line within a program in place of a symbol.
mainframe computers	These are computers with large memory capacity and can processes large volumes of data and can solve complicated, scientific and mathematical problems.
magnetic tape	A low cost secondary storage device used for storing data and files.
malloc	It stands for *"memory allocation"*, which allocates a block of memory (in bytes) at run time and it returns a pointer variable. The block of memory is not initialized.
MAR	Memory Address Register.The register, which stores the address of the location from where a word is to be retrieved or to be stored.
Mask-programmed ROM	Type of ROM in which contents are written at the time of its manufacture.
matrix	A two-dimensional array in which the position of a data element must be specified by giving two co-ordinates (i.e., row and column.)

MDR	Memory Data Register. The register, which stores the data retrieved from memory or stored in memory.
microcomputers	These are small computers with built on a single silicon chip (IC/ microprocessor) as its CPU.
merge sort	It is the most commonly used external sorting. A file is divided into two subfiles. These files are compared, one pair of records at a time, and merged by writing them to other files for further comparisons.
Modularity	Process of splitting a large program / project in to smaller modules to perform the operations fast and, which helps in easy error checking.
multidimensional array	It is an array of rank higher than one. Each data element must be specified by giving one co-ordinate for each dimension.
nibble	It is a collection of bits on 4-bit boundary.
next address field	Also referred as the forward link field, which stores the address of its next node.
node	Each data item in a data structure.
non-impact printers	Type of printers that produce their output without making contact (such as electro sensitive, electrostatic, ink jet, and laser for printing) on the paper to be printed.
non-terminal nodes	All intermediate nodes that traversing the given tree from its root node to the terminal nodes.
non-volatile cell	A memory cell, which does not loose the information stored in it when power is switched off.
null address	It is the address stored by the NULL pointer of the last node of the list, which indicates the end of the list.
null character	It is used for terminating strings. It has the ASCII code of zero.
octal	A numbering system with a base of 8. The octal digits are 0 1 2 3 4 5 6 and 7.
OCR	Optical Character Recognition. An input device commonly referred as optical scanner recognizes the printed characters directly from printed documents.
OMR	Optical Mark Reader. Recognizes characters that have been darkened by a dark pencil or ink.
operand	An object in a statement on which an operation is performed.
operating system	Collection of programs that acts as an interface between the user and the computer by coordinating the operations of hardware and software.
outdegree	It is the number of edges exiting from the vertex in a digraph.
output restricted deque	It is another form of a deque which allows deletion at one end (it can be either front / rear) only.
parameter	It is a value or a variable passed to a function.
path (tree)	It is a sequence of distinct nodes in which successive nodes are connected by edges in the tree.
path (graph)	It is a sequence of distinct vertices each adjacent to the next, except possibly the first vertex and last vertex is different in a graph.

piles	see stack.
pointer	It is a variable that represents the location (not the value) of a data item, such as a variable or an array element.
plotter	An output device, which is used to draw graphs and figures using a computer.
pop	It is an operation in stack which is used to delete an existing element from the stack at the top.
postfix notation	Also called as reverse polish notation, in which the operator is represented after its operands.
precedence	It determines the order in which operations are performed.
precision	It specifies the number of meaningful digits in value that must be represented in a fixed number of binary digits.
prefix notation	Also called as polish notation, in which the operator is represented before its operands.
preprocessor	It is the part of the compiler that manipulates the program text before any further compiling is done.
previous address field	Also referred as the backward link field, which stores the address of its previous node.
printf	It stands for *"print function"* or *"print formatted"*. This function is used to output any combination of digits, characters and strings according to the format string on the standard output device.
program	It is the series of instructions provided to the computer that directs the computer in executing a task.
programming	It is the process of preparing and feeding the instructions into the computer for execution.
PROM	Programmable Read Only Memory. Type of memory where the information can be written only once and then onwards it can only be read. Its contents cannot be altered in future.
pseudocode	It is an informal language written in English, particularly used for developing algorithms.
push	It is an operation in stack which is used to add a new element in the stack at the top.
push-down lists	see *stack*.
puts	It stands for *"put string"*. This function is used to output a null-terminated string on the standard output device.
queue	It is an ordered collection of elements in which insertions are made at one end and deletions are made at the other end.
quick sort	This sort divides the initial unsorted list in to two parts, such that every element in the first list is less than all the elements present in the second list. The procedure is then repeated recursively for both the parts, up to relatively short sequences, which can be sorted until the sequences, reduces to length one.

radix sort	It derives its name from the fact that while sorting the data, it considers the radix (digit position) of elements from lower order byte to higher order byte and sorts the data items in the list.
RAM	Random Access Memory. It is a read write memory, in which the contents can be read and can be used to write any information on it. It forms the major part of the main memory.
realloc	It stands for *"re allocation"*. This function increases or decreases the size of dynamically allocated block of memory.
rear end	The end at which insertions are made in a queue.
recursion	It is a process by which a function calls itself repeatedly until some specified condition has been satisfied.
register variable	It is similar to an auto variable which is used to store variables in the CPU register in order to access them as quickly as possible.
register	It is a storage class specifier.
return	It is used to mark the end of a function execuetion and to transfer the control back to the calling function.
return value	It is the value sent by the called function to the calling function.
right sub trees	All the nodes to the right of a given node in a binary tree.
right-skewed binary tree	It is a binary tree, which has only right child nodes.
ROM	Read Only Memory. It is a RAM fabricated with permanently stored information, which cannot be erased. It is a non-volatile memory.
root	See root node.
root node	It is the highest level or the first level in a tree.
sorting	Rearranging a group or sequence of elements or data items in either ascending or descending order depending upon the relationship among the data items present in the group.
scanf	It stands for *"scan function"* or *"scan formatted"* This function is used to get input any combination of digits, characters or strings according to the format string from the standard input device.
searching	It is a programming technique that determines whether an element or a data item is present in the given list or not.
secondary memory	The memory, which is not directly connected to the CPU and stores the data and instructions permanently.
selection sort	Successive elements are selected from a file or from an array and placed in their proper position.
sequential search	The value to be found is searched, from left to right one by one until the element is found or until the end of the list is reached. This search doesn't require that the data items in the list to be in sorted order.
short	It is a data type modifier that can be applied to some of the basic data types to decrease their size.
siblings	The child nodes of a given parent node.

signed	It is a data type modifier used with character and integer data type variables to indicate that the variables are stored with the sign.
singly linked list	It is a linked list in which each node contains only one link field pointing to the next node in the list.
sink vertex	It is a vertex whose outdegree is 0.
sizeof	It is an operator used to get the size of a data type in bytes.
SRAM	Static Random Access Memory. Type of physical read/write memory used in personal computers and holds data, as long as power is supplied to the circuit.
source vertex	It is a vertex whose indegree is 0.
stack	It is an ordered collection of elements in which insertions and deletions are restricted to one end. Stacks are also referred as "piles" and "push-down lists".
static variable	It is a variable whose value persists between different function calls until the end of the program.
static	It is a storage class specifier.
static data structure	A data structure formed when the number of data items are known in advance.
stderr	It is the standard error files operand automatically for displaying the error messages.
stdin	It is the standard input file opened automatically for getting the input for the program.
stdout	It is the standard output file opened automatically for displaying the output of the program.
string	It is an array of characters terminated by a null character.
strongly connected graph	It is a directed graph in which for every pair of distinct vertices their is a **graph** directed path from every vertex to every other vertex.
structure (struct)	It is a collection of one or more variables possibly of different data types, grouped together a single name for convenient handling. They are also used to create user-defined datatypes.
sub tree	A subset of a tree that is itself a tree.
super computers	Large sophisticated, expensive computers, using the latest state-of-the-art technology.
switch	It is a control statement which provides multi-way branching that matches its argument against each of the pattern arguments in order.
syntax	The rules that govern the order of words and relationships in a language.
system software	It is a collection of programs written to service other programs, such as the basic input-output system (bios), device drivers, an operating system, typically a graphical user interface and so on.
terminal node	A node that has no children (i.e., with zero degree).
top	The end from which elements are added and/or removed from a stack.
traverse	Moving through all the nodes in the binary tree, visiting each node in the tree exactly once.

tree	It is a non-linear, two-dimensional data structure, which represents hierarchical relationships between individual data items.
tribonacci numbers	Numbers that are the sum of the preceding three numbers in a sequence. For example : 1 , 1 , 2 , 4 , 7 , 13 , 24...
touch screen	Input device that allows users to operate a computer by simply touching the display screen.
type cast operator	It is the operator that changes the type of an expression.
typedef	It stands for *"type definition"* which allows the user to define new data types that are equivalent to existing data types.
unary minus	It is the operator that complements the value of an expression.
undirected graph	It is a graph in which an edge between any two nodes is not directionally oriented.
union	It is similar to struct in declaration which is used to allocate storage for several data items at the same location.
UNIVAC	UNIVersal Automatic Computer.
unqualified graph	*See undirected graph.*
unsigned	It is a data type modifier used with character and integer data type variables to indicate that the variables are stored without the sign.
User-programmed ROM	Type of ROM introduced before two decades, which are used by users to store their own data / information.
variable	It refers to a quantity which may vary during execuetion of a program
VDU	Visual display unit. Most commonly used output device, which is also referred, as monitor is a device resembling a television that displays what you have typed or otherwise entered in to the computer on the screen in front of you.
void	It is a data type which is used to indicate that the function returns nothing.
volatile	It is a data type modifier used in variable declarations. It indicates that the variable may be modified by factors outside the control of the program.
weakly connected graph	It is a directed graph if any vertex doesn't have a directed path to any other vertices.
weighted graph	It is a graph in which every edge in the graph is assigned some weight or value. The weight of an edge is a positive value that may be representing the distance between the vertices or the weights of the edges along the path.
while	It is a control statement used to execute a set of statements repeatedly depending on the outcome of a test.

INDEX